BEYOND QUANTUM

BEYOND QUANTUM

Andrei Khrennikov

PAN STANFORD PUBLISHING

Published by

Pan Stanford Publishing Pte. Ltd.
Penthouse Level, Suntec Tower 3
8 Temasek Boulevard
Singapore 038988

Email: editorial@panstanford.com
Web: www.panstanford.com

British Library Cataloguing-in-Publication Data
A catalogue record for this book is available from the British Library.

Beyond Quantum

ISBN 978-981-4411-73-8 (Hardcover)
ISBN 978-981-4411-74-5 (eBook)

Printed in the USA

Contents

Preface

The present wave of interest in quantum foundations is caused by the tremendous development of quantum information science and its applications to quantum computing and quantum communication. Nowadays this interest has increased even more because *it has become clear that some of the difficulties encountered while realizing quantum information (especially while creating quantum computers and design of new quantum algorithms) are not simply technicalities, but instead have roots at the very fundamental level.* To solve such difficult problems, quantum theory has to be reconsidered, and some prejudices must be discarded, particularly, the prejudice on completeness of quantum mechanics (QM), which Albert Einstein strongly criticized throughout his life.

This book is dedicated to Einstein's vision of physics and specifically his hope for what quantum theory could and, in his view, should be. Two of Einstein's dreams about the future of quantum theory are particularly realized in this book: reduction of quantum randomness to classical ensemble randomness and the total elimination of particles from QM, that is, the creation of a classical field model of quantum phenomena. Thus, contrary to a number of the so-called no-go arguments and theorems that advanced throughout the history of quantum theory (such as those of von Neumann, Kochen–Specker, and Bell), quantum probabilities and correlations can be described in a classical manner.

There is, however, a crucial proviso. While this book argues that QM can be interpreted as a form of classical statistical mechanics (CSM), this classical statistical theory is not that of particles, but of fields. This means that the mathematical formalism of QM must be translated into the mathematical formalism of CSM on the infinite-dimensional phase space. The infinite dimension of the phase space of this translation is a price of classicality. From the mathematical viewpoint, this price is very high, because in this case the theories of measure, dynamical systems, and distributions are essentially more complicated than in the case of the finite-dimensional phase space found in a CSM of particles. However, at the model level (similar to quantum information theory), one can proceed with the finite-dimensional phase space by approximating physical prequantum fields by vectors with finite number of coordinates. To simplify the presentation and by taking into account that usage of infinite-dimensional analysis in the rigorous mathematical framework is still not common in the quantum community, the basic constructions in finite-dimensional Hilbert spaces are presented. Special sections are devoted to generalization to the case of fields.

From the physical and philosophical viewpoints, by considering QM as a CSM of fields, the basic interpretational problems of QM can be resolved. For example, quantum correlations of entangled systems can particularly be reduced to correlations of classical random fields. From this perspective, quantum entanglement is not mysterious at all, since quantum correlations are no longer different from the classical ones. In fact, the situation is more complicated. Averages and correlations provided by the classical field theory are related to continuous signals, but the quantum ones are based on statistics of discrete clicks of detectors. However, it is possible to discretize continuous signals with the aid of the *threshold-type detectors* and transform probabilistic quantities for continuous signals into statistics of discrete clicks that coincide with quantum ones. The main part of this book is devoted to the representation of quantum averages and correlations with the aid of continuous random signals, although this step is nontrivial both from physical and mathematical viewpoints. The corresponding discretization model is presented in Chapter 9. We call this model the *threshold detection model* (TSD).

The main difficulty is that the classical situation is very tricky by itself. All correlations contain the irreducible contribution of a background field (vacuum fluctuations). Roughly speaking, quantum systems are classical random signals that are measured against a sufficiently strong random background. The data of QM is a result of our ignorance concerning the contribution of this random background. Thus quantum probabilities and correlations are not simply classical quantities. They are obtained from classical quantities by means of a renormalization procedure: a subtraction of the contribution of the background field. Accordingly, the reduction of quantum randomness to classical is not totally straightforward. Nevertheless, it is possible. For example, the otherwise mysterious nature of the Heisenberg's uncertainty principle can be resolved in the following way: Quantum dispersions are not simply classical dispersions, but the results of the subtraction of the dispersion of vacuum fluctuations. There is nothing mysterious in the fact that renormalized quantities satisfy this type of inequality. This is not so unusual from the viewpoint of the classical probability theory.

Max Planck had already emphasized the role of a random background field, a concept that was widely used in stochastic electrodynamics. (In his letter to Einstein he pointed out that spontaneous emission can be easily explained by taking into account the background field.) This field also plays an essential role in our model, CSM of classical fields, which is also termed as *prequantum classical statistical field theory* (PCSFT). This model is purely that of the field type. A classical random field is associated with each type of quantum particles. There are, for example, the electronic, neutronic, and protonic fields. The photonic field is simply the classical electromagnetic field. We speculate that all these classical fields can be unified through a single prequantum field. However, this is still the open problem. This "grand unification" can include the gravitational field. Thus, instead of quantization of gravity, this book advertises "dequantization" of QM as the first step toward unification of QM with gravity.

There is also a deep-going analogy between the present approach and the *classical theory of random signals.* Quantum measurements can be described as measurements for classical random signals with a noisy background. This analogy between QM and classical signal theory has already been explored from the reverse perspective, for example, while using quantum information theory in the theory of classical Gaussian random signals (the model of Ohya–Watanabe). Indeed, restricting the present discussion to Gaussian signals alone would significantly simplify the presentation of PCSFT. However, the present book attempts to proceed by considering arbitrary random signals as much as possible.

I hope that this book will stimulate research that aims to demystify QM and to create a pure field model of quantum reality and to even go beyond QM and find classical wave phenomena behind the basic laws of QM, such as Born's rule for probabilities. In PCSFT, prequantum random fields fluctuate on a time scale that is essentially finer than the time scale of quantum measurements. The fundamental question whether this time scale is physically approachable remains open. In particular, if the prequantum time scale were the Planck scale, the PCSFT-level would be inapproachable. There would be no hope to monitor prequantum waves and show experimentally how the quantum statistics of clicks of detectors is produced through interaction of such waves with threshold-type (macroscopic) detectors. In such a situation, PCSFT can be considered as a metaphysical theory, like the string theory. However, if the prequantum time scale is essentially coarser than the Planck scale, one might dream of finding experimental confirmation of emergence of quantum laws (which are fundamentally probabilistic) from behavior of prequantum (random) waves. Either way, however, PCSFT provides an adequate theoretical model of reduction of QM to CSM of fields.

Andrei Khrennikov
Växjö, Sweden
Winter 2013

Acknowledgments

I discussed the basic ideas of this book with world's leading experts in quantum foundations, both theory and experiment. I am especially thankful to experimenters for discussions about the process of detection and other experimental technicalities which, in my opinion, play a crucial role in transition from microworld to quantum mechanics. (The latter is considered as the operational formalism describing results of measurements.) I would like to thank Alain Aspect, Catalina Curceanu, Francesco De Martini, Marco Genovese, Philippe Grangier, Daniel Greenberger, Paul Kwiat, Alan Migdal, Sergey Polyakov, Helmut Rauch, Stig Stenholm, Rupert Urzin, Gregor Weihs, and Anton Zeilinger.

I also would like to thank Luigi Accardy, Guillaume Adenier, Ana Maria Ceto, Luis de la Pena, Hans de Raedt, Thomas Elze, Gerhard Grössing, Karl Hess, Werner Hofer, Gerard't Hooft, Marian Kupczynski, Kristel Michielsen, Theo Nieuwenhuizen, Arkady Plotnitsky, and Arnt Vistnes for debates on possible ways beyond quantum.

This book was written with the support of grants "Mathematical Modeling and System Collaboration" and "Mathematical Modeling of Complex Hierarchic Systems" of the Faculty of Natural Science and Engineering of Linnaeus University, Sweden. A part of this book was written during my visits to Japan and supported by visiting professorships of Tokyo University of Science. I would like to thank Masanori Ohya and Noboru Watanabe for their hospitality and discussions on usage of quantum entropy in theory of classical Gaussian signals with infinite-dimensional state space.

Chapter 1

Introduction

1.1 The Author's Views on Quantum Foundations

This section contains the author's reminiscences of meetings in Växjö, philosophical and historical remarks on views of Hertz, Boltzmann, Planck, Einstein, Bohr, Schrödinger, Fock, von Neumann, de Broglie, Kolmogorov, von Mises, Lamb and Scully, Lande, Bohm and Hiley, Mackey, Marshall and Brafford, Klyshko, Vorob'jev, Boyer, de la Pena and Cetto, Emch, Cole, Ballentine, Mermin, Gudder, Lahti, Ozawa, Vidman, Peres, Ohya, Accardi, Gill, Larsson, Fuchs, Plotnitsky, Zeilinger, Genovese, Jaeger, Hardy, Zurek, D' Ariano, Aspect, Rauch, Weihs, Volovich, Belavkin, Ozawa, De Muynck, De Baere, Elitzur, Peres, Nieuwenhuizen, Greenberger, Migdal, Polyakov, Hess and Philipp, 't Hooft, Adenier, De Raedt,... This section may be of interest to some readers as it contains details of history of quantum foundations; or the reader can jump directly to Section 1.2, which contains a short introduction to my "beyond quantum model"—*prequantum classical statistical field theory* (PCSFT).

1.1.1 *Debates in Växjö*

To better understand what this book is about, it may be useful to know a little bit about the author's views. In the quantum

Beyond Quantum
Andrei Khrennikov

ISBN 978-981-4411-73-8 (Hardcover), 978-981-4411-74-5 (eBook)
www.panstanford.com

foundations community the author is well known as the organizer of a series of conferences held in Växjö, a small town in the South-East part of Sweden. This town surrounded by woods and lakes is really a good place for contemplations of kinds, in my case, specifically, about quantum theory, one of the most exciting theories ever created. The theory is exciting not only because of its tremendous advances, but also because of its paradoxical claims and conclusions.

The series of Växjö conferences on foundations of quantum mechanics (especially, probabilistic foundations) combines two subseries: *Foundations of Probability and Physics*: 2000, 02, 04, 06, 08, 11 [7, 10, 141, 148, 159]; and *Quantum Theory: Reconsideration of Foundations*: 2001, 03, 05, 07, 09, 12 [8, 12, 144, 156, 200]. A new series *Advances in Quantum Theory* started in 2010. All the conferences have been notable not only for the original contributions, but also for several exciting debates that took place there. These debates offered a great diversity of perspectives on foundations of quantum mechanics (QM) and its future developments: from the orthodox Copenhagen interpretation (which rejects realism and causality) at one end of the spectrum, to more realistic views as advocated by Einstein at the other end.

During the last ten years, I have been lucky to meet the world's leading experts in quantum foundations and discuss with them the most intriguing problems. What surprised me (at least at the first conferences) was the huge diversity of opinions and views on the very fundamental and old problems. My expectation that by inviting great quantum gurus I can get clear answers was naive. The first conference, *Bohmian Mechanics 2000*, was the total fiasco: two leading representatives of Bohmian school, Shelly Goldstein and Basil Hiley, presented two totally different interpretations of Bohmian mechanics. Finally, they accused each other in misunderstanding of Bohm's views (both had very close connections to David Bohm). My students whom I invited to learn Bohmian mechanics from its creators were really confused. The only useful information which I extracted from *Bohmian Mechanics 2000* was that Bohmian mechanics does not give new experimental predictions comparing to conventional QM. Thus, although formally (mathematically) Bohmian mechanics provides a finer description of micro processes, it is impossible to design experiments that will distinguish Bohmian mechanics and QM.

Similar stories have repeated quite a few times with various fundamental problems. Only in some cases was I lucky to learn something. For example, I got the answer to the question: "What is crucial in quantum computing: superposition or entanglement?" I learned that superposition plays a subsidiary role, the crucial is entanglement; the classical wave computer (e.g., optical) cannot beat the classical digital computer. However, yet another simple question has never been clarified: "Do pure states provide better quantum computational resource than mixed?" Opinions of quantum computing gurus did not converge to a common point. And, I can mention a series of similar questions, e.g., "How dangerous for quantum cryptography is detectors inefficiency?" People who spontaneously answered me that the impact of the detectors inefficiency can be easily taken into account a few years later applied for grants to study this "very important problem."

I can mention a series of simple Växjö questions without answers, e.g., "What is electron? What is the origin of discreteness of the electric charge? What is the essence of vacuum fluctuations? Can the mathematical formalism of QM be applied outside of physics, e.g., in cognitive science?"

1.1.1.1 Interpretations of wave function

The most exciting spectacle started each time when the question of *interpretations of the wave function* attracted the attention. Finally, I understood that the number of different interpretations is in the best case equal to the number of participants. If you meet two people who say that they are advocates of, e.g., the Copenhagen interpretation of QM, ask them about the details. You will see immediately that their views on what is the Copenhagen interpretation can differ very much. The same is true for other interpretations. If two scientists tell that they are followers of Albert Einstein's *ensemble interpretation,* ask them about the details. At one of the round tables (after two hours of debates with opinions for and against completeness of QM) we had decided to vote on this problem. Incompleteness advocates have won, but only because a few advocates of completeness voted for incompleteness. The situation is really disappointing: the basic notion of QM has not yet been properly interpreted (after 100 years of exciting, but not very productive, debates).[a] I specifically appreciate the activity of Arkady Plotnitsky, a philosopher studying Bohr's views (see, e.g., Refs. [262–264]). He taught us (participants of Växjö conferences) a lot. First of all, we got to know that the Copenhagen interpretation cannot be rigidly coupled with Bohr's views. On many occasions Niels Bohr emphasized that QM is not about physical processes in microworld, but about our measurements [38]: "Strictly speaking, the mathematical formalism of quantum mechanics and electrodynamics merely offers rules of calculation for the deduction of expectations pertaining to observations obtained under well-defined experimental conditions specified by classical physical concepts." The basic postulate of the Copenhagen interpretation of QM—"the wave function describes completely the state of a quantum system" (i.e., a concrete system, not an ensemble)—cannot be assigned to Bohr. Then we learnt (again from Plotnitsky) that Bohr's views have been crucially changed a few times during his life. Thus, there can be found many different Bohr's interpretations of QM. Bohr was definitely the father of the *operational interpretation* of QM. As was already pointed out, Bohr emphasized that the formalism of QM does not provide the intrinsic description of processes in microworld, it describes only results of measurements. Bohr can also be considered as the father of the so-called *information interpretation* of QM: the QM formalism describes information about microsystems extracted by means of macroscopic measurement devices. Heisenberg (and to some extent Schrödinger) shared this viewpoint. Nowadays the information interpretation of QM became very popular (see, e.g., Refs. [86, 144, 264]). I can mention Anton Zeilinger [300] and Christopher Fuchs [87–89] among the active promoters of this interpretation. I can also mention Mermin's paper [242].

[a]"When I speak with somebody and get to know his interpretation, I understand immediately it is wrong. The main problem is that I do not know whether my own interpretation is right." (Theo Nieuwenhuizen) This is the standard problem of participants of Växjö conferences.

1.1.1.2 Debates on Bell's inequality: Gill against Accardi

One of the most commemorative debates, on *Bell's inequality,* was ignited by Luigi Accardi (see, e.g., Refs. [2, 5] for his views), and Richard Gill (see, e.g., Ref. [92]). They had put 1000 euros for (Luigi) and against (Richard) a possibility to simulate the Einstein–Podolsky–Rosen (EPR) correlations in purely classical local framework, later the sum reached 3000 euro. I, Slava Belavkin, and Inge Helland agreed to be in the jury. (Very soon we realized that it was a wrong decision.) Although the positions of both parties involved in this great debate have not changed much after 10 years of discussions, nor the jury was able to make a well-grounded decision, this debate had an impact in the quantum community (see, e.g., Ref. [147]). The main message of this debate was that Bell's arguments were not so well justified as it was commonly believed. A part of Bell's critique [31] against von Neumann's no-go theorem can be redirected against Bell's own theorem. And, it is not as easy as it was believed to defend the Bell's position.

As an expert in probability, in general, I agree with Luigi Accardi. (But I am not sure that computer simulation can play the role of a crucial argument.) This inequality is a general statistical test to check a possibility to fit the data collected in a few experiments into a single *Kolmogorov probability space*, see monograph [192] for the complete analysis of this problem (see also Refs. [9, 133, 135–137, 139, 143, 147, 227, 235–238]). This is a test of the possibility to use one special model of probability theory, the Kolmogorov's model [218], for the collected data. I recall that already Lobachevsky and Gauss planned experiments to check applicability of the Euclidean model of geometry (see Section 1.4 for more detail). They proposed concrete tests. Bell's inequality is a similar test for Kolmogorov's probability model. We recall that, in fact, Bell's inequality was invented many years ago by Boole [40, 41]: to check that statistical data can be described by a single Boolean algebra (see also Ref. [260] for a detailed analysis of the problem). General statistical tests of Kolmogorovness of data were found by Soviet mathematician Vorob'ev [297]. Where does non-Kolmogorovness come from? It is a separate problem. Two possibilities were mentioned by Bell: violation of locality or (and) violation of realism. But, in principle, non-Kolmogorovness can come from various sources different from those mentioned by Bell, see my monograph [192]. Thus the honest position would be that violation of Bell's inequality can only tell us that the Kolmogorov's model did not pass the test. We cannot derive the definite conclusion on concrete sources of non-Kolmogorovness.

1.1.1.3 Around loopholes: Aspect, Kwiat, Zeilinger, Hess and Philipp, Larsson and Gill, Adenier, Khrennikov and Volovich, Raedt and Michielsen

It is not easy to understand why physicists do not like to consider violation of Bell's inequality as just a sign of non-Kolmogorovness of statistical data (with one exception: the famous Soviet experimenter in quantum optics Klyshko, see Ref. [215]). Theoretical physics is about creation of mathematical models of reality.[a] Elaboration and experimental verification of the statistical test, the Bell–Boole inequality, which shows the boundary of applications of one of mathematical models, Kolmogorov's one, is a great success! If one does not like to proceed in such a way, then he should take other possible sources of non-Kolmogorovness not less seriously than Bell's sources. For example, my graduate student Guillaume Adenier studied the impact of the so-called *unfair sampling* for the Clauser–Horne–Shimony–Holt (CHSH) inequality [11–21],[b] impossibility to guarantee reproducibility of statistical properties of ensembles used in experiments with incompatible pairs of orientations of polarization beam splitters (PBSs).

It is clear that for CHSH inequality unfair sampling implies non-Kolmogorovness and, hence, violation of Bell's inequality. From our viewpoint, unfair sampling is not a less important source of non-Kolmogrovness than, e.g., nonlocality. In turn, there are various sources of unfair sampling. One of the most well known is the inefficiency of detectors. I spoke a few times with Alain Aspect about this problem. He was not even sure that experimenters should concentrate efforts to close this "loophole." For him, this source of

[a]One of the sides of this problem is that physicists often identify mathematical models in use with reality.

[b]I remark that for the Clauser-Horne (CH) inequality and the Eberhard inequality the situation is very different. These inequalities are derived without the fair sampling assumption.

violation of Bell's inequality cannot be considered on the same level of importance as "Bell's sources." Contrary to him, I think that closing of the "efficiency of detectors loophole" is not less important than, e.g., locality loophole; without this it is totally meaningless to try to restrict sources of non-Kolmogorovness to "Bell's sources." From a purely probabilistic viewpoint, it is even more natural to expect that non-Kolmogorovness is induced by the cutoff of ensembles as the result of detectors inefficiency. Therefore, during the last few years I and Adenier advertised the EPR–Bell experiment with *tungsten-based superconducting transition-edge sensors* (W-TESs)—the ultrasensitive microcalorimeters (see, e.g., Ref. [194] for an experimental proposal).

Paul Kwiat tried to design such an experiment in his lab a few years ago. He promised me a talk on such an experiment at Växjö conferences 2006, 07, but, as I understood, his group did not overcome technical problems. It was a risky project. One PhD student spent a few years struggling with experimental devices and without any result. (I guess that for him it was a terrible experience. However, real research in science can be done only in this way.) Nowadays the group of Anton Zeilinger (a few PhDs and postdocs) work hard to overcome the aforementioned technical problems. However (if I understood rightly the explanations of Marco Genovese, in a private communication), from the experimental viewpoint the problem is so complex that its solution may take many years. I remark that in 2013 both aforementioned groups made important steps toward a loophole free Bell test.[a,b] However, the chance that the loophole-free and clean experiment with W-TESs would never be performed is very high, in spite of the tremendous efforts of world's leading experimental groups. This is a good place to remark that the author of this book and Igor Volovich elaborated an analog of Heisenberg's uncertainty principle for the Bell–Bohm experiment: two loopholes, the locality loophole and the detectors efficiency

---

[a] M. Giustina, Al. Mech, S. Ramelow, B. Wittmann, J. Kofler, J. Beyer, A. Lita, B. Calkins, Th-Gerrits, S. Woo Nam, R. Ursin, and A. Zeilinger, Bell violation using entangled photons without the fair-sampling assumption. *Nature* 497, 227–230 (2013).

[b] B. G. Christensen, K. T. McCusker, J. Altepeter, B. Calkins, T. Gerrits, A. Lita, A. Miller, L. K. Shalm, Y. Zhang, S. W. Nam, N. Brunner, C. C. W. Lim, N. Gisin, P. G. Kwiat, Detection-Loophole-Free Test of Quantum Nonlocality, and Applications. *Phys. Rev. Lett.*, 111, 1304–06 (2013).

(more generally unfair sampling) loophole, are complementary and, hence, they would be never closed in a single experiment [147], see also Ref. [192]. Up to today (September 25, 2012), the experimental situation confirms our loopholes-complementarity principle. (We remark that Richard Gill, with whom we discussed our principle via email exchanges, wrote a paper in which he called this principle "Bell's fifth position" [92]). In fact, in Ref. [147] we advertised the viewpoint that loopholes are not just "experimental technicalities," but the fundamental issues of the Bell's test of local realism (cf. Chapter 9).

Surprisingly, some leading experimenters in quantum optics cannot exclude a possibility that if the EPR–Bohm experiment with W-TESs be finally performed it might happen that *no violation of Bell's inequality* would be observed. At the same time they do not believe in hidden variables. This is an interesting psychological phenomenon.

In any event it is clear that this problem, closing the unfair sampling loophole for the Bell's tests with photons, cannot be solved with the aid of threshold-type detectors, photomultiplier tubes (PMTs), avalanche photodiodes (APDs), and visible light counters (VLPCs). Nevertheless, even such detectors give a possibility to detect the presence of unfair sampling,[a] at least as it is described by the most natural models with hidden variables exploring unfair sampling to violate Bell's inequality [13]. This can be done with a test of fair sampling [14]. Unfortunately, we were not able to convince experimenters to do this test (in spite of a number of promises, it has never been done). I also point to the really important contribution of Jan-Ake Larsson [226] to the study of the probabilistic structure of the problem of the detectors efficiency and unfair sampling.

Another source of non-Kolmogorovness is the temporal structure of the EPR–Bohm experiments and, in particular, the use of the *time window*, see Refs. [71, 72, 92, 106–108, 185, 228] for different viewpoints on this problem. Non-Kolmogorovness need not be reduced to ensemble cutoff, as due to inefficiency of detectors or time window. Unfair sampling can take place for 100% efficient detectors and practically zero time window. It appears naturally in

[a]Thus we do not try to close the detectors efficiency loophole. However, even with inefficient detectors we may find the trace of unfair sampling.

models of the EPR–Bell experiment taking into account parameters of measurement devices [186, 192]. In such models randomness induced by preparation procedure is combined with randomness induced by the measurement device.

1.1.1.4 Is it worth trying to go beyond quantum mechanics?

The majority of participants of Växjö conferences believed that QM, as it stands now, will, for a long time to come or even indefinitely, remain a correct, or indeed the correct, theory within its proper scope. On this view, improvements in our experimental technology would not alter the essential features of QM in its present form, although new exciting experimental and theoretical findings, including of foundational nature, are possible. Indeed, even in this group, there was no consensus, not only on whether there is a single correct interpretation of QM but also on whether we really have a proper foundation of the theory, say, of the type we have in relativity theory. Several papers presented in Växjö explored this question and possible new foundational approaches to the standard version of QM.[a]

On the other hand, a number of participants actively promoted alternative models, which might enable us to *go beyond the standard QM.* In particular, some of them considered QM as an emergent theory and envisioned, "dreamed of," the possibility of a reduction of quantum randomness (viewed as irreducible by the orthodox interpretation) to classical randomness, for example, of the type found in classical statistical physics.[b] The author belongs to this (minority) group and considers *QM as an approximation, possibly a very good one, of a more fundamental theory of microscopic reality.*

I emphasize that the reduction of quantum randomness to classical randomness, e.g., consideration of quantum systems as classical random signals (as in this book), does not imply a kind

---

[a] I can mention V. Belavkin, B. Coecke, W. De Muynck, C. Fuchs, R. Schack, M. Appleby, A. Plotnitsky, L. Hardy, M. D'Ariano, A. Elitzur, W. Zurek, P. Busch, D. Greenberger, R. Balian, K. Svozil, J. Smolin, M. Ozawa, A. Peres, D. Mermin, S. Stenholm, J. Summhammer, P. Lahti, I. Volovich.

[b] I can mention G. 't Hoot, T. Elze, C. Garola, M. Davidson, T. Boyer, T. Nieuwenhuizen, S. Gudder, G. Emch, B. Hiley, C. Wetterich, D. Cole, L. de la Pena, G. Adenier, H. D. Doebner, A. F. Kracklauer, Ch. Roychoudhuri, G. Groessing.

of comeback to Laplacian determinism. Consider the standard Brownian motion. Suppose, we are not able to take into account the effect of the random background for an individual particle. In such a situation we are not able to predict the trajectory of this particle, even if the initial conditions are determined with high precision. Nevertheless, the impossibility to predict the trajectory is not interpreted as the absence of the trajectory. (In Ref. [19] it was shown that classical Brownian motion exhibits even such a "fundamentally quantum property" as entanglement.)

More generally, it may be argued that strategically there are two main ways *to go beyond QM.* The first is to stimulate the development of the conventional QM, especially by designing new experiments, in a hope that, sooner or later, QM will reach the limit of its validity, even within its proper experimental scope (i.e., as a nonrelativistic theory of quantum phenomena), "peacefully," as it were, as the result of its own internal development. The second approach is to pursue a critique of the conventional approach in order to find its weak points, which would require an alternative theory for handling properly.

I support both approaches; and, as I said, I believe, with Einstein, that the standard QM will ultimately prove to be an *approximation of a more fundamental theory,* perhaps based on a prequantum mathematical model of the type described above. At the moment, however, I do not think that any available model of this type provides a viable possibility in this regard. Mathematics offers great opportunities to explore, to "play with," various prequantum models. However, such mathematical games are not the same as real physical models, which must rigorously relate their mathematics to experimentally observed phenomena. The main problem of prequantum approaches, including my own, is the lack of proposals for realistic experiments that could demonstrate that the QM formalism describes micro-reality only by a way of approximations[a] that can be superseded by a better theory. Nevertheless, these approaches offer a possible new trajectory for future development of quantum theory, and the main aim of Växjö conferences was to explore such trajectories.

---

[a] My classical field-type model predicts that even the basic rule of QM, Born's rule, is an approximate rule.

Overall, Växjö conferences have played an important role in the ongoing investigation of quantum foundations and possibilities of new discoveries in QM and beyond it,[a] for example, in quantum field theory (QFT). Several additional aspects and high points of the conferences are worth mentioning here. The *interaction of a large number of experimenters* at the conferences (e.g., A. Aspect, H. Rauch, G. Weihs, M. Genovese, C. Roychoudhuri, M. Zukowski, F. De Martini, A. Zeilinger, F. Sciarrino, B. C. Hiesmayr, C. Roos, A. Migdal, S. Polyakov, Yu. Hasegawa) will, hopefully, facilitate the aim of designing new foundational experiments, especially other tests of possible violations of Bell's inequalities, the approach that dominated this field for a long time. The discussions and debates on the foundations of quantum information theory (especially quantum cryptography and computing) that took place during the conference will, undoubtedly, contribute to further developments in this important field of research. Indeed, quantum information theory can be considered as a great experiment to test the validity of the main principles of quantum mechanics and its interpretation. Finally, Växjö conferences offered significant new insights into the question of quantum probability, which remains essential to quantum foundations, whichever way one pursues them.

1.1.2 *The Role of Probability*

Main message: *Probability in QM should be taken seriously; this is a tricky notion; even "classical probability" can induce rather counterintuitive (paradoxical) conclusions. Therefore a rigorous mathematical presentation of probabilistic statements of QM is very important.*

In fact, *quantum probability* or, to be more precise, the difference between quantum and classical probabilistic models was the starting point of my interest in quantum foundations. As I graduated from the Department of Mechanics and Mathematics of Moscow State University I was well-trained in probability theory; on a few occasions I was lucky to meet the founder of the

[a] See, e.g., Refs. [1–6, 17, 18, 20, 21, 23, 24, 28–30, 34, 47, 52–60, 62, 67–69, 86–89, 92, 99, 103, 105–111, 119, 146, 147, 152, 176, 214, 226, 227, 231, 242, 249, 259, 262, 271, 283, 286, 292, 298].

modern probability theory Andrei Nikolaevich Kolmogorov [218]. In particular, he communicated to Doklady USSR my paper [280] (written in collaboration with Oleg Georgievich Smolyanov) devoted to a generalized model with complex valued probabilities: an analog of the central limit theorem and the law of large numbers and so on. This was an example of non-Kolmogorovian probability model.

My PhD thesis [127] was about *probability on infinite-dimensional spaces* (including distributions on Hilbert spaces) with applications to mathematical problems of QFT [128–131]. This pathway to quantum theory, namely, through theory of continual integral, secured me for a long time from terrible problems and prejudices related to quantum foundations which everybody meet immediately in standard textbooks. The continual integral approach, especially its Euclidean version, induced an illusion that quantum theory, at least QFT, is about integration on infinite-dimensional spaces. For quite long time I was completely sure that the infinite number of the state space dimensions is the main point of departure of QM from classical mechanics.

The main surprise for me in von Neumann's book [296] was the notion of *irreducible quantum randomness*. Von Neumann sharply distinguished classical and quantum randomness. The first one, which is exhibited everywhere, besides QM (in classical statistical mechanics, economics, finances, biology, engineering), is a consequence of the impossibility to take into account all parameters describing a system and its interaction with a measurement device (or environment). Although it is difficult, and sometimes even really impossible, to specify the values of these parameters (so to say "hidden variables"), there are no doubts of their existence. For example, in statistical mechanics the Liouville equation describes dynamics of the probability distribution on a phase space. It is very difficult to solve the corresponding system of Hamiltonian equations describing trajectories of millions of individual particles. Nevertheless, there are no doubts that these particles really move in physical space. The easiest way to describe diffusion (including the Brownian motion) is to solve the Fokker–Planck (and, in general, the forward Kolmogorov) equation for the probability distribution on configuration space. Still, it is also possible to solve the corresponding stochastic differential equation and obtain the description of dynamics by means of trajectories of individual particles, the classical stochastic process.

In contrast, it has been claimed that quantum randomness cannot be reduced to our lack of knowledge of "hidden variables." It became rather fashionable to claim that QM has demonstrated the violation of laws of classical probability theory, see, e.g., Richard Feynman [85], p. 2: "But far more fundamental was the discovery that in nature the laws of combining probabilities are *not* those of the classical probability theory of Laplace."

Such a viewpoint supports the illusion that something mystical goes on in QM, since the probability laws which have been valid everywhere are violated in the microworld. In my first studies I clarified this problem.[a] One cannot speak about probability without specifying a mathematical model of probability. In particular, the notion of "classical probability" is related to a variety of mathematical models: Kolmogorov's measure-theoretic, von Misses frequency, subjective probability, and so on [135, 293–295]. Quantum probability is nothing else than a probability described by Dirac–von Neumann model [74, 296]: a complex Hilbert space; wave functions and, more generally, density matrices as states; self-adjoin operators as observables; Born's rule. Therefore, when one speaks about matching or mismatching of classical and quantum probability, the classical model and *matching rules should be specified.*

At the first stage of my exciting journey into the probabilistic foundations of QM, I found that there is no contradiction between the QM probabilistic model and von Misses frequency probability theory [135, 137, 140, 142, 143]. And this is not surprising. The von Mises approach [135, 293–295] is a very general empiric approach providing the probabilistic model for statistical data collected in experiments. In fact, von Mises (in 1919) elaborated the probabilistic model based on the empiricist ideology: statistics of outcomes of experiments is described by frequencies of realizations. Thus, *the classical probabilistic model of R. von Mises does not contradict QM.* This was already well known to von Neumann, who proposed to consider von Mises theory as the probabilistic ground

---

[a] See Refs. [133, 135, 137, 139, 140, 142, 143, 149, 150, 152–155, 160, 163, 164, 192, 293–295].

of QM.[a] The approach of von Mises is especially useful to describe statistical data collected in quantum experiments. It reflects the temporal structure of an experimental data stream. Also, a typical quantum experiment has the well-defined temporal structure. This is a run of preparations and corresponding detections.

The main difference between von Mises' and von Neumann's views was that von Mises' did not claim that additional variables responsible for randomness could not be introduced.

Most of this book is based on the Kolmogorov's model of probability. We show that continuous prequantum random fields (Kolmogorovian random variables which are in general valued in infinite-dimensional space) can reproduce all quantum averages and correlations. However, to proceed to measurements describing individual events, clicks of detectors, the frequency interpretation of probability will be used. In Chapter 9 we shall derive relative frequencies predicted by QM for experiments with microsystems. Hence, the complete prequantum model is based on the hybrid of the Kolmogorov (measure-theoretic) and von Mises (frequency) approaches. The former describes "reality as it is" and the latter describes results of measurements.

1.1.3 *To Hilbert Space from Probability*

Main message: *The Hilbert space model provides a linear space representation of probabilistic data.*

In his famous book [233] Mackey started in purely empiricist manner, i.e., with probabilities collected in various (in general incompatible) experiments, and tried to find conditions inducing representation of data by complex probability amplitudes, or in the abstract framework by normalized vectors in the complex

---

[a] "However, the investigation of the physical quantities related to a single object S is not the only thing which can be done—especially if doubts exist relative to the simultaneous metastability of several quantities. In such cases it is also possible to observe great statistical ensembles which consist of many systems $S_1, \ldots, S_N$ (i.e., N models of S, N large). (Such ensembles, called collectives, are in general necessary for establishing probability theory as the theory of frequencies. They were introduced by R. von Mises, who discovered their meaning for probability theory, and who built up a complete theory on this foundation.)" See Ref. [296], p. 298.

Hilbert space.[a] Unfortunately, he did not succeed completely. Finally, the complex Hilbert space representation was postulated and this decreased the value of Mackey's studies. I spent a few years by working to complete Mackey's program and I developed the general contextual approach based on families of probabilities obtained by measuring of various observables for various contexts. I found an algorithm representing probabilities (under special, but natural conditions) by complex amplitudes—*quantum-like representation algorithm* (QLRA); probabilities and amplitudes are coupled by Born's rule, i.e., the squared amplitude coincides with the probability. Thus, the complex Hilbert space was not postulated (as was done by everybody, from Dirac [74] and von Neumann [296] to Mackey [233]), but appeared in the constructive way [192].[b]

1.1.4 *Against Completeness*

Main message: *In the past, various models of reality have been claimed to be final (complete). The most known examples are Euclidean geometry (see, especially Kant* [123]*) and Newtonian mechanics. However, sooner or later such scientific myths died. QM is the latest myth of a complete theory.*

My next action toward demystification of the QM formalism was against Bohr's thesis on *completeness of QM.* I remark that von Neumann's statement of irreducibility of quantum randomness is nothing else than the probabilistic performance of Bohr's statement of *completeness of QM.* Bohr's statement provides the finest possible description of micro phenomena; a finer description of the state of a quantum system than that given by the wave function is totally

[a] I called this problem the "inverse Born problem" [177, 192, 251]. Born's rule obtains probability from the wave function. We are interested in production of a complex amplitude from probabilistic data. This amplitude has to satisfy to the "direct Born's rule."

[b] Moreover, it was found that the formalism of QM is too restrictive for a general empiricist model: not all probabilities are represented by complex amplitudes. Some of them are represented by the so-called hyperbolic amplitudes [151, 154, 161, 192, 206] (taking values of the form $z = x + jy, x, y \in \mathbf{R}, j^2 = +1$) or mixed hyper-complex amplitudes. Thus the complex Hilbert space is too restrictive for the linear representation of the general empiricist (contextual) model; the hypercomplex Hilbert space provides the proper base.

impossible. To justify completeness of QM, Bohr need not any "no-go theorem" (such as von Neumann's, Kochen–Specker's, or Bell's theorems [31, 217, 296]). He was completely fine with Heisenberg's uncertainty relation.

Einstein had never accepted Bohr's thesis on completeness of QM. All his life he dreamed of creation of a new fundamental theory of micro phenomena. He was sure that the wave function does not provide the complete description of the state of an individual quantum system. Einstein was the father of the *ensemble interpretation* of the wave function as describing statistical properties of an ensemble of systems created by some preparation procedure.[a] Part of the quantum community borrowed Einstein's ensemble interpretation of the wave function and combined it peacefully with Bohr's thesis of completeness of QM. They support the *operational or empiricist interpretation* of QM [46, 59, 61, 115, 116, 216, 232, 255, 256] and consider QM formalism as a story on preparation and measurement procedures. At this stage the positions of Bohr and Einstein coincide. However, Einstein did not see any fundamental barrier to complete QM (to know what goes on behind rough macroscopic preparation and measurement procedures. Majority of people using the operational interpretation do not support Einstein's views. In contrast to Einstein, they do not dream of new, more sensitive preparations and measurements which will show that QM provides only an approximate description of phenomena. But even this interpretation is better than the orthodox Copenhagen interpretation, according to which the wave function describes completely the state of a microsystem.

The first dream of Albert Einstein was *to reduce quantum randomness to classical randomness by creating a finer description of microsystem's state than given by the wave function.* It was the dream of creation of a kind of classical statistical mechanics for microsystems. Einstein did not specify prequantum, so to say hidden, variables. They need not be the canonical pair of

[a]This interpretation was later elaborated by Leslie Ballentine [25, 26], who used the term *statistical interpretation.* Unfortunately, this terminology is rather misleading, since it had been used by von Neumann, too: the wave function, although assigned to the state of an individual system, expresses statistics of measurements (but this statistics is coupled to irreducible randomness).

(q, p), position and momentum. Classical prequantum statistical mechanics has to be based on some sort of deterministic dynamics. For example, one may expect to find a micro-analog of the Liouville equation and the underlying Hamilton equations. We emphasize that Einstein, whose contribution to theory of Brownian motion is well known, did not dream of the comeback to the Laplacian determinism. (An essential part of book [79] was devoted to critique of mechanical determinism.) The prequantum deterministic dynamics is definitely influenced by classical randomness of the initial conditions and random background. (By discussing the views of Einstein one has to be very careful, since Einstein's views changed drastically during his life, see Section 1.1.10.)

1.1.5 *'t Hooft's Dream of Prequantum Determinism*

Gerard 't Hooft (who got the Nobel Prize for his contribution to QFT) plays an important role in the quantum foundational community. The fact that such a big name in conventional quantum theory questioned all basic interpretation issues of QM is very supporting for researchers going beyond QM. In a few words, his viewpoint on the interrelation of classical and quantum physics can be represented in the following way, see Refs. [287–289] for details. (The presentation of 't Hooft's *deterministic subquantum mechanics* is based on my personal understanding of his ideas. I could not guarantee that he would agree with such a picture. In fact, reading the works of 't Hooft reminds me reading the works of Bohr. For me, it took 15 years to start to understand Bohr...)

At the very deep level (the Planck's scale of space and time) *the world is totally deterministic.* State's evolution can be described by an ordinary differential equation. In particular, it can be the Newton's equation. And, quantum uncertainty appears as the result of identifying a huge cluster of subquantum states into one quantum state.

't Hooft claims that the comeback to Laplacian determinism is possible, but at the Plank scale. And this position is very rigid: there is no need even in classical randomness of, e.g., the Brownian motion type (a private discussion between the author and G."t Hooft). This position is close to the position of "early

Einstein" (see Section 1.1.10). Although "late Einstein" [79] criticized the Laplacian determinism of *particle mechanics*, it seems that "early Einstein" dreamed of a purely deterministic particle model completing QM. Hence, one may say that 't Hooft's deterministic subquantum mechanics is the dream of "early Einstein." Later Einstein rejected Laplacian determinism of particles in favor of the purely field model (Section 1.1.10). However, it seems that 't Hooft would also support such an viewpoint to prequantum reality. In a private discussion with the author he accepted that subquantum states can be considered as the states of fields rather than of particles, but at the same time he was not much in favor of *classical fields*. His subquantum fields are *ontological beables.* They are used for construction of the operator representation via the procedure of canonical quantization. However, subquantum Hamiltonian functions are of the very special form: they are linear with respect to beable momentum. (Hence, 't Hooft does not quantize the classical Hamiltonian function.) 't Hooft's position is not so straightforward as the position of Einstein and Infeld [79] and from my viewpoint it is based on the mixture of ideas and notions of classical Newtonian mechanics and QFT. Although in general 't Hooft's subquantum mechanics is totally deterministic,[a] to explain the violation of Bell's inequality, vacuum fluctuations have to be involved. The latter are understood as fluctuations of space–time itself. This is an interesting question whether the total determinism on fluctuating space–time can indeed be treated as determinism.

1.1.6 *Einstein's Dream of the Pure Field Model*

Main message: *Particles are illusions and fields are reality.*

It is less known that Einsteinian's "beyond quantum world" was not imagined as a microcopy of our macroscopic world populated by particles. Einstein's greatest dream was to *eliminate particles totally from coming fundamental theory.* New theory should be a purely field model of reality: no particles, but only fields (or may be just one field). He considered particles as the relict of the old mechanistic model of reality. It may be not so well known, but for Einstein QM

[a]In particular, there is even no free will of an experimenter changing experimental setting.

was not a theory too novel (so that he even could not understand it, as some people claim), but, in contrast, it was too old fashioned to be considered a new fundamental theory.

In Ref. [79] he discussed a lot on Bohr's principle of complementarity, so to say, wave–particle duality. He was not happy with the quantum jargon of mixing waves and particles. Einstein was sure that the wave–particle duality will be finally resolved in favor of a purely wave model:

> But the division into matter and field is, after the recognition of the equivalence of mass and energy, something artificial and not clearly defined. Could we not reject the concept of matter and build a pure field physics? What impresses our senses as matter is really a great concentration of energy into a comparatively small space. We could regard matter as the regions in space where the field is extremely strong. In this way a new philosophical background could be created. Its final aim would be the explanation of all events in nature by structure laws valid always and everywhere. A thrown stone is, from this point of view, a changing field, where the states of greatest field intensity travel through space with the velocity of the stone. There would be no place, in our new physics, for both field and matter, field being the only reality. This new view is suggested by the great achievements of field physics, by our success in expressing the laws of electricity, magnetism, gravitation in the form of structure laws, and finally by the equivalence of mass and energy. Our ultimate problem would be to modify our field laws in such a way that they would not brake down for regions in which the energy is enormously concentrated. But we have no so far succeeded in fulfilling this program convincingly and consistently. The decision, as to whether it is possible to carry it out, belongs to the future. At present we must still assume in all our actual theoretical constructions two realities: field and matter. (See Ref. [79], pp. 242–243.)

Then they discussed QM and a possibility to interpret the wave function, the probability wave, as a physical field:

> For one elementary particle, electron or photon, we have probability waves in a three-dimensional continuum, characterizing the statistical behavior of the system if the experiments are often repeated. But what about the case of not one but two interacting particles, for instance, two electrons, electron and

> photon, or electron and nucleus? We cannot treat them separately and describe each of them through a probability wave in three dimensions, just because of their mutual interaction. Indeed, it is not very difficult to guess how to describe in quantum physics a system composed of two interacting particles. We have to descend one floor, to return for a moment to classical physics. The position of two material points in space, at any moment, is characterized by six numbers, three for each of the points. All possible positions of the two material points form a six-dimensional continuum and not a three-dimensional one as in the case of one point. If we now again ascend one floor, to quantum physics, we shall have probability waves in a six-dimensional continuum and not in a three-dimensional continuum as in the case of one particle. Similarly, for three, four, and more particles the probability waves will be functions in a continuum of nine, twelve, and more dimensions. This shows clearly that the probability waves are more abstract than the electromagnetic and gravitational field existing and spreading in our three-dimensional space. (See Ref. [79], pp. 290, 291; cf. the recent work of Hofer [113].)

This discussion is very important for our further studies of a possibility of creation of a purely wave picture of physical reality. It was directly emphasized that one of the main problem is the impossibility to realize quantum "waves of probability" for composite quantum systems on physical space. Recently this problem was solved by the author of this book, see Refs. [189, 190, 197–199, 202–205], and the solution will be presented in Section 3.2. Finally, Einstein and Infeld concluded [79], p. 293:

> But there is also no doubt that quantum physics must still be based on the two concepts: matter and field. It is, in this sense, a dualistic theory and does not bring our old problem of reducing everything to the field concept even one step nearer realization.

As we have seen, Einstein's picture of electron or neutron is very simple: these are fields densely concentrated in small areas of space. These are classical fields (not quantum!). Hence, Einstein was sure that the classical space–time picture of reality could be combined with QM. (We state again that classical means classical field theory and not classical mechanics of particles).

1.1.7 *Antiphoton*

Main message: *Photon is a pulse of classical electromagnetic field. Discreteness is an illusion produced by detectors.*

Now we discuss the notion of photon. It is well known that Albert Einstein invented the notion of the *quantum of light,* which was later called photon.[a]

Bohr was not happy with the invention of light quanta. In particular, the *Bohr–Kramers–Slater theory* [39] was an attempt to describe the interaction of matter and electromagnetic radiation without using the notion of photon. We also mention the strong opposition to the notion of photon from two pioneers of QM: Alfred Lande [223, 224] (in particular, this name is associated with Lande g-factor and the first explanation for the anomalous Zeeman effect) and Willis E. Lamb [222] (e.g., Lamb shift). Their views on electromagnetism differ crucially from the view of Albert Einstein (at least, young Einstein, see Section 1.1.9 for the evolution of the Einstein views). The latter written in 1910 contained the following:

> What we understand by the theory of "light quanta" may be formulated in the following fashion: a radiation of frequency ν can be emitted or absorbed only in a well-defined quantum of magnitude $h\nu$. The theoreticians have not yet even come to an agreement in regard to the following question: Can the light quanta be accounted for entirely by a characteristic of the emitting or absorbing substance, or should the electromagnetic radiation itself be assigned, besides a wave structure, such that the energy of the radiation itself is already divided in definite quanta? I believe that I have proven that this latter view should be adopted. (Ref. [78], p. 207.)

Both Lande and Lamb rejected the existence of discrete quanta of electromagnetic field. They were sure that one can proceed in the so-

[a]Originally the concept of photon was invented by physical chemist G. N. Lewis, who really considered photons as light particles that transmit radiation from one atom to another. Wave-like properties of photon were attributed to guiding ghost field. See Lamb's "Anti-photon" [222], pp. 201–211, for more details. We underscore the difference, the photon terminology, unlike the quantum of light terminology, is not so innocent as one may think. By calling the quantum of light the photon, people emphasized the role of a particle picture of light.

called *semiclassical approach*, describing the interaction of classical electromagnetic field with quantum matter, see, e.g., Refs. [230, 234, 275] and recently Refs. [219, 220, 268, 269]. We cite Lamb:

> It is high time to give up the use of the word "photon," and of a bad concept which will shortly be a century old. Radiation does not consist of particles... ([222], p. 211.)

For adherents of the semiclassical approach quantization of the electromagnetic field is done by detectors; it is not present in electromagnetic field propagating in the vacuum.[a] We also recall that Max Planck opposed Einstein's idea of quantum of light from the beginning, and remained a champion of the unquantized Maxwell field throughout his life. In 1907 in a letter to Einstein, he wrote:

> I am not seeking the meaning of the quantum of action (light quantum) in the vacuum but rather in places where emission and absorption occur, and I assume that what happens in the vacuum is rigorously described by Maxwell equations. (See, e.g., Ref. [247]).

We also mention *stochastic electrodynamics* (SED)—a variant of classical electrodynamics which postulates the existence of a classical Lorentz-invariant radiation field (zero point field, Marshall and Brafford, Boyer, de la Pena, Cetto, and Coli, see, e.g., Refs. [43, 53, 63, 67], see also Refs. [44, 249]). The presence of this field plays the crucial role in SEDs description of quantum effects. We recall that already in 1911 Planck introduced the hypothesis of the *zero point electromagnetic field* in an effort to avoid Einstein's ideas about discontinuity in the emission and absorption processes.

It is important for our further considerations that neither the semiclassical approach nor SED resolve the wave–particle dualism. Neither was it resolved by Bohmian mechanics, the modern version of de Broglie's double solution approach. Bohmian mechanics reduces the quantum randomness to classical ensemble randomness, and particles are the basic objects of this theory.

[a]The semiclassical approach can describe a number of quantum effects, e.g., the photoelectric effect (G. Wentzel and G. Beck, 1926; see Ref. [221] for more detailed calculations).

1.1.8 *Schrödinger's Wave Mechanics*

Main message: *Wave mechanics is an alternative to the Laplacian deterministic model of particles' motion.*

By now, you might be wondering that Schrödinger's name has not yet been mentioned in the discussion of classical and quantum wave mechanics. I refrained from it till this chapter to have enough place to consider not only Schrödinger's wave mechanics, but also his philosophic doctrine, which played an important role in my own theory.

In the beginning, Schrödinger considered the squared wave function of the electron (multiplied by its electric charge e) as the density of its charge:

$$p(t, x) = -e|\psi(t, x)|^2.$$

The solution $\psi(t, x)$ of Schrödinger equation for, e.g., hydrogen atom describes oscillations of such electronic cloud, which induce electromagnetic radiation with frequencies and intensities matching the experiment.[a] This picture of a quantum particle as a field, in this case electronic field, coincides with the picture from Einstein's field dream. Unfortunately, Schrödinger was not able to proceed in this way. He understood, as well as Einstein, that already for two electrons the wave function cannot be interpreted as a field on a physical space: $\psi(t, x, y)(x = (x_1, x_2, x_3), y = (y_1, y_2, y_3))$ is defined on $\mathbf{R}^6$. Although formally Schrödinger gave up and accepted Born's interpretation of the wave function, he did not like the Copenhagen interpretation, as Einstein did neither, especially, Bohr's thesis on completeness of QM. Schrödinger dreamed to go beyond QM, and to refined purely wave resolution of wave–particle duality. Nor, however, did Schrödinger accept Einstein's ensemble interpretation of the wave function. No wonder why (!) Schrödinger would rather have a wave associated with an individual quantum system, and the wave function was the best candidate for such wave. Therefore, he rejected Einstein's idea to associate the wave function with an ensemble of quantum systems, see their correspondence in Ref. [84].

[a] Unfortunately, I was not able to find in Schrödinger's papers any explanation of the impossibility to divide this cloud into a few smaller clouds, i.e., no attempt to explain the fundamental discreteness of the electric charge.

As I mentioned, Einstein did not want the comeback to the Laplacian determinism, and Schrödinger did neither: Schrödinger's views on scientific description of physical reality were based on a well-elaborated approach, the so-called *Bild-conception tradition*, see Ref. [57], p. 351, for details:

Schrödinger called "the classical ideal of uninterrupted continuous description," at both observables and theoretical levels, an "old way," meaning, of course, that this ideal is no longer attainable. He acknowledged that this problem was at the center of the scientific debate in the nineteenth and twentieth centuries as well:

> Very similar declarations... (were) made again and again by competent physicists a long time ago, all through the Nineteenth Century and the early days of our century... they were aware that the desire for having a clear picture necessarily led one to encumber it with unwarranted details [35], p. 24.

I would like now to cite a rather long passage from D'Agostino [57], pp. 351–352, presenting philosophic views of Schrödinger on two levels of description of reality: observational (empiricist) and theoretical.

> The competent physicists are almost certainly Hertz, Boltzmann and their followers. One can thus argue that Schrödinger's two-level conception above is, at bottom and despite its "amazing" appearance, part of the tradition of the nineteenth-century Bild-conception of physics, formulated by Hertz in his 1894 Prinzipien der Mechanik, and also discussed by Boltzmann, Einstein et altri. He partially inherited this tradition from his teacher Exner and he deepened this conception through his intense study of Boltzmann's work. One of the main features of the above tradition is its strong anti-inductionism. If theory is not observation-depended—in the sense that it is not constructed on (or starting from) observations—it consequently possesses a sort of distinction as regards observations. This distinction may be pushed to various degrees of independence. Hertz implied that a term-to-term correspondence between concepts and observables was not needed when he introduced hidden quantities among the theory's visible ones. In his often quoted dictum, Boltzmann

asserted that only one half of our experience is ever experience. At bottom, Schrödinger was thus orthodox in his assertions that theory and observations are not necessarily related in a term-to-term correspondence and a certain degree of independence exists between them. However, when he added the further qualification that a repugnancy might exist between them, he stretched this independence to its extreme consequences, introducing a quasidichotomy between a pure theory and an observational language.

This extreme position was not acceptable to the majority of his contemporaries and to Einstein in particular. Causal gaps, even if limited to the observables level, could not be accepted by Einstein and other scientists. In fact, Einstein's completeness implied a *bi-univocal* correspondence between concepts and observables. It followed from Einstein's premises that, if Schrödinger's wave function did not correspond to a complete description of the system, the reason was to be sought in its statistical (in Einstein's sense!) features: i.e., Schrödinger's wave function refers to an ensemble not to an individual system. Differently, Schrödinger thought that incompleteness in description was generated by an illegitimate (due to indistinguishability) individualization of classical or quasi-classical particles in microphysics. On the other hand, Schrödinger could not accept Heisenberg's and Bohr's Copenhagenism, because, for him, their position represented a concession to an old conception of the theory-observations relation, implying that causality-gaps and discontinuities on the observation-level would forbid the construction of a complete theory (a complete model). One can thus argue that Schrödinger considered the fundamental defect of the Copenhagen view its missing the distinction between the two levels of language, the descriptive and the purely theoretical level. From the QM impossibility of a continuous descriptive language on the observable level, the Copenhagenists would have rushed to conclude the uselessness of a continuous purely theoretical language.

In this book I present a theoretical (causal and continuous) model of physical reality—PCSFT.[a] Since my starting point was

[a] In principle, causality is approachable in the PCSFT framework, but the situation is quite complicated, because of the presence of vacuum fluctuations; we shall come back to this problem in Section 1.2.7.

not the observations, my model does not rely completely on the descriptive language of QM. This is in total accordance with views of Boltzmann, Hertz, Exner, and Schrödinger on relation between theoretical and observational models. The correspondence between concepts of PCSFT and QM is not straightforward, see Chapter 9 for coupling of PCSFT and QM through a measurement theory for PCSFT.

Let us return to the views of Schrödinger on QM and physical reality. I cite from Lockwood [229], pp. 385–386:

> Two possibilities then present themselves. One possibility (a) is that individual physical systems do, after all, possess determinate states in essentially the classical sense. That is to say, the classical dynamical variables do have well-defined values at every moment, arbitrary precise simultaneous knowledge of which is, however, in principle unattainable. Consequently, we have to fall back on statistical statements. The assertions of quantum mechanics should accordingly be understood to refer, as in statistical mechanics, to the distribution of values of these variables within an ideal ensemble of similarly prepared systems. Schrödinger assumed this to be Einstein's position. The other possibility (b) is that the quantum-mechanical description, as embodied in the ψ-function, is a complete specification of an objectively "fuzzy" state. On this conception, quantum mechanics does offer a model of reality; but the model it presents us with is of an objectively "blurred" reality. The difference between these two interpretations, Schrödinger regards as analogous to the difference between an out-of-focus photograph of something with perfectly sharp outlines, and a properly focused photograph of something lacking sharp outlines, such as a patch of fog. Having set up these alternatives, Schrödinger then, disconcertingly, proceeds to argue that neither is tenable.

In fact, the viewpoint to QM generated by PCSFT in combination with the corresponding threshold detection (TSD) model matches neither with (a) the classical statistical mechanical viewpoint to QM, nor with (b) the completeness viewpoint to QM. The aforementioned two-level description of reality based on the combination of theoretical and observational models is sufficiently close to the one given by PCSFT/TSD. However, opposite to Schrödinger and other adherents of the Bild concept, I think that all basic features of observational model have to be derivable from the theoretical model. This is a good place to point to a general scientific methodology which was advertised during many years by Atmanspacher and Primas [23]. Any scientific theory is based on two levels of description of reality: ontic (reality as it is) and epistemic (the image of reality obtained with the aid of a special class of observables). The QM formalism is an example of an epistemic model. In this framework PCSFT can be considered as the ontic model and TSD as the epistemic model which is equivalent to QM. However, I am not fine with the notion of the ontic model as a model of objective reality, i.e., reality existing independently of our observational abilities. I rather prefer a two-level description with two collections of variables, so to say fine and coarse variables. The coarse variables are already approachable and the fine ones are yet not, but in future they will be. The coarse variables are determined by the fine ones. However, in PCSFT/TSD approach this realization is very tricky.

1.1.9 *Bohr–Kramers–Slater Theory*

Bohr, Kramers, and Slater (BKS) tried to treat the interaction of matter and electromagnetic radiation without photons [39]. According to their model atoms produce a virtual field (induced by virtual oscillators), which induces the emission and absorption processes. This virtual field contains the contributions of all atoms and hence each transition in a single atom is determined by processes in all atoms nearby. The BKS theory can be coupled with PCSFT. In the latter, any "quantum particle" is represented by a classical random field. In particular, any atom is nothing else than an atomic field. A group of atoms induces a collective atomic field. Therefore we might try to interpret the virtual BKS field as the real atomic field of PCSFT. Any transition in atom (by the QM terminology from one level to another) is a completely causal process of evolution of this field. Fields of various types of "quantum particles" can interact with each other or better to say there is a single fundamental prequantum field which have various configurations: photonic (electromagnetic), electronic, atomic, ... In PCSFT we stress a role of the background field—vacuum fluctuations. This field is present even in the absence of "quantum particles." The presence of the background field may solve one of the main problems of the BKS theory, namely, possible violation of the law of conservation of energy on the individual level: the impossibility to account for conservation of energy in a process of de-excitation of an atom followed by excitation of a neighboring one. In PCSFT the energy of the fundamental prequantum field is not changed in this process.

The BKS theory was an attempt to unify wave and particle pictures on the basis of the classical field theory. This was an attempt of *causal continuous description* of quantum jumps in the processes of absorption and emission. We remark that Bohr elaborated his principle of complementarity only because he was not able to construct a satisfactory causal field-type model. Later he advertised completeness of QM [37, 38]. Roughly speaking he tried to stop studies similar to his own in 1924th. (The Freudian background of such behavior is evident.)

1.1.10 *On the Evolution of Einstein's Views: From Classical Electrodynamics to Photon and Back*

Einstein's views, as presented in Sections 1.1.6 and 1.1.7, appear to be in conflict. On the one hand, as discussed in Section 1.1.6, he was the discoverer of the particle of light, the photon, as it eventually became known, and thus advocated the particle-like model of the behavior of light in certain circumstances, a view confirmed by the Compton scattering experiment in 1923, shortly before the discovery of quantum mechanics. On the other hand, as discussed in Section 1.1.7, he championed the classical-like field model as the best, if not the only model, for fundamental physics, which, given the continuous character of the classical field theory, is difficult to reconcile with the concept of photon. This discrepancy leads one to suspect that these two positions reflect the views of two different "Einsteins," especially given that they correspond to two different periods of Einstein's work: the first, roughly between 1905 and 1920, and second, from roughly 1920 till his death in 1955. His book with Infeld, discussed in Section 1.1.7, was originally published in 1938 and, thus, it might be added, was written shortly after the EPR argument, which solidified Einstein's critical assessment of quantum mechanics. The evolution of Einstein's views is instructive and one might sketch this evolution roughly as follows.[a]

It may be argued that Einstein's primary model for doing fundamental physics had always been Maxwell's electrodynamics as a field theory, which was the ground for special relativity, introduced in 1905, the same year he introduced the idea of photon. It also appears, however, that his thinking at the time was more flexible regarding what type of physical theory one should or should not use. His approach was determined more by the nature of the experimental phenomena with which he was concerned, or in his own later words, his attitude was more "opportunistic" [272], p. 684, rather than guided by a given set of philosophical preferences, as in his later works. In this respect, the term "opportunistic" may no longer easily apply to his later thinking, or at least his opportunism was conditioned by his philosophical inclinations toward a classical-like field-theoretical approach to fundamental physics. Einstein appears to have introduced the concept of photon under the pressure of experimental evidence, such as that reflected in Planck's law or the law of photoeffect (for which Einstein was actually awarded his Nobel prize). He went further than Planck by proposing that the photon was a real particle (rather than a mathematical convenience), the idea that took a while, until the 1920s and much additional experimental evidence, most especially, again, Compton's scattering experiments, to accept. Intriguingly, not only Planck but also Bohr was among the skeptics, and Bohr only accepted the idea in view of these experiments. Planck, who, as discussed earlier, strongly resisted Einstein's introduction of the concept of the photon, had never reconciled himself to the idea. Thus, it appears that until roughly 1920, Einstein did not have a strongly held philosophical position of the type he developed later on, first, following his work on general relativity (a classical-like field theory) and, secondly and most especially, in the wake of quantum mechanics. It is worth noting in this connection that he

[a]The account of this evolution sketched here is courtesy of Arkady Plotnitsky [private communication]. See also Pais [258] for a discussion of the development of Einstein's views on fundamental physics, from his earlier work to his work on general relativity and beyond; and for Einstein's earlier views, see Ref. [118] and also Ref. [78]. For Einstein's later views, see especially both of his contributions to the Schilpp volume [272].

initially resisted Minkowsky's concept of spacetime as insufficiently physical, but eventually came to appreciate its significance, again, especially in view of its effectiveness in general relativity. It is true, however, that theoretical physics at the time, including quantum theory (the "old" quantum theory), was still more classically oriented, as against quantum mechanics in the Heisenbergian approach. In addition, given that, in some circumstances, light would still exhibit wave behavior, Einstein also believed at the time (until even 1916 or so), that a kind of new synthesis of the particle-like and the wave-like theory of radiation would be necessary. However, this hope had not materialized in any form that he found acceptable, and he was especially dissatisfied with Heisenberg's approach [104], developed into the matrix mechanics by Born and Jordan, or related schemes, such as Dirac's one [74]. The success of general relativity as a classical-like field theory was significantly responsible for strengthening Einstein's field-theoretical predilections, and shaped his program of the unified field theory (with a unification of gravity and electromagnetism as the first task), which he pursued for the rest of his life. The problems of quantum mechanics and his debate on the subject with Bohr continued to preoccupy him as well, as reflected in particular in his persistent thinking concerning the EPR experiment, on which he commented virtually until his death. His view of fundamental physics following his work on relativity was also more mathematically oriented than the earlier one. In particular, he came to believe that it is a free mathematical conceptual construction, such as those of Riemann's geometry and tensor calculus in the case of general relativity and indeed of a similar classical-like field-theoretical type, that should and, he even argued, will allow us to come closest to capturing, in a realist manner, the ultimate character of physical reality. He expressly juxtaposed this approach to that of the Copenhagen–Göttingen approach in quantum mechanics [272], pp. 83–85. In sum, Einstein had come to be convinced that a strictly field-like theory unifying the fundamental forces of nature should be pursued. He saw this kind of theory as the best and even the only truly acceptable program for the ultimate theory of nature, while he believed quantum mechanics to be a provisional theory, eventually to be replaced by a field theory of the type he envisioned.

It may be remarked that the idea of particle poses difficulties for this view, especially the particle nature of radiation, initially represented in the idea of photon. This is why Einstein preferred and saw Schrödinger's wave mechanics as more promising (than matrix mechanics), or why earlier he liked de Broglie's approach (which he used in his work on the Bose–Einstein theory). It is true that the latter does retain the concept of particle and, as such, represents an attempt at a synthesis of the wave and the particle pictures, which, as noted above, Einstein contemplated initially. Later on, however, he did not like Bohmian mechanics, which pursued a similar line of thinking, although his negative attitude appears to have been determined by a complex set of factors. Eventually it became apparent that Schrödinger's formalism could not quite be brought under the umbrella of Einstein's unified field-theoretical program, a la Maxwell, although in his later years (in 1940s and 1950s) Schrödinger return to his initial ideas concerning wave mechanics. Quantum electrodynamics and then other quantum field theories appeared even more difficult to reconcile with this approach. Even general relativity posed certain significant problems for Einstein's vision, such as singularities, eventually leading to ideas such as black holes, although the full measure of these difficulties became apparent only later on, after Einstein's death.

There is thus quite a bit of irony to this history. While Einstein was fundamentally responsible for several theoretical ideas that eventually led others to quantum mechanics, he had developed grave doubts about quantum mechanics as a "useful point of departure for future development" [272], p. 83. Since, however, our fundamental physics remained incomplete at the time, Einstein thought that his vision might ultimately be justified. It might yet be, since our fundamental physics still remains incomplete, and in particular, is defined by a manifest conflict between relativity and quantum mechanics or higher level quantum theories. It would be curious to contemplate whether Einstein would have liked something like the string and brane theories, or any other currently advanced programs for fundamental physics and cosmology.

1.2 Prequantum Classical Statistical Field Theory: Introduction

Now I turn to my model, PCSFT, which is based on the unification of Einstein's two dreams: to reduce quantum randomness to classical randomness and to create a purely field model of physical reality. I emphasize from the very beginning that the majority of PCSFT structures are already present in QM, but in PCSFT they obtain a new (classical random field) interpretation. Therefore the introduction in PCSFT presented in this section can be considered as a short dictionary that establishes a correspondence between terms of QM and PCSFT. However, PCSFT not only reproduces QM, but provides a possibility to go beyond it. Therefore, advanced structures of PCSFT do not have counterparts in QM.

1.2.1 *Classical Fields as Hidden Variables*

Main message: *Quantum randomness is reducible to randomness of classical fields.*

Classical fields are selected as the hidden variables.[a] Mathematically, they are functions $\phi : \mathbf{R}^3 \to \mathbf{C}$ (or, more generally, $\to \mathbf{C}^k$) that are square-integrable, i.e., elements of the L_2-space. The latter condition is standard in the classical signal theory.

In particular, for electromagnetic field, this is just the condition of the finiteness of energy

$$\int_{\mathbf{R}^3} (E^2(x) + B^2(x))dx = \int_{\mathbf{R}^3} |\phi(x)|^2 < \infty, \tag{1.1}$$

where

$$\phi(x) = E(x) + iB(x) \tag{1.2}$$

is the *Riemann–Silbertstein vector* (the complex representation of the electromagnetic field).

Thus, the state space of our prequantum model is $H = L_2(\mathbf{R}^3)$. Formally, the same space is used in QM, but we couple it with the

[a] PCSFT is not a deterministic-type model with hidden variables. By fixing a classical "prequantum" field we cannot determine the values of observables. These values can be predicted with probabilities which are determined by the prequantum field, see Chapters 8 and 9 for a measurement theory in the PCSFT framework.

classical signal theory. For example, the quantum wave function satisfies the normalization condition

$$\int_{\mathbf{R}^3} |\phi(x)|^2 = 1, \tag{1.3}$$

but any vector ϕ in H can be selected as a PCSFT state. These prequantum waves evolve in accordance with Schrödinger equation; formally, the only difference is that the initial condition ϕ_0 is not normalized by 1, see Section 1.2.4, Eq. (1.13). Thus, these PCSFT waves are closely related to Schrödinger's quantum waves. However, opposite to Schrödinger and to the orthodox Copenhagen interpretation, the wave function of the QM formalism is not a state of a quantum system. In the complete accordance with Einstein's dream of reducibility of quantum randomness, wave function is associated with an ensemble. The ensemble, however, not of quantum systems, but the ensemble of classical fields, or, more precisely, a *classical random field*, random signal. It is appropriate to say that, although our model supports Einstein's views on the origin of quantum randomness, it also matches von Neumann's views [296] on individual quantum randomness. By using ergodicity (see Section 3.1.5) we can switch from the ensemble description to the individual signal description and vice versa. We state again that such a possibility of the peaceful combination of Einstein's and von Neumann's views on quantum randomness is a consequence of the rejection of the corpuscular model in the complete accordance with the views of "late Einstein." (It seems that at first he wanted to reduce quantum randomness to randomness of ensembles of particles.)

1.2.1.1 Random fields

A *random field* (at a fixed instance of time) is a function $\phi(x, \omega)$, where ω is a random parameter[a] and x is the coordinate vector.

For each fixed value of the random parameter $\omega = \omega_0$, we obtain the classical field, $x \mapsto \phi(x, \omega_0)$. The random field is the H-valued random variable,[b] each fixed ω_0 determines a vector $\phi(\omega_0) \in H$.

---

[a]A short introduction to the mathematical formalism of classical probability theory can be found in Section 2.5.

[b]For simplicity, we can consider a finite-dimensional Hilbert space instead of $L_2(\mathbf{R}^3)$ (as people often do in quantum information theory). In this case, PCSFT considers

By using the Riemann–Silberstein representation the electromagnetic random field can be represented in the complex form

$$\phi(x, \omega) = E(x, \omega) + iB(x, \omega).$$

In the classical signal theory one can move from the ensemble description of randomness to the time series description—under the *ergodicity hypothesis* (see Section 3.1.5 and Chapter 9). Random signals are widely used, e.g., in radiophysics [265]; these are electromagnetic fields depending of a random parameter.

Remark 2.1. Einstein used to make a point that the wave function Ψ is a label for an ensemble of identically prepared quantum systems. However, it was far from clear which statistical characteristics of an ensemble are encoded in Ψ. Obviously, not all of them, since Einstein lamented that QM is not complete. Our model, PCSFT, specifies the statistical characteristics are encrypted in Ψ, these are correlations between components of the field. The correlations are described by the covariance operator of the probability distribution of hidden variables of the field type. This is an important improvement of the statistical interpretation of QM. Instead of the Einstein's vague statement (see also Margenau [240] and Ballentine [25–29]) about statistical characteristics of an ensemble, we discovered the classical statistical variable, the covariance operator, which was formally used in the QM formalism under the name "wave function." Finally, we remark that people using the operational interpretation of QM (e.g., Ludwig, Davis, D'Ariano, Holevo, Busch, Grabowski, Lahti, Ozawa [46, 59, 61, 70, 115, 116, 215, 232, 255, 256]) typically proceed with the ensemble interpretation, too. In contrast to Einstein, Margenau, and Ballentine, they are sure that Ψ encodes all possible statistical characteristics of an ensemble, because they believe in completeness of QM. At the first sight, PCSFT presents a strong argument against such a viewpoint (introducing a new statistical characteristic): the covariance operator does not determine a probability distribution uniquely. Therefore a random field contains essentially more information than given by the covariance operator. However, if a prequantum field is Gaussian, it is completely determined by its covariance operator. (We shall consider only random fields with

H-valued random vectors, where $H = \mathbf{C}^n$. However we strongly emphasize the role of the physical state space $H = L_2(\mathbf{R}^3)$, see also Refs. [147, 292].

zero average.) Thus, for Gaussian prequantum fields, views of the adherents to the "orthodox ensemble interpretation" can be easily combined with views of the adherents to the operational approach to QM. (As we see, surprisingly many contradictions between different interpretations of QM can be resolved by PCSFT.)

1.2.2 *Covariance Operator Interpretation of Wave Function*

Main message: *The wave function is not a field of probabilities or a physical field. It encodes correlations between degrees of freedom of a prequantum random field.*

For simplicity, in this introductory section we considered the case of a single, i.e., noncomposite, system, e.g., the electron (nonrelativistic, since the present PCSFT is a nonrelativistic theory)[a] and we neglect for a moment (again for simplicity) fluctuations of vacuum that will play an important role in our further consideration.

In our model the wave function Ψ of the QM formalism encodes a class of prequantum random fields having the same covariance operator (determined by Ψ and determining a unique Gaussian random field.) We state again that we consider the case of a noncomposite quantum system; for composite systems, e.g., for a pair of photons or electrons, the correspondence between the wave function of QM and the covariance operator of PCSFT is more complicated, see Chapter 3.

In this situation the covariance operator (normalized by dispersion) is given by the orthogonal projector on the vector Ψ:

$$D_\Psi = \Psi \otimes \Psi, \tag{1.4}$$

i.e., $D_\Psi u = \langle u, \Psi \rangle \Psi, \ \ u \in H$. Thus,

$$D_\Psi = |\Psi\rangle\langle\Psi|$$

in Dirac's notation, i.e.,

$$D_\Psi u = \langle u|\Psi\rangle \ |\Psi\rangle.$$

We also suppose that all *prequantum fields have zero average*

$$E\langle y, \phi\rangle = 0, y \in H, \tag{1.5}$$

---

[a]It seems that there are no problems (neither physical nor mathematical) to develop a relativistic variant of PCSFT. I plan to do this in future.

where E denotes the classical *mathematical expectation* (average, mean value). By applying a linear functional y to the random vector ϕ we obtain the scalar random variable. In the L_2-case we get a family of scalar random variables:

$$\omega \mapsto \xi_y(\omega) \equiv \int_{\mathbf{R}^3} \phi(x, \omega)\overline{y(x)}dx, \; y \in L_2.$$

We recall that the covariance operator D of a random field (with zero average) $\phi \equiv \phi(x, \omega)$ is defined by its bilinear form

$$\langle Du, v\rangle = E\langle u, \phi\rangle\langle\phi, v\rangle, \; u, v \in H. \tag{1.6}$$

Under the additional assumption that the prequantum random fields are *Gaussian*, the covariance operator uniquely determines the field. Although this assumption seems to be quite natural both from the mathematical and physical viewpoints, we should be very careful. In the case of a single system we try to proceed as far as possible without this assumption. However, the PCSFT description of composite systems is based on Gaussian random fields (Section 3.2, see also Section 3.12 for general discussion of a possible physical origin of Gaussian probability distributions on the prequantum level.) Let $H = \mathbf{C}^n$ and $\phi(\omega) = (\phi_1(\omega), \ldots, \phi_n(\omega))$, then the zero average condition (1.5) is reduced to

$$E\phi_i \equiv \int_{\Omega} \phi_k(\omega)dP(\omega) = 0, \; k = 1, \ldots, n;$$

the covariance matrix $D = (d_{kl})$, where

$$d_{kl} = E\phi_k\bar{\phi}_l \equiv \int_{\Omega} \phi_k(\omega)\overline{\phi_l(\omega)}dP(\omega).$$

We also recall that the dispersion of the random variable ϕ is given by

$$\sigma_\phi^2 = E\,\|\phi(\omega) - E\phi(\omega)\|^2 = \sum_{k=1}^{n} E\,|\phi_k(\omega) - E\phi_k(\omega)|^2.$$

In the case of zero average we simply have

$$\sigma_\phi^2 = E\,\|\phi(\omega)\|^2 = \sum_{k=1}^{n} E\,|\phi_k(\omega)|^2.$$

Here it is always possible to select Ω (the set of random parameters) as $\mathbf{C}^n$. Then, the above integrals will be over $\mathbf{C}^n$. In particular, by selecting Gaussian, complex-valued, random fields we obtain Gaussian integrals over $\mathbf{C}^n$.

The case $H = L_2$ is more complicated from the measure-theoretic viewpoint, since this space is infinite dimensional. In the case of noncomposite systems (i.e., a single photon or electron) it is also possible to select $\Omega = H$, i.e., to integrate with respect to all fields of the L_2 class. For composite systems, the situation is more complicated. Here we cannot proceed without taking into account the background field, that is, of the white noise type. And the well-known fact is that the probability distribution of white noise cannot be concentrated on L_2, one has to select Ω as a space of distributions, i.e., to integrate with respect to singular fields.

We also remark that the random field $\phi(x, \omega)$ corresponding to a pure quantum state is not L_2-normalized. Its L_2-norm

$$\|\phi\|^2(\omega) \equiv \int_{\mathbf{R}^3} |\phi(x, \omega)|^2 dx \tag{1.7}$$

fluctuates depending on the random parameter ω.

By assigning to the squared field the physical dimension of the *energy density*, i.e., energy per volume, we interpret the quantity

$$\pi_2(\phi) \equiv \|\phi\|^2$$

as the *total energy* of the random field ϕ. This quantity will play a crucial role in the measurement theory corresponding to PCSFT, see Chapters 8 and 9.

1.2.3 *Quantum Observables from Quadratic Forms of the Prequantum Field*

Main message: *In spite of all no-go theorems (e.g., the Kochen–Specker theorem), a natural functional representation of quantum observables exists.*

In PCSFT, quantum observables are represented by corresponding quadratic forms of the prequantum field.[a] A self-adjoint operator $\widehat{A}$ is considered as the symbolic representation of the PCSFT variable

$$\phi \mapsto f_A(\phi) = \langle \widehat{A}\phi, \phi \rangle. \tag{1.8}$$

---

[a]This is true for a part of PCSFT reproducing QM, see Section 3.1.4 and Chapter 5 for the PCSFT "beyond quantum model."

This is a map from the L_2 space of classical prequantum fields into real numbers, a quadratic form.

We remark that f_A can be considered as a function on the phase space of classical fields: $f_A \equiv f_A(q, p)$, where $\phi(x) = q(x) + ip(x)$, i.e., it is possible to move from the complex representation to the phase space representation and vice versa, see Chapter 3. A crucial point is that the *prequantum phase space* is infinite dimensional (and the "post-quantum phase space," i.e., the phase space of ordinary classical mechanics is finite-dimensional).

The average of this quadratic form with respect to the random field determined by the wave function Ψ coincides with the corresponding quantum average:

$$\langle f_A \rangle = \langle \widehat{A}\Psi, \Psi \rangle \tag{1.9}$$

or

$$\langle f_A \rangle = \langle \Psi | \widehat{A} | \Psi \rangle$$

in Dirac's notation. Here

$$\langle f_A \rangle = E f_A(\phi) = \int_H f_A(\phi) d\mu_\Psi(\phi)$$

is the classical average and μ_Ψ is the probability distribution of the prequantum random field $\phi \equiv \phi_\Psi$ determined by the pure quantum state Ψ. In the real physical case H is infinite dimensional; the classical average is given by the integral over all possible classical fields; probabilistic weights of the fields are determined, in general, nonuniquely, by the Ψ. Thus, the quantum formula for the average of an observable was demystified:

$$\langle \widehat{A} \rangle_\Psi \equiv \langle \widehat{A}\Psi, \Psi \rangle = \int_H f_A(\phi) d\mu_\Psi(\phi) \tag{1.10}$$

It can be obtained via the classical average procedure.

1.2.4 *Quantum and Prequantum Interpretations of Schrödinger Equation*

Main message: *Schrödinger equation with random initial conditions describes dynamics of the physical random field.*[a]

[a] In the biparticle case Schrödinger equation describes dynamics of the two-points correlation function for field components (see Section 3.4).

Before going to the PCSFT dynamics, we consider the Schrödinger equation in the standard QM formalism:

$$ih\frac{\partial \Psi}{\partial t}(t, x) = \widehat{\mathcal{H}}\Psi(t, x), \tag{1.11}$$

$$\Psi(t_0, x) = \Psi_0(x), \tag{1.12}$$

where $\widehat{\mathcal{H}}$ is the Hamiltonian, the energy observable. We recall that Schrödinger tried to interpret $\Psi(t, x)$ as a classical field (e.g., the electron field; the distribution of electron charge in space). However, he gave up and, finally, accepted the conventional interpretation, the probabilistic one, due to Max Born.

We recall that a time-dependent random field $\phi(t, x, \omega)$ is called a *stochastic process* (with the state space $H = L_2$). Dynamics of the prequantum random field is described by the simplest stochastic process which is given by *deterministic dynamics with random initial conditions.*

In PCSFT the Schrödinger equation, but with the random initial condition, describes dynamics of the prequantum random field, i.e., the prequantum stochastic process can be obtained from the mathematical equation used in QM for dynamics of the wave function:

$$ih\frac{\partial \phi}{\partial t}(t, x, \omega) = \widehat{\mathcal{H}}\phi(t, x, \omega), \tag{1.13}$$

$$\phi(t_0, x, \omega) = \phi_0(x, \omega), \tag{1.14}$$

where the initial random field $\phi_0(x, \omega)$ is determined by the quantum pure state Ψ_0. Standard QM gives the covariance operator of this random field.

Roughly speaking, we combined Schrödinger's and Born's interpretations: the Ψ-function of QM is not a physical field, but for each t it determines a random physical field, i.e., the H-valued stochastic process $\phi(t, x, \omega)$.

PCSFT dynamics matches standard QM dynamics by taking into account the PCSFT interpretation of the wave function, see (1.4). Denote by $\rho(t)$ the covariance operator of the random field $\phi(t) \equiv \phi(t, x, \omega)$, the solution of Eqs. (1.13) and (1.14). Then

$$\rho(t) \equiv \rho_{\Psi(t)} = \Psi(t) \otimes \Psi(t),$$

where $\Psi(t)$ is a solution of Eqs. (1.11) and (1.12).

Such simple description can be used only for a single system and in the absence of fluctuations of vacuum. In the general case of a composite system, e.g., a biphoton system, in the presence of vacuum fluctuations Schrödinger dynamics of the Ψ-function encodes only dynamics of the covariance operator of the prequantum stochastic process (see Section 3.2). The situation is essentially more complicated than in the case of a single system. We found that it is possible to construct a few different prequantum dynamics which match (on the level of correlations) QM dynamics (see Section 3.3).

We considered the simplest situation—the covariance operator of the initial random field $\phi_0(x, \omega)$ has the unit trace. In general, this is not the case.

1.2.5 *Toward Prequantum Determinism*

Main message: *The background field is everywhere.*

From the PCSFT viewpoint, the source of quantum randomness is the randomness of initial conditions (if one neglects vacuum fluctuations), i.e., impossibility to prepare a nonrandom initial prequantum field $\phi_0(x)$.

We expect that in future very stable and precise preparation procedures will be created. The output of such a procedure will be a deterministic field $\phi(x)$, i.e., random fluctuations will be eliminated.

However, this dream of creating supersensitive "subquantum" technologies which would recover determinism on the microlevel may never come true. In such a case PCSFT will play the role of classical statistical mechanics of prequantum fields.[a] Unfortunately, there are a few signs that it really might happen. First of all, it might be that the scale of prequantum fluctuations is very fine, e.g., the *Planck scale.* In this case it would be really impossible to

[a]In ordinary classical statistical mechanics the existence of Hamiltonian dynamics has merely a theoretical value. In real applications we operate with probability distributions on the phase space. Corresponding dynamics is described by Liouville equation.

prepare a deterministic prequantum field. Also, there is another reason. The PCSFT model presented up to now has been elaborated for noncomposite quantum systems, e.g., a single electron. The extension of PCSFT to composite systems, e.g., a pair of entangled photons or electrons, is based on a more complicated model of prequantum randomness (see Section 3.2). We should complete the present model by considering fluctuations of the background field (zero-point field, vacuum fluctuations), in the same way as in SED. In reality, these are always present. Therefore Einstein's dream of determinism cannot be peacefully combined with the presence of the background field. If this field is irreducible (as a fundamental feature of space), then deterministic prequantum fields will never be created. If this background field is simply noise,[a] which can be eliminated, then we can dream of the creation of deterministic prequantum fields. However, a possibility to prepare such fields does not imply deterministic reduction of QM. As was already pointed out, the interrelation between prequantum fields and quantum observables given by TSD (measurement theory of classical waves with threshold detectors) is really tricky, see Chapter 9.

1.2.6 *Random Fields Corresponding to Mixed States*

Main message: *A density matrix is the normalized covariance operator of a prequatum random field.*

We now consider the general quantum state given by a density operator ρ. (We still work with noncomposite quantum systems.) According to PCSFT, ρ determines the covariance operator of the corresponding prequantum field (under normalization by its dispersion)

$$D_\rho = \rho. \tag{1.15}$$

Dynamics of the corresponding prequantum field $\phi(t, x, \omega)$ is also described by the Schrödinger equation, see Eqs. (1.13) and (1.14), with the random initial condition $\phi_0(x, \omega)$. The initial random field has the probability distribution μ_{ρ_0} having zero mean value and the

[a]Hence the completely empty physical space can be really prepared, "distilled from noise."

covariance operator

$$D(t_0) = \rho_0.$$

Under the assumption that all prequantum random fields are Gaussian, the initial probability distribution is determined in the unique way. In the general (non-Gaussian) case we lose the solid ground. The $\phi_0(x, \omega)$ can be selected in various ways, i.e., it can be any distribution having the covariance $D(t_0)$. We could not exclude such a possibility. It would simply mean that macroscopic preparation procedures are not able to control even the probability distribution (only its covariance operator).

Denote by $\rho(t)$ the covariance operator of the random field $\phi(t) \equiv \phi(t, x, \omega)$ given by Eqs. (1.13) and (1.14) with ϕ_0 having the covariance operator $\rho(t_0) = \rho_0$. Then $\rho(t)$ satisfies the von Neumann equation. However, $\rho(t)$ has the classical probability interpretation as the covariance operator $D(t)$. In the Gaussian case $D(t)$ determines completely the prequantum probability distribution.

In this chapter we consider the simplest situation: the covariance operator of the initial random field has the unit trace. In general, to couple the covariance operator (the term of the prequantum theory) with the density operator (the term of the quantum theory), we have to normalize the covariance operator by its trace. We shall discuss this normalization throughout this book.

1.2.7 *Background Field*

Main message: *QM is a formalism of measurement with filtering vacuum fluctuations.*

In the general PCSFT framework the randomness of the initial conditions has to be completed by taking into account fluctuations of vacuum (to obtain a consistent PCSFT that works both for one-particle systems and biparticle systems). In our model the background field (vacuum fluctuations) is of the white noise type. It is a Gaussian random field with zero average and the covariance operator

$$D_{\text{background}} = \varepsilon I, \ \ \varepsilon > 0.$$

It is a stationary field, so its distribution does not change with time.

Consider (by using the QM language) a quantum system in the mixed state ρ_0. It determines the prequantum random field $\phi_0 \equiv \phi_0(x, \omega)$ with the covariance operator

$$\tilde{D}(t_0) = \rho_0 + \epsilon I.$$

The value of $\varepsilon > 0$ is not determined by PCSFT, but it could not be too small for a purely mathematical reason (see Section 7.6.1). (Hence QM is a theory of filtration of strong noise.) Now consider the solution $\phi(t)$ of the Schrödinger equation (1.13) and (1.14) with the initial condition ϕ_0. Its covariance operator can be easily found:

$$\tilde{D}(t) = D(t) + \varepsilon I,$$

where $D(t)$ is the covariance operator of the process in the absence of the background field, $D(t) = \rho(t)$. Here $\rho(t)$ satisfies the QM equation for evolution of the density operator, i.e., the von Neumann equation. Thus on the level of dynamics of the covariance operator the contribution of the background field is very simple: an additive shift. However, on the level of the field dynamics the presence of vacuum fluctuations changes the field behavior crucially.

Consider the prequantum random field $\phi_0(x, \omega)$ corresponding to a pure quantum state Ψ_0. Now (in the presence of the background field) the prequantum random field $\phi_0(x, \omega)$ is not concentrated on a one-dimensional subspace[a]

$$H_{\Psi_0} = \{\phi = c\Psi_0 : c \in \mathbf{C}\};$$

the vacuum fluctuations smash it over H.

In canonical QM, the background field of the white noise type is neglected; in fact, it is eliminated "by hand" using detectors of the threshold type (see Chapter 9). And it is the right strategy for a formalism describing measurements on the random background. However, in an ontic model, i.e., a model of reality as it is, this background field should be taken into account. Neglecting it induces a rather mystical picture of quantum randomness.

[a]In the absence of vacuum fluctuations the covariance operator of the random field $\phi_\Psi(x, \omega)$ corresponding to a pure state Ψ is given by the orthogonal projector on Ψ (1.4); the corresponding Gaussian measure is concentrated on a one-dimensional subspace generated by Ψ. Of course, the latter is valid only for Gaussian prequantum fields.

We shall see that in the PCSFT formalism the background field plays the fundamental role in the derivation of Heisenberg's uncertainty relation (see Section 3.10). Roughly speaking, Heisenberg's uncertainty is a consequence of vacuum fluctuations.[a]

1.2.8 *Coupling between Schrödinger and Hamilton Equations*

Main message: *The Schrödinger equation is a complex form of the Hamilton equation for a special class of quadratic Hamilton functions on an infinite-dimensional phase space.*

We remark that the Schrödinger equation can be written as the system of Hamilton equations on the (infinite-dimensional) phase space $Q \times P$, where $Q = P$ is the real Hilbert space and $H = Q \oplus i P$ is the corresponding complex Hilbert space. The prequantum field $\phi(x) = q(x) + ip(x)$, where $q(x)$ and $p(x)$ are real-valued fields (or more generally, they take values in $\mathbf{R}^m$). Consider the Hamilton function

$$\mathcal{H}(q, p) = \frac{1}{2}\langle \widehat{\mathcal{H}}\phi, \phi\rangle, \tag{1.16}$$

or, in Dirac's notation,

$$\mathcal{H}(\phi) = \frac{1}{2}\langle \phi | \widehat{H} | \phi \rangle.$$

see Chapter 4 for details; in PCSFT $\mathcal{H}(q, p)$ is the energy of the prequantum field $\phi(x) = q(x) + ip(x)$. The Schrödinger equation (1.13) can be written as the system [284]

$$\dot{q} = \frac{\partial \mathcal{H}}{\partial p}, \quad \dot{p} = -\frac{\partial \mathcal{H}}{\partial q}. \tag{1.17}$$

From the PCSFT viewpoint, there is no reason (at least mathematical) to use only quadratic Hamiltonian functions. By considering nonquadratic Hamilton functions we obtain Hamilton systems connected with the nonlinear Schrödinger equation, cf. Refs. [32, 75, 76, 93, 94, 299]. PCSFT naturally induces a nonlinear extension of QM (see Sections 5.7 and 6.1).

---

[a] A similar viewpoint on Heisenberg's uncertainty relation can be found in Hofmann's PhD thesis [114].

1.2.9 *Nonquadratic Functionals of the Prequantum Field and Violation of Born's Rule*

Main message: *Nonlinear, of order higher than two, contribution of the prequantum field induces violation of Born's rule.*

As was pointed out, there is no reason to restrict PCSFT variables to quadratic functionals of the prequantum fields, see (1.9). Let us consider an arbitrary smooth functional $f(\phi)$, $\phi \in H$, which maps the field $\phi \equiv 0$ into zero, $f(0) = 0$. Let us also consider a random field $\phi = \phi(x, \omega)$ corresponding to a quantum density operator ρ. We can find the classical average

$$\langle f \rangle_\mu = \int_H f(\phi) d\mu(\phi), \tag{1.18}$$

where μ is the probability distribution of the random field. We shall show (see Chapter 5) that this classical average can be approximated by the quantum average

$$\langle \widehat{A} \rangle_\rho = \mathrm{Tr} \rho \widehat{A}, \tag{1.19}$$

where

$$\widehat{A} = f''(0)/2 \tag{1.20}$$

is the second derivative of the field functional $f(\phi)$ at the point $\phi = 0$ (divided by the factor 2 which arises from the Taylor expansion). If a Hilbert state space is finite-dimensional, then this is the usual second derivative. Its matrix (*Hessian*) is symmetric. If a Hilbert space is infinite dimensional (of the L_2 type), then the derivatives are the so-called variations. In the rigorous mathematical framework they are Frechet derivatives, that are used, e.g., in optimization theory. In the latter case the second (variation) derivative is given by a self-adjoint operator. This is the PCSFT origin of the representation of quantum observables by self-adjoint operators.

Quantum observables are represented by self-adjoint operators, since they correspond to Hessians of smooth functionals of the prequantum field.

Thus the QM formalism gives approximations of classical averages with respect to the prequantum random fields by approximating field functionals $f(\phi)$ by the quadratic terms of their Taylor expansions.

If the functional $f(\phi)$ is linear, $f(\phi) = \langle \phi, y \rangle, y \in H$, then its QM image, the second derivative, is equal to zero. Linear field effects are too weak and they are completely ignored by QM. However, such functionals and their correlations are well described by PCSFT. Observation of such effects can be the first step beyond QM (see Section 3.4).

1.2.10 *Wave Comeback: A Solution Too Cheap?*

Main message: *Physical space exists!*[a] *Hence waves propagating in this space are basic entities of nature.*

It is well known that Einstein was not happy with Bohmian mechanics. He considered this solution of the problem of completion of QM as cheap. Recently Anton Zeilinger mentioned (in his lecture at the Växjö conference, 2010, "Advances in Quantum Theory") that QM may be not the last theory of micro processes and in future a new fundamental theory may be elaborated. And looking me in the face, he added that those who nowadays criticize QM and dream of a prequantum theory will be terrified by this coming new theory, by its complexity and extraordinariness. They will recall the old QM formalism, i.e., the present one, with great pleasure, since it was so close to classical mechanics. A similar viewpoint on a coming prequantum theory was presented by Claudio Garola during our dialog on possible ways to proceed beyond QM [90].

PCSFT is a comeback to classical field theory; roughly speaking, in the spirit of early Schrödinger and late Einstein: the Maxwell classical field theory is extended to "matter waves." Of course, this comeback is not the dream of the majority of those who nowadays are not afraid to speculate on prequantum models and criticize the Copenhagen QM. Nevertheless, I do not think that PCSFT is a cheap completion of the standard QM. I hope that, in contrast to Bohmian mechanics, Einstein might accept PCSFT as one of the possible ways beyond QM. In any event the Laplacian mechanistic determinism was totally excluded from PCSFT; reality became blurred in the sense

[a]By this statement Igor Volovich has started his talks at Växjö conferences for ten years criticizing quantum information theory which practically ignores this fact.

of Schrödinger [273, 274]. This is reality of fields and not particles, but still reality.

1.3 Where Is Discreteness? Devil in Detectors?

Main message: *All detectors have thresholds.*

Prequantum variables $f_A(\phi) = \langle \widehat{A}\phi, \phi \rangle$ have continuous ranges of values. On the other hand, in QM some observables have discrete spectra. Thus, although PCSFT matches precisely probabilistic predictions of QM,[a] it violates the *spectral postulate* of QM. How can one obtain discrete spectra?

The continuous field model supports the viewpoint that "ontic reality," i.e., reality as it is, is continuous.[b] *Discreteness of some observable data is created by our macroscopic devices,* which split a prequantum signal in a number of discrete channels. Take a PBS and consider first a classical signal. Suppose, the PBS is oriented at an angle θ. Then the classical signal is split into two channels. We can label these channels as "polarization up," $S_\theta = +1$, and "polarization down," $S_\theta = -1$, (for θ-direction). The only problem is that the classical signal is present in both channels. Thus we cannot assign to a classical signal (even to a short pulse) a concrete value of S_θ. On the contrary, for a "quantum signal" (photon), detectors never click in both channels; we get either $S_\theta = +1$ or $S_\theta = -1$. This is a standard example of quantum discreteness.

The first comment of this common description is that the situation "no double clicks" is never occurred in real experiments

---

[a] In fact, the situation is more complicated. By considering quadratic prequantum variables $f_A(\phi)$ we obtain the coincidence of prequantum classical and quantum averages, see Chapters 3 and 4. However, by considering nonquadratic functionals of prequantum fields we find that the quantum probability given by Born's rule is just the main contribution to the prequantum (classical) average.

[b] In fact, my viewpoint on a proper mathematical model of reality is more complicated. Of course, the usage of continuous space–time based on real numbers is just a way to unify a huge hierarchy of scales of space and time. In this book we do not criticize this model, cf. p-adic models, Vladimirov, Volovich, Witten, Freund, Dragovich, Aref'eva, Frampton, Parisi, Khrennikov, Zelenov, Kozyrev, see, e.g., Refs. [132, 133, 291]. At the moment we "just" criticize Bohr's postulate, the existence of the fundamental quant of action given by the Planck constant. We predict splitting of values of quantum observables at finer (" prequantum") time scales.

[9, 11, 22, 95, 96]. There are always *double clicks,* clicks in both channels of PBS (or simply beam splitter). However, as was shown in the experiment of Grangier [95], their relative number is small. The latter is known as the *coefficient of second-order coherence,*

$$g^{(2)}(0) = \frac{P_{12}}{P_1 P_2}, \tag{1.21}$$

where P_{12} is the probability of coincidence of clicks in both channels and P_i is the probability for ith channel, $i = 1, 2$. For a single photon source, quantum theory predicts that $g^{(2)}(0) = 0$. Of course, in all experiments on measurement of second-order coherence this coefficient is essentially larger than zero. Nevertheless, $g^{(2)}(0) < 1$. At the same time known models of reduction of quantum light to classical light, including the semiclassical model, predict that $g^{(2)}(0) \geq 1$. Grangier [95] obtained $g^{(2)}(0) < 1$. Thus classical and semiclassical models of light were rejected. In particular, the interest to the semiclassical model decreased drastically. Nowadays this model is considered as just a computational tool which can essentially simplify calculations [275].

However, this problem of mismatching of classical and quantum optics on the level of second-order coherence can be solved by taking into account the *threshold functioning* of all known quantum detectors. The first detailed study of the impact of thresholds was performed in the PhD thesis [13] of my student Guillaume Adenier. He performed numerical simulation for the TSD model of classical signals. He reproduced quantum probabilities of detection. For classical bi-signals interacting with two PBSs (oriented at angles θ_1 and θ_2), the EPR–Bohm correlations were reproduced and Bell's inequality was violated [13]. (In principle, Adenier's model can be applied even to the problem of second-order coherence.)

Later (motivated by the numerical study of Adenier) I created a theoretical model of threshold detection of classical random fields which reproduces quantum probabilities, gives $g^{(2)}(0) < 1$, and violates Bell's inequality [209–213] (see Chapter 9). In particular, according to this model, the electromagnetic field is quantized only in the process of interaction with matter. This viewpoint matches well with views of Lamb [222], Lande [223, 224], Kracklauer [219, 220], Roychoudhuri [268, 269], Adenier [13], people working in SED, e.g., Marshall and Brafford, Boyer, de la Pena, Ceto, Coli, ...

[43, 53, 63–67]. However, PCSFT differs essentially from a rather popular idea that the electromagnetic field is continuous, but matter is quantized, see, e.g., the books of Lande [223, 224]. PCSFT does not quantize the matter, the latter also composed of continuous fields fluctuating on very fine space–time scales. These scales are not yet approachable. In future we expect to get a possibility to monitor these fields and not only their averaged images.

Since the main issues of PCSFT have already been presented, the reader who is interested in connection of theoretical prequantum models with experiment can jump directly to Chapters 8 and 9. Chapter 8 contains a phenomenological model of detection for PCSFT. This model couples the number of clicks in a detector with the average energy of the prequantum random signal in the corresponding channel. The concrete mechanism of transformation of energy into clicks, threshold-type detection, is presented in Chapter 9.

1.4 On Experiments to Test the Euclidean Model

One of the most famous stories about Gauss depicts him measuring the angles of the great triangle formed by the mountain peaks of Hohenhagen, Inselberg, and Brocken for evidence that the geometry of space is non-Euclidean. He tested the inequality:

$$\alpha_{12} + \alpha_{23} + \alpha_{13} < 2\pi, \tag{1.22}$$

where α_{ij} is the angle between the corresponding sides of the triangle. Gauss understood how the intrinsic curvature of the Earth's surface would theoretically result in slight discrepancies when fitting the smaller triangles inside the larger triangles, although in practice this effect is negligible, because the Earth's curvature is so slight relative to even the largest triangles that can be visually measured on the surface. Still, Gauss computed the magnitude of this effect for the large test triangles because, as he wrote, "the honor of science demands that one understand the nature of this inequality clearly."

On the other hand, if the curvature of space was actually great enough to be observed in optical triangles of this size, then presumably Gauss would have noticed it, so we may still credit him with having performed an empirical observation of geometry, but in this sense every person who ever lived has made such observations. The first person to publicly propose an actual test of the geometry of space was apparently Lobachevsky, who suggested that one might investigate a stellar triangle for an experimental resolution of the question. The stellar triangle he proposed was the star Sirius and two different positions of the Earth at 6-month intervals. This was used by Lobachevsky as an example to show how we could place limits on the deviation from the flatness of actual space based on the fact that, in a hyperbolic space of constant curvature, there is a limit to how small a star's parallax can be, even for the most distant star. The first definite measurement of the parallax for a fixed star was performed by Friedrich Bessel (a close friend of Gauss) in 1838, on the star 61 Cygni. Shortly thereafter he measured Sirius (and discovered its binary nature). Lobachevsky's first paper on the new geometry was presented as a lecture in Kasan in 1826 followed by publications in 1829, 1835, 1840, and 1855 (a year before his death). He presented his lower bound for the characteristic length of a hyperbolic space in the later editions based on the still fairly recent experimental results of stellar parallax measurements.

The inequality

$$\alpha_{12} + \alpha_{23} + \alpha_{13} = 2\pi \tag{1.23}$$

is a geometric analog of Bell's inequality. Violation of (1.23), e.g., in the form of (1.22) implies impossibility to use the Euclidean model. In the same way violation of Bell's inequality implies impossibility to use the Kolmogorov's model.

Chapter 2

Conventional Quantum Theory: Fundamentals

The aim of this chapter is to present (in a compact way) the fundamentals of conventional quantum theory. We select the formulation of QM as an axiomatic theory in the complete accordance with the Hilbert's program on axiomatization of physical theories. We remark that the first step toward the present formulation of QM was done in the seminal work of Hilbert, Nordheim, and von Neumann [112]. This work was substantially extended, completed, and clarified in the book of von Neumann [296]. Another version of the rigorous mathematical approach to QM was presented in the book of Dirac [74]. From the mathematical viewpoint, the main difference between the approaches of von Neumann and Dirac is that the first one is based on the theory of self-adjoint operators in complex Hilbert space and the second one is based on the theory of self-adjoint operators acting (by using the modern mathematical terminology) in rigged Hilbert spaces. This theory operates with generalized eigenvectors belonging to the spaces of distributions. Although the von Neumann's formalization of QM became dominant in quantum mathematical physics, the Dirac's formalization is widely used in the works on the physical level

Beyond Quantum
Andrei Khrennikov

ISBN 978-981-4411-73-8 (Hardcover), 978-981-4411-74-5 (eBook)
www.panstanford.com

of rigorousness. Moreover, the Dirac's terminology, bra- and ket-vectors, became the standard for quantum information theory. However, since quantum information is based on the castrated version of QM, namely, QM in *finite-dimensional Hilbert spaces*, the so-called n-qubit spaces, the basic tool of the Dirac's formalism, usage of generalized eigenvectors, is useless (since in the finite-dimensional case everything is reduced to manipulation with matrices). During the last 80 years, there were written hundreds of books on QM enlightening its foundational aspects. Unfortunately, the majority of these writings advertise exotic (and some of them really esoteric) views on QM and, in particular, on interpretation of the wave function. They are full of such terms as the collapse of the wave function, quantum nonlocality, many worlds, the wave function of universe, the preparation of Schrödinger cats, influence of the future on the past, quantum teleportation. Of course, manipulations with such terminology create an atmosphere of mystery and a feeling that something unusual and impossible can happen. And this atmosphere plays a positive role in attracting young people to QM and especially to quantum foundations.

Among the books on quantum foundations without mysteries, I can recommend the books by Lande [223, 224], Ballentine [26, 27], and my own book [192]. An interesting book was written by Accardi [3], but he was not able to publish it in English; there exist only versions in Italian and Russian. Quantum foundations without the mystical dimension are also discussed in the book by Ohya and Volovich [253], especially those related to quantum information theory, e.g., the real analysis of the degree of security of quantum cryptography.

2.1 Postulates

We state again that H denotes complex Hilbert space with the scalar product $\langle \cdot, \cdot \rangle$ and the norm $\| \cdot \|$ corresponding to the scalar product.

Postulate 1. (The mathematical description of quantum states). *Quantum (pure) states (wave functions) are represented by normalized vectors* ψ *(i.e.,* $\|\psi\|^2 = \langle \psi, \psi \rangle = 1$*) of a complex Hilbert space*

H. Every normalized vector $\psi \in H$ may represent a quantum state. If a vector ψ corresponding to a state is multiplied by any complex number c, $|c| = 1$, the resulting vector will correspond to the same state.[a]

The physical meaning of "a quantum state" is not defined by this postulate. It must be provided by a separate postulate; see Postulates 6, 6a, 6b (respectively, the ensemble, Copenhagen, and Växjö interpretations).

Postulate 2. (The mathematical description of physical observables). *A physical observable a is represented by a self-adjoint operator $\widehat{a}$ in complex Hilbert space H. Different observables are represented by different operators.*

Postulate 3. (Spectral) *For a physical observable a which is represented by the self-adjoint operator $\widehat{a}$ we can predict (together with some probabilities) values $\lambda \in Spec(\widehat{a})$.*

We restrict our considerations by simplest self-adjoint operators which are analogous to discrete random variables in classical probability theory. We recall that a self-adjoint operator $\widehat{a}$ has *purely discrete spectrum* if it can be represented as

$$\widehat{a} = \alpha_1 P^a_{\alpha_1} + \cdots + \alpha_m P^a_{\alpha_m} + \cdots, \quad \alpha_m \in \mathbf{R}, \tag{2.1}$$

where $P^a_{\alpha_m}$ are orthogonal projection operators related to the orthonormal eigenvectors $\{e^a_{km}\}_k$ of $\widehat{a}$ corresponding to the eigenvalues α_m by

$$P^a_{\alpha_m}\psi = \sum_k \langle \psi, e^a_{km}\rangle e^a_{km}, \quad \psi \in H. \tag{2.2}$$

Here k labels the eigenvectors e^a_{km} which belong to the same eigenvalue α_m of $\widehat{a}$.

Postulate 4. (Born's rule—in formalization of Dirac and von Neumann) *Let a physical observable a be represented by a self-adjoint operator $\widehat{a}$ with purely discrete spectrum. The probability $\mathbf{P}_\psi(a = \alpha_m)$ to obtain the eigenvalue α_m of $\widehat{a}$ for measurement of a in a state ψ is given by*

$$\mathbf{P}_\psi(a = \alpha_m) = \| P^a_m \psi \|^2. \tag{2.3}$$

---

[a] Thus states are given by elements of the unit sphere of the Hilbert space H.

If the operator $\widehat{a}$ has nondegenerate (purely discrete) spectrum, then each α_m is associated with one-dimensional subspace. The latter can be fixed by selecting any normalized vector, say e^a_m. In this case orthogonal projectors act simply as

$$P^a_{\alpha_m}\psi = \langle\psi, e^a_m\rangle e^a_m. \tag{2.4}$$

The formula (2.3) takes a very simple form

$$\mathbf{P}_\psi(a = \alpha_m) = |\langle\psi, e^a_m\rangle|^2. \tag{2.5}$$

It is Born's rule in the Hilbert space formalism.

To obtain original Born's rule, one should choose H as the L_2 space of square-integrable functions, $\psi : \mathbf{R} \mapsto \mathbf{C}$. (We consider one-dimensional particle). The position observable x is represented by the multiplication operator $\widehat{x}$ as

$$\widehat{x}(\psi)(x) = x\psi(x). \tag{2.6}$$

This operator has continuous spectrum. It coincides with the whole real line. So, this operator is unbounded. Its eigenvectors do not belong to the L_2-space. They are given by Dirac's δ-functions, i.e., these are generalized eigenfunctions [74]

$$\widehat{x}(e_\alpha)(x) = \alpha e_\alpha(x), \ \ \alpha \in \mathbf{R}, \tag{2.7}$$

where $e_\alpha(x) = \delta(x - \alpha)$. One can reasonably define paring[a]

$$\langle\psi, e_\alpha\rangle = \psi(\alpha). \tag{2.8}$$

Then the rule (2.5) gives

$$\mathbf{P}_\psi(x = \alpha) = |\psi(\alpha)|^2. \tag{2.9}$$

This rule was invented in the following way. Originally Schrödinger considered the ψ-function as a classical field. The quantity $E(\alpha) = |\psi(\alpha)|^2$ was interpreted as the density of the electron's charge. Born invented the rule (2.9) by criticizing

[a]In the rigorous mathematical framework elements of the L_2-space are given by equivalent classes of functions. Two functions belong to the same class if measure of points where they are distinct equals to zero. To proceed rigorously, one should select a subspace in the L_2-space and consider Dirac's delta function and its shifts $e_\alpha(x) = \delta(x - \alpha)$ as continuous linear functionals on this subspace. This can be done in the framework of distribution theory. Here paring (2.8) is nothing else than action of the functional e_α to the test function ψ. However, physicists typically do not pay attention to such mathematical problems.

Schrödinger's interpretation. Instead of the charge density, he considered this quantity as the probability density. The latter induces automatically the normalization condition

$$1 = \int_{-\infty}^{+\infty} |\psi(\alpha)|^2 = \langle \psi, \psi \rangle,$$

which was absent in Schrödinger's model. After a few years of struggle for the classical field interpretation of the wave function, Schrödinger gave up and he kept to the Born's probabilistic interpretation. In our model which is based on classical random fields the wave function also has the probabilistic interpretation. However, it encodes probabilities not directly, but in a rather tricky way. By itself the wave function encodes the covariance operator of the prequantum random field. (We state again that by PCSFT each quantum system is a symbolic representation of the purely classical object, the random field.) The Born's representation of detection probabilities through the wave function is not fundamental. It is related to usage of the special class of detectors, see Chapter 9 for the TSD model for PCSFT. Thus the PCSFT/TSD interpretation does not coincide neither with the Born's probabilistic interpretation nor with the Schrödinger's physical field interpretation. We point that the element of the Schrödinger's interpretation is also present in PCSFT: the dynamics of the prequantum random field is described by the Schrödinger equation. Finally, we stress once again that the wave function of QM (the operational quantity used for the Born's representation of probabilities) should not be identified with the corresponding prequantum random field. The latter is the fundamental entity of the prequantum theory and the former can be derived from it by using the classical probabilistic rule for calculation of the covariance operator. As we shall see in Chapter 9, in the PCSFT/TSD approach the Born's rule for the detection probabilities is not postulated (as it is in QM), but derived from properties of prequantum random fields and detectors.

In the same way one can consider momentum measurement. Schrödinger defined the momentum operator as

$$\widehat{p}(\psi)(x) = -i\frac{d}{dx}\psi(x). \tag{2.10}$$

(We eliminate the Planck constant from consideration by choosing corresponding system of units.) It is easy to see that its spectrum is also continuous and it coincides with **R**. Its generalized eigenfunctions can be easily found from the equation

$$-i\frac{d}{dx}f_\beta(x) = \beta f_\beta(x),\ \ \beta \in \mathbf{R}.$$

Thus $f_\beta(x) = e^{i\beta x}$. Thus by (2.5)

$$\mathbf{P}_\psi(p = \beta) = |\langle\psi,\ f_\beta\rangle|^2. \tag{2.11}$$

By taking into account that

$$\langle\psi,\ f_\beta\rangle = \int_{-\infty}^{+\infty} \psi(x)e^{-i\beta x}dx = \tilde{\psi}(\beta)$$

is the Fourier transform of ψ, we write Born's rule for the momentum measurement as

$$\mathbf{P}_\psi(p = \beta) = |\tilde{\psi}(\beta)|^2, \tag{2.12}$$

cf. (2.9).

We point out that, for any fixed quantum state ψ, each quantum observable $\widehat{a}$ can be represented as a classical random variable. In the discrete case the corresponding probability distribution (see Section 2.5) is defined as

$$\mathbf{P}(A) = \sum_{\alpha_m \in A} \mathbf{P}_\psi(a = \alpha_m),$$

where $\mathbf{P}_\psi(a = \alpha_m)$ is given by Born's rule.

Thus each concrete quantum measurement can be described by the classical probability model.

Problems (including deep interpretational issues) arise only when one tries to describe classically data collected for a few incompatible observables. We remark that such attempts contradict to Kolmogorov's ideology [218]. Kolmogorov emphasized that each probability space (see Section 2.5) is determined by corresponding complex of experimental conditions (context). The same message we got from Bohr who pointed out that the whole experimental arrangement should be taken into account and whose principle of complementarity supports the Kolmogorov's ideology. For example, impossibility to embed the collection of probabilities for the position and momentum measurements (for all possible quantum states) into a single probability space is often considered as a new astonishing probabilistic situation. However, Kolmogorov's approach implies that attempts of such an embedding are not justified, since the position and momentum measurements for quantum systems cannot be performed in a single experimental setting (context).

By using the Born's rule (2.3) and the classical probabilistic definition of average, see, e.g., (2.25) and (2.26), it is easy to see that the average value of the observable a in the state ψ (belonging to the domain of definition of the corresponding operator $\widehat{a}$) is given by

$$\langle a \rangle_\psi = \langle \widehat{a}\, \psi, \psi \rangle. \tag{2.13}$$

For example, for an observable represented by an operator with the purely discrete spectrum, we have:

$$\langle a \rangle_\psi = \sum_m \alpha_m \mathbf{P}_\psi(a = \alpha_m) = \sum_m \alpha_m \| P_m^a \psi \|^2 = \langle \widehat{a}\, \psi, \psi \rangle.$$

Postulate 5. (Time evolution of wave function). *Let $\widehat{\mathcal{H}}$ be the Hamiltonian of a quantum system, i.e., the self-adjoint operator corresponding to the energy observable. The time evolution of the wave function $\psi \in H$ is described by the Schödinger equation*

$$i \frac{d}{dt} \psi(t) = \widehat{\mathcal{H}} \psi(t) \tag{2.14}$$

with the initial condition $\psi(0) = \psi$.

2.2 Quantization

We remark that the operators of position and momentum, $\widehat{x}$ and $\widehat{p}$, see (2.6) and (2.10), do not commute and they satisfy Heisenberg's *canonical commutation relation*

$$[\widehat{x}, \widehat{p}] = i. \tag{2.15}$$

Consider any real-valued function on the classical phase space, i.e., a function of classical coordinate and momentum, $f(x, p)$. The quantization procedure is the map

$$f \mapsto \widehat{f} = f(\widehat{x}, \widehat{p}). \tag{2.16}$$

In general, it is a tricky mathematical problem to define a function of two noncommuting operators. It is typically done by using the calculus of pseudo-differential operators.[a]

However, in the simplest case the operator of energy $\widehat{\mathcal{H}}$ can be easily defined. Consider a classical particle with the mass m moving in the potential $V(x)$. Its Hamiltonian function (representing classical energy of this particle) is given by

$$\mathcal{H}(x, p) = \frac{p^2}{2m} + V(x). \tag{2.17}$$

Quantization gives us the operator

$$\widehat{\mathcal{H}} = \mathcal{H}(\widehat{x}, \widehat{p}) = \frac{\widehat{p}^2}{2m} + V(x). \tag{2.18}$$

2.3 Interpretations of Wave Function

Now we are going to discuss one of the most important and complicated notions of quantum mechanics: the notion of a quantum state. There are two main points of view which are formulated in the following postulates.

Postulate 6. (The ensemble interpretation). *A wave function provides a description of certain statistical properties of an ensemble of similarly prepared quantum systems.*

This interpretation is upheld, for example by Einstein, Popper, Blokhintsev, Margenau, Ballentine, Klyshko, and in recent years by, e.g., De Muynck, De Baere, Holevo, Santos, Khrennikov, Nieuwenhuizen, Adenier, Groessing, and many others (see, e.g., the recent works of Groessing on the interplay between classical statistical mechanics and QM [97, 98]; we also point to works of Raedt *et al.* demonstrating that the ensemble interpretation is compatible with statistical data from the two-slit experiment and from the EPR–Bohm experiment [71, 72, 122]).

Postulate 6a. (The Copenhagen interpretation). *A wave function provides a complete description of an individual quantum system.*

[a]See Refs. [132, 133] for operator quantization over non-Archimedean (in particular, p-adic) number fields.

This interpretation was supported by a great variety of members, from Schrödinger's original attempt to identify the electron with a wave function solution of his equation to the several versions of the Copenhagen interpretation; for example, Heisenberg, Bohr, Pauli, Dirac, von Neumann, Landau, Fock, and in recent years by, e.g., Greenberger, Mermin, Lahti, Peres, Summhammer.[a] Nowadays the individual interpretation is extremely popular, especially in quantum information and computing.

Instead of Einstein's terminology *ensemble interpretation*, Ballentine [25, 26] used the terminology *statistical interpretation*. However, Ballentine's terminology is rather misleading, because the term statistical interpretation was also used by von Neumann [296] for individual randomness! For him "statistical interpretation" had the meaning which is totally different from the Ballentine's ensemble-statistical interpretation. von Neumann wanted to emphasize the difference between deterministic (Newtonian) classical mechanics in that the state of a system is determined by values of two observables (position and momentum) and quantum mechanics in that the state is determined not by values of observables, but by probabilities. We shall follow Albert Einstein and use the terminology *ensemble interpretation*.

2.4 Växjö Interpretation of Quantum Mechanics

The Växjö interpretation [145] was born (in 2001) as an alternative to the Copenhagen interpretation. The basic assumption of the latter that the wave function describes completely the state of a quantum system is the main source of all quantum mysteries. (We remind that Schrödinger elaborated the "Schrödinger cat" experiment to show the absurdness of this assumption. Nowadays the origin of the Schrödinger cat illustration of the absurdness of the

[a] There is an interesting story about the correspondence between Bohr and Fock on the individual interpretation. This story was told me by one of former students of Fock who pointed out that one of the strongest supporters of this interpretation was Vladimir A. Fock, and that even though Bohr himself had doubts about its consistency, Fock demonstrated to Bohr inconsistency in the Einsteinian ensemble interpretation. Thus interpretation which is commonly known as the *Copenhagen interpretation* might be as well called the "*Leningrad interpretation*."

Copenhagen interpretation is practically forgotten. The presence of such cat-states is often considered as one of natural features of QM.) Therefore the easiest way to resolve the majority of interpretation problems of QM is to assume that the wave function description is not the final description of micro phenomena. By proceeding in this way one has to develop prequantum models inducing the QM formalism as an operational formalism ignoring details of processes in the microworld and describing only results of measurements. Measurements are operationally described by Hermitian operators or more generally by positive operator valued measures (POVM). However, there are known many no-go statements which prohibit any motion beyond QM (under the natural assumption of locality[a]). The first version of the Växjö interpretation [145] was created as the result of analysis of the "impossibility statements" (see, e.g., Refs. [31, 217, 296]) playing the crucial role in the contemporary debates on quantum foundations. Surprisingly for myself, I found that *all no-go statements* contain some unphysical assumptions which are not valid for real experimental situations, see Ref. [192] for details; cf. with Bell's and Ballentine's critical analysis of assumptions of von Neumann no-go theorem [25, 26, 31]. I understood that usage of the operational quantum formalism for results of measurements does not contradict with the possibility of creating prequantum classical models. These models can be both realistic and local. Here *realism (objectivity) is defined as a possibility to assign the values of quantum observables to quantum systems before measurement.* This viewpoint to realism is common in discussions related to Bell's inequality [31] and in general to interrelation of classical and quantum physics.

The Växjö interpretation 2001 was the (local) realistic interpretation [145].

In 2004 I visited Beckman Institute for Advanced Science and Technology (University of Illinois at Urbana-Champaign). After my talk (May 3, 2004), I discussed with A. Leggett a role of no-go theorems in QM. In particular, I wondered why Einstein had never

---

[a]We remind that Einstein, Podolsky, and Rosen considered nonlocality as the *absurd alternative* to incompleteness of QM [80]. This is also practically forgotten (ignored?). One can often find just the statement that quantum nonlocality was originally discussed by Einstein, Podolsky and Rosen [80]. In his reply to the EPR paper [37] Bohr did not mention nonlocality at all. He also was sure that QM is local.

mentioned the von Neumann's no-go theorem. (After Einstein's death, the book of von Neumann [296] was found in Einstein's office.) A. Leggett remarked that Einstein was mainly interested in the real physical situation rather than in formal mathematical statements. I started to think about the real physical situation described by QM. I understood that *even if some classical prequantum model is formally possible, in spite of all no-go "theorems," this does not imply that this model matches with the real physical situation.* Hence, although local realism cannot be rejected as a consequence of, e.g., violation of Bell's inequality, one has to carefully analyze matching of local realism with the real physical situation. Since, as for Einstein and Bohr, I did not take seriously quantum nonlocality, for me the questionable point was the possibility of realistic representation of quantum observables.

By reading Bohr's works [37, 38] I understood the fundamental role of *measurement context* in quantum measurements. Bohr stressed that the result of measurement is the sum of impacts of the quantum system and the measurement device. And it is impossible to distill the contribution of the system from the integral measurement result. Thus quantum observables are not objective. We could not assign a value of, e.g., position to a quantum system before measurement. We remark that we discuss measurement of a single quantum observable and not of a pair of observables. Hence, the essence of this discussion is not in the impossibility of the joint measurement of some quantum observables, the so-called incompatible observables, e.g., position and momentum (the Bohr's principle of complementarity [38]). We discuss *contextuality* of even a single quantum observable: the properties of measurement context cannot be separated from intrinsic properties of a quantum system.[a]

Unfortunately, in recent discussions related to Bell's theorem [31] the notion of contextuality is coupled to the joint measurement of compatible observables. In this book (following Bohr), we always

---

[a] May be Bohr would not accept that a quantum system has intrinsic properties at all. (Bohr's works are very difficult for understanding; try to read, e.g., Ref. [37].) At least it is clear that he was not an idealist. So, he might agree that, e.g., an electron has some intrinsic properties. However, for Bohr, these properties were fundamentally inapproachable in measurements.

use the notion of contextuality as the irreducible dependence of the result of measurement (of even a single observable) on the whole experimental arrangement.

Under influence of Bohr I rejected *Einstein's realism.*[b] Here two circumstances of totally different nature were very important.

One of my main scientific interests is the application of the mathematical formalism of QM outside of physics: to study statistical data from cognitive science, psychology, and finances [206]. In these applications observables, "mental observables," are nonobjective by their nature; contexts play a crucial role. And this is well known for psychologists and researchers working in cognitive science, sociology, and economics. This out-physics activity improved my understanding of the role of contextuality.

Another strong motivation to reject realism of quantum observables was related to the development of a measurement theory for PCSFT.[c] In Chapter 9 we shall develop a TSD model for measurements of prequantum random fields. TSD describes production of clicks by detectors and it reproduces the quantum probabilities of detection. It couples quantum observables with prequantum random fields. And this coupling is not objective. The value of a quantum observable is contextual, it depends on parameters of the measurement device in use and on the noisy background field. It could not be assigned to the prequantum field before fixing context of measurement.

For example, consider the position of a quantum particle. In PCSFT this particle is a wave, it is distributed over space (may be over whole space). By TSD to determine "its position," we put

---

[b]I also mention an intermediate version of the Växjö interpretation [152] in which I tried to combine realism with contextuality, see also Ref. [192]. This interpretation might be useful in some applications of quantum information outside of physics.

[c]We state again that in this discussion realism is regarded as objectivity of quantum observables, i.e., the possibility to assign the value of a quantum observable to the quantum system *before measurement.* We do not reject realism in its general philosophical meaning. In particular, quantum systems have their own properties, properties of objects. However, these objective properties could not be approached by the measurement devices in use. The problem is that the class of measurement devices is not large enough to approach the subquantum level. Of course, this theoretical discussion does not imply that it would be possible (at least for our civilization) to elaborate novel measurement devices to approach subquantum spatial and temporal scales.

barrier-type detectors in different places. And if in some place, say at the point x, this wave overcomes the barrier, the click is produced—the sign that the particle's position is given by x. In such a wave model, the particle's position depends on the experimental arrangement (context).

We also point to another source of nonobjectivity. The wave is random and moreover it is mixed with the random background field; in general, the wave's amplitude is low, but occasionally in some points it becomes high enough to overcome the barriers. TSD describes a class of prequantum random fields for which the probability that the random wave overcomes the barriers in a few places simultaneously is very small.[a] Nevertheless, it is nonzero. Thus the position observable is not only contextual (in the sense of the fundamental dependence on the experimental arrangement), but even *multivalued.* Thus nonobjectivity has even more fundamental character than the dependence on experimental context. Suppose, we consider not a single (idealized) position observable, but a class of observables corresponding to all existing detectors used to measure position. (In particular, they have different detection thresholds and operate in different noisy backgrounds.)[b] And let us discuss objectivity of the position observable corresponding to a fixed collection of detectors located in various points of space. The experimental arrangement is fixed! (In particular, all detection thresholds are fixed.) Suppose, we might totally eliminate the noisy background field. (In reality this is impossible.) Nevertheless, even in this situation we will not recover objectivity. Sometimes a few detectors would click simultaneously, producing the "multivalued position" of a quantum "particle."

Thus the new Växjö interpretation combines Einstein's and Bohr positions. As well as Einstein, we do not believe that the quantum state describes completely the physical state of a quantum

---

[a] This is in fact the problem of the magnitude of second-order coherence (see Section 1.3).

[b] By TSD, although the appearance of each individual click depends on the detector's parameters, especially on its threshold, the probability of detection is independent of these parameters. This is the result of the normalization procedure for detection probabilities: not by the total number of pulses (quantum systems) emitted by a source, but by the sum of clicks of the detectors in all channels. This is the natural operational normalization procedure (see Chapter 9).

system. The quantum state describes only statistical features. As well as Bohr, we do not consider quantum observables as objective quantities (intrinsic properties of objects). Quantum observables are fundamentally contextual.

Postulate 6c. (The Växjö interpretation). *A wave function describes correlations in prequantum random fields (which are symbolically represented as quantum systems).*

2.5 A Short Introduction to Classical Probability Theory

As we have seen, the main aim of QM is to provide probabilistic predictions about the results of measurements. Moreover, we have seen, that statistics of the results of measurements of a single quantum observable can be described by classical probability theory. We now present an elementary introduction to classical probability theory. This short section is written for readers who want to proceed on the mathematical level of rigorousness. In principle, one can proceed even without the mathematical formalization and jump to the next section devoted to quantum conditional probability.

The *Kolmogorov probability space* [218] is any triple

$$(\Omega, \mathcal{F}, \mathbf{P}),$$

where Ω is a set of any origin and $\mathcal{F}$ is a σ-algebra of its subsets, $\mathbf{P}$ is a probability measure on $\mathcal{F}$. The set Ω represents random parameters of the model. Kolmogorov called elements of Ω *elementary events*. This terminology is standard in mathematical literature. Sets of elementary events are regarded as *events*. The key point of Kolmogorov's axiomatization of probability theory is that not any subset of Ω can be treated as an event. For any stochastic model, the system of events $\mathcal{F}$ is selected from the very beginning. The key mathematical point is that $\mathcal{F}$ has to be a σ-algebra. (Otherwise it would be very difficult to develop a proper notion of integral. And the latter is needed to define average of a random variable.)

We remind that a σ-algebra is a system of sets which contains Ω and empty set, it is closed with respect to the operations of countable union and intersection and to the operation of taking the complement of a set. For example, the collection of all subsets of Ω is a σ-algebra. This σ-algebra is used in the case of finite or countable sets:

$$\Omega = \{\omega_1, \ldots, \omega_n, \ldots\}. \tag{2.19}$$

However, for "continuous sets," e.g., $\Omega = [a, b] \subset \mathbf{R}$, the collection of all possible subsets is too large to have applications. Typically it is impossible to describe a σ-algebra in the direct terms. To define a σ-algebra, one starts with a simple system of subsets of Ω and then consider the σ-algebra generated from this simple system with the aid of aforementioned operations. In particular, one of the most important for applications σ-algebras, the so-called *Borel σ-algebra*, is constructed in this way by staring with the system consisting of all open and closed intervals of the real line. In a metric space (in particular, in a Hilbert space), the Borel σ-algebra is constructed by starting with the system of all open and closed balls.

Finally, we remark that in American literature the term σ-field is typically used, instead of σ-algebra.

The probability is defined as a *measure*, i.e., a map from $\mathcal{F}$ to nonnegative real numbers which is σ-additive:

$$\mathbf{P}(\cup_j A_j) = \sum_j \mathbf{P}(A_j), \tag{2.20}$$

where $A_j \in \mathcal{F}$ and $A_i \cap A_j = \emptyset, i \neq j$. The probability measure is always normalized by one:

$$\mathbf{P}(\Omega) = 1. \tag{2.21}$$

In the case of a discrete probability space, see (2.19), the probability measures have the form

$$\mathbf{P}(A) = \sum_{\omega_j \in A} p_j, \;\; p_j = \mathbf{P}(\{\omega_j\}).$$

In fact, any finite measure μ, i.e., $\mu(\Omega) < \infty$, can be transformed into the probability measure by normalization:

$$\mathbf{P}(A) = \frac{\mu(A)}{\mu(\Omega)}, \; A \in \mathcal{F}. \tag{2.22}$$

A (real) random variable is a map $\xi : \Omega \to \mathbf{R}$, which is measurable with respect to the Borel σ-algebra $\mathcal{B}$ of $\mathbf{R}$ and the σ-algebra $\mathcal{F}$ of Ω. The latter means that, for any set $B \in \mathcal{B}$, its pre-image $\xi^{-1}(B) = \{\omega \in \Omega : \xi(\omega) \in B\}$ belongs to $\mathcal{F}$. (in fact, it is sufficient to check that pre-images of all intervals in $\mathbf{R}$ belong to $\mathcal{F}$.) This condition provides the possibility to assign the probability to the events of the type "values of ξ belong to a (Borel) subset of the real line." The probability distribution of ξ is defined as

$$\mathbf{P}_\xi(B) = \mathbf{P}(\xi^{-1}(B)), \ B \in \mathcal{B}. \tag{2.23}$$

In the same way we define the real (and complex) vector valued random variables, $\xi : \Omega \to \mathbf{R}^n$ ($\xi : \Omega \to \mathbf{C}^n$). Moreover, in this book we shall operate with random variables valued in complex Hilbert spaces (endowed with the Borel σ-algebras), $\xi : \Omega \to H$. The corresponding probability distributions are measures on H, see the books of Skorohod [279] and Daletsky and Fomin [58] for theory of such measures.

A random variable is called discrete if its image consists of finite or countable number of points, $\xi = \alpha_1, \ldots, \alpha_n, \ldots$. In this case its probability distribution has the form

$$\mathbf{P}(B) = \sum_{\alpha_j \in B} P_{\alpha_j}, \ P_{\alpha_j} = \mathbf{P}(\omega \in \Omega : \xi(\omega) = \alpha_j). \tag{2.24}$$

The mean value (average) of a real-valued random variable is defined as its integral (the Lebesgue integral)

$$E\xi = \int_\Omega \xi(\omega) dP(\omega). \tag{2.25}$$

For a discrete random variable, its mean value has the simple representation:

$$E\xi = \sum_{\alpha_j \in B} \alpha_j P_{\alpha_j}. \tag{2.26}$$

In the Kolmogorov's model the conditional probability is *defined* by the *Bayes formula*

$$\mathbf{P}(B|A) = \frac{\mathbf{P}(B \cap A)}{\mathbf{P}(A)}, \ \mathbf{P}(A) > 0. \tag{2.27}$$

We stress that other axioms are independent of this definition. In principle, one can proceed with the measure-theoretic model of probability, but change the definition of conditional probability, see Ref. [135] for discussions.

We also present the *formula of total probability* (FTP) which is a simple consequence of the Bayes formula. Consider the pair, a and b, of discrete random variables. Then

$$\mathbf{P}(b = \beta) = \sum_{\alpha} \mathbf{P}(a = \alpha)\mathbf{P}(b = \beta | a = \alpha). \qquad (2.28)$$

Thus the b-probability distribution can be calculated from the a-probability distribution and the conditional probabilities $\mathbf{P}(b = \beta | a = \alpha)$. These conditional probabilities are known as *transition probabilities.*

We remark that the Kolmogorov's measure-theoretic formalization of classical probability theory is just a special mathematical model. This model was created not by God, but by a man. (In fact, for Kolmogorov, one of the main motivations to write his book [218] was a possibility to earn money from its publication in German.) Comparing classical and quantum probabilities one should not forget that he compares just two *mathematical models.*

Some Kolmogorov's axioms reflect the properties of relative frequencies of realizations of events in a random experiment:[a] positivity and additivity of probability, its normalization by one. However, its countable additivity, i.e., the possibility to consider series and not only finite sums in (2.20), as well as usage of the σ-algebra as the domain of definition of probability are the purely mathematical conditions. They were borrowed from the already existing theory of the Lebesgue measure. Kolmogorov stressed this point in his book [218]. In reality, it is impossible to repeat a random experiment infinitely many times. Therefore, the condition of countable additivity is metaphysical. At the very beginning, i.e., before the final formulation of the axiomatics of the modern (classical) probability theory, Kolmogorov even did not exclude a possibility that, for some random experiments, the domain of definition of probability need not be even a set algebra (see Ref. [135] for the history of axiomatization of probability), i.e., it need not be closed even with respect to finite unions and intersections

[a]This part of Kolmogorov's axiomatics is simply the measure-theoretic counterpart of the von Mises frequency theory of probability in which the probability was directly defined as the limit of relative frequencies in a long series of measurements [293–295], see also Ref. [135].

of sets. The latter is, in fact, one of the basic properties of quantum probability [135].

We state again, see also Section 2.1, that Kolmogorov pointed to contextuality of his approach [218], also see Ref. [135]. Each experiment determines its own probability space; whether for some experiments their probability spaces can be embedded in a *single probability space* is a separate question. Thus the impossibility to describe quantum statistical data collected in a few experiments, measurements of incompatible observables, matches well with the Kolmogorov's ideology. In fact, he studied the problem of the existence of a single probability space in his seminal book [218]. He proved the famous theorem on the existence of a probability space for a stochastic process, say $\xi(t)$, starting with its finite-dimensional distributions,

$$\mathbf{P}_{t_1 \cdots t_n}(A_1 \times \cdots \times A_n) = \mathbf{P}((\xi(t_1), \cdots, \xi(t_n)) \in A_1 \times \cdots \times A_n).$$

He formulated the conditions on the family of finite-dimensional distributions which are necessary and sufficient for the existence of the single probability measure (on the space of all possible trajectories). This probability measure induces all these finite-dimensional probabilities.

However, Kolmogorov did not discuss examples of incompatible probability spaces. (Although he clearly understood that they exist.) It is not clear whether he understood that quantum probability and quantum incompatibility are related to incompatibility of classical probability spaces. Once he was asked why he did not study the probabilistic structure of QM. His answer was that "this is a too difficult problem" (private communication of prof. O. G. Smolyanov). May be this answer was related to the recognition of the probability multispace structure of QM.

Later in 1960s, when Vorob'jev [297] formulated the most general necessary and sufficient conditions of the existence of the joint probability space and proposed to work with the probability multispace structures, the reaction of Kolmogorov and the Soviet probabilistic school to his paper was in general negative; the paper was practically forgotten. We stress that the famous Bell's inequality [31] which made so much noise in quantum physics [2, 14, 69, 71, 72, 92] can be found in Vorob'jev's paper [297]. In fact, he considered all possible Bell-type inequalities; some of them have already been derived by physicists struggling with locality and realism, but some still did not appear in physical literature. In physical literature the first signal that Bell's inequality is related to the existence of the joint probability space was sent by Fine [84]. Unfortunately, it and a huge series of similar signals from mathematically thinking experts in quantum foundations (see Ref. [182] for review) were practically ignored by physicists.

We emphasize that quantum probability theory is not reduced to operation with a family of incompatible probability spaces. For the fixed quantum state Ψ, probability spaces corresponding to quantum measurements of even incompatible observables are fundamentally coupled through this state. One can transfer predictions on statistical data from one probability space to another by using the transition from one orthogonal basis (corresponding to one quantum observable) to another (corresponding to another quantum observable). Gudder [99] compared the structure of quantum probability with the structure of a manifold: the possibility to move from one chart of a manifold to another is a geometric equivalent of transition from one orthogonal basis to another. He also developed a very general model of *probabilistic manifolds* with applications to quantum probability.

We remark that the classical FTP (2.28) also provides the possibility of transition from the probabilities for the random variable a and the probabilities of transition from a to b to the probabilities for the random variable b. As we shall see in Section 2.7, the essence of QM is in the generalization of the classical FTP, a more general transition rule based on the quantum version of FTP, the FTP with an additional (interference) term corresponds to the transition from one basis to another.

How do incompatible probability spaces arise in the PCSFT framework? In fact, PCSFT by itself is based on the classical probability theory. By fixing the prequantum random field we fix the probability space. All prequantum physical variables (given by the quadratic forms of a random field) are realized on this single probability space. Even if two operators do not commute, the corresponding quadratic forms can still be realized as random variables on the same probability space. However, measurements disturb this classical harmony! By considering the measurement theory for PCSFT, namely, TSD, we found that one is not able to proceed with a single probability space. In fact, the situation is even worse (from the classical probabilistic viewpoint): we are not able to use the Kolmogorov's model even for a single quantum observable. We are not able to represent it as a classical random variable. This is a delicate point and we discuss it in more detail. As was pointed in Section 2.1, the statistical data from measurements of a single quantum observables can be described by a classical probability space. And it seems that the latter statement contradicts to the former (on "total nonclassicality" of TSD). However, in Section 2.1 we discussed a possibility of the formal mathematical representation of quantum observables considered as *operational entities.* In TSD, quantum observables are not just operational entities, they describe transformations of classical random fields into clicks of detectors (Chapter 9). And we found that such transformations cannot be represented by classical random variables. Here the temporal structure of statistical data plays a crucial role. We describe this structure by using the frequency von Mises approach to detection probability for a quantum observable [293–295], see also [135] for its modern presentation. The formulas (9.19) and (9.20) have the meaning only in the frequency probabilistic approach. We remark that in the frequency approach for each fixed sequence of observations, the so-called collective, it is possible to define the classical probability distribution on the set of all possible results of the measurement. This distribution was discussed in Section 2.1. However, this is not at all the probability distribution which is directly based on the set of random parameters of the prequantum random field (see Chapter 9 for details).

2.6 Quantum Conditional Probability

As in the classical Kolmogorov's probabilistic model, Born's postulate should be completed by a definition of conditional probability. We present the contemporary definition which is conventional in quantum logic and quantum information theory.

Definition 2.6.1. *Let physical observables a and b be represented by self-adjoint operators with purely discrete (may be degenerate) spectra:*

$$\widehat{a} = \sum_m \alpha_m P^a_{\alpha_m}, \ \widehat{b} = \sum_m \beta_m P^b_{\beta_m} \tag{2.29}$$

Let ψ be a pure state and let $P^a_{\alpha_k}\psi \neq 0$. Then the probability to get the value $b = \beta_m$ under the condition that the value $a = \alpha_k$ was observed in the preceding measurement of the observable a on the state ψ is given by

$$\mathbf{P}_\psi(b = \beta_m | a = \alpha_k) \equiv \frac{\| P^b_{\beta_m} \ P^a_{\alpha_k} \psi \|^2}{\| P^a_{\alpha_k} \ \psi \|^2} \tag{2.30}$$

Let the operator $\widehat{a}$ has nondegenerate spectrum, i.e., for any eigenvalue α the corresponding eigenspace (i.e., generated by eigenvectors with $\widehat{a}\psi = \alpha\psi$) is one dimensional. We can write:

$$\mathbf{P}_\psi(b = \beta_m | a = \alpha_k) = \| P^b_{\beta_m} e^a_k \|^2 \tag{2.31}$$

(here $\widehat{a} e^a_k = \alpha_k e^a_k$). Thus the conditional probability in this case does not depend on the original state ψ. We can say that the memory about the original state was destroyed. If also the operator $\widehat{b}$ has nondegenerate spectrum then we have: $\mathbf{P}_\psi(b = \beta_m | a = \alpha_k) = |\langle e^b_m, e^a_k \rangle|^2$ and $\mathbf{P}_\psi(a = \alpha_k | b = \beta_m) = |\langle e^a_k, e^b_m \rangle|^2$. By using symmetry of the scalar product we obtain the following proposition.

Proposition 5.1. *Let both operators $\widehat{a}$ and $\widehat{b}$ have purely discrete nondegenerate spectra and let $P^a_k \psi \neq 0$ and $P^b_m \psi \neq 0$. Then conditional probability is symmetric and it does not depend on the original state ψ :*

$$\mathbf{P}_\psi(b = \beta_m | a = \alpha_k) = \mathbf{P}_\psi(a = \alpha_k | b = \beta_m) = |\langle e^b_m, e^a_k \rangle|^2.$$

We remark that classical (Kolmogorov–Bayes) conditional probability is not symmetric, besides very special situations. Thus *QM is described by a very specific probabilistic model.*

Consider two nondegenerate observables. Set $p_{\beta|\alpha} = \mathbf{P}(b = \beta | a = \alpha)$. The matrix of transition probabilities $\mathbf{P}^{b|a} = (p_{\beta|\alpha})$ is not only *stochastic,* i.e.,

$$\sum_\beta p_{\beta|\alpha} = 1$$

but it is even *doubly stochastic.*:

$$\sum_\alpha p_{\beta|\alpha} = \sum_\alpha |\langle e^b_\beta, e^a_\alpha \rangle|^2 = \langle e^b_\beta, e^b_\beta \rangle = 1.$$

In the Kolmogorov's model, stochasticity is the general property of matrices of transition probabilities. However, in general classical matrices of transition probabilities are not doubly stochastic. Hence, double stochasticity is a very specific property of quantum probability.

2.7 Interference of Probabilities in Quantum Mechanics

We shall show that the quantum probabilistic calculus violates the conventional FTP (2.28).

Let $H_2 = \mathbf{C} \times \mathbf{C}$ be the two-dimensional complex Hilbert space and let $\psi \in H_2$ be a quantum state. Let us consider two dichotomous observables $b = \beta_1, \beta_2$ and $a = \alpha_1, \alpha_2$ represented by self-adjoint operators $\widehat{b}$ and $\widehat{a}$, respectively (one may consider simply Hermitian matrices). Let $e^b = \{e^b_\beta\}$ and $e^a = \{e^a_\alpha\}$ be two orthonormal bases consisting of eigenvectors of the operators. The state ψ can be represented in the two ways

$$\psi = c_1 e^a_1 + c_2 e^a_2, \;\; c_\alpha = \langle \psi, e^a_\alpha \rangle; \tag{2.32}$$

$$\psi = d_1 e^b_1 + d_2 e^b_2, \;\; d_\beta = \langle \psi, e^b_\beta \rangle. \tag{2.33}$$

By Postulate 4 we have

$$\mathbf{P}(a = \alpha) \equiv \mathbf{P}_\psi(a = \alpha) = |c_\alpha|^2. \tag{2.34}$$

$$\mathbf{P}(b = \beta) \equiv \mathbf{P}_\psi(b = \beta) = |d_\beta|^2. \tag{2.35}$$

The possibility to expand one basis with respect to another basis induces connection between the probabilities $\mathbf{P}(a = \alpha)$ and $\mathbf{P}(b = \beta)$. Let us expand the vectors e^a_α with respect to the basis e^b

$$e^a_1 = u_{11} e^b_1 + u_{12} e^b_2 \tag{2.36}$$

$$e^a_2 = u_{21} e^b_1 + u_{22} e^b_2 \tag{2.37}$$

where $u_{\alpha\beta} = \langle e^a_\alpha, e^b_\beta \rangle$. Thus $d_1 = c_1 u_{11} + c_2 u_{21}$, $d_2 = c_1 u_{12} + c_1 u_{22}$. We obtain the *quantum rule* for transformation of probabilities:

$$\mathbf{P}(b = \beta) = |c_1 u_{1\beta} + c_2 u_{2\beta}|^2. \tag{2.38}$$

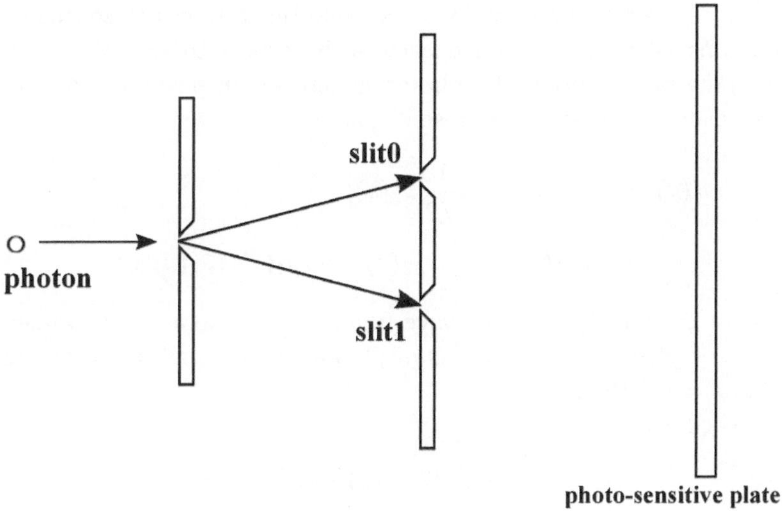

Figure 2.1 Experimental setup.

On the other hand, by the definition of quantum conditional probability, see (2.30), we obtain

$$\mathbf{P}(b = \beta | a = \alpha) \equiv \mathbf{P}_{\psi}(b = \beta | a = \alpha) = |\langle e^a_\alpha, e^b_\beta \rangle|^2. \tag{2.39}$$

By combining (2.34), (2.35) and (2.38), (2.39) we obtain the *quantum FTP—the formula of the interference of probabilities*:

$$\mathbf{P}(b = \beta) = \sum_\alpha \mathbf{P}(a = \alpha)\mathbf{P}(b = \beta | a = \alpha) \tag{2.40}$$

$$+2\cos\theta\sqrt{\mathbf{P}(a = \alpha_1)\mathbf{P}(b = \beta | a = \alpha_1)\mathbf{P}(a = \alpha_2)\mathbf{P}(b = \beta | a = \alpha_2)}$$

In general $\cos\theta \neq 0$. Thus the quantum FTP does not coincide with the classical FTP (2.28), which is based on the Bayes' formula.

2.8 Two-Slit Experiment

The two-slit experiment is the basic example demonstrating that QM describes statistical properties in microscopic phenomena, to which the classical probability theory seems to be not applicable, see, e.g., Ref. [85].

In this experiment (with the symmetric setting: the source of photons is located symmetrically with respect to two slits, Fig. 2.1), the probability that a photon is detected at position x on the photosensitive plate[a] is represented as

$$P(x) = \left| \frac{1}{\sqrt{2}} \psi_1(x) + \frac{1}{\sqrt{2}} \psi_2(x) \right|^2$$
$$= \frac{1}{2} |\psi_1(x)|^2 + \frac{1}{2} |\psi_2(x)|^2 + |\psi_1(x)|\,|\psi_2(x)| \cos\theta,$$

where ψ_1 and ψ_2 are two wave functions, whose absolute values $|\psi_k(x)|^2$ give the distributions of photons which pass through the slits $a = 1, 2$ (Fig. 2.2).

The term

$$|\psi_1(x)|\,|\psi_2(x)| \cos\theta$$

implies the interference effect of two wave functions. Let us denote $|\psi_k(x)|^2$ by $P(x|k)$, and then the above equation is represented as

$$P(x) = P(x|1)P(1)+P(x|2)P(2)+2\sqrt{P(x|1)P(1)P(x|2)P(2)} \cos\theta, \quad (2.41)$$

where $P(1) = P(2) = 1/2$. In the above form, it seems that a classical probability law, FTP,

$$P(x) = P(x|1)P(1) + P(x|2)P(2), \quad (2.42)$$

is violated, and the term of interference $2\sqrt{P(x|1)P(1)P(x|2)P(2)} \cos\theta$ specifies the violation. For this violation of FTP, we give the following interpretation.

2.8.1 *Contextuality of Data*

The statistical property in context that a photon can pass through slit 1 or slit 2 (we denote this context by $C_{1\cup 2}$, Fig. 2.2) is not a simple sum of contributions of contexts C_1 and C_2 (Figs. 2.3 and 2.4), in

[a] To match with notation of Section 2.7, we select a as the "slit passing variable," i.e., $a = 1, 2$, and b as the position on the photosensitive plate. The main mathematical difference from Section 2.7 is that the b-variable has the continuous range of values. Hence, its eigenvectors are given by Dirac δ-functions: $\psi(x) = \langle \psi | x \rangle$, where $|x\rangle$ is the generalized eigenvector corresponding to the position x on the photosensitive plate.

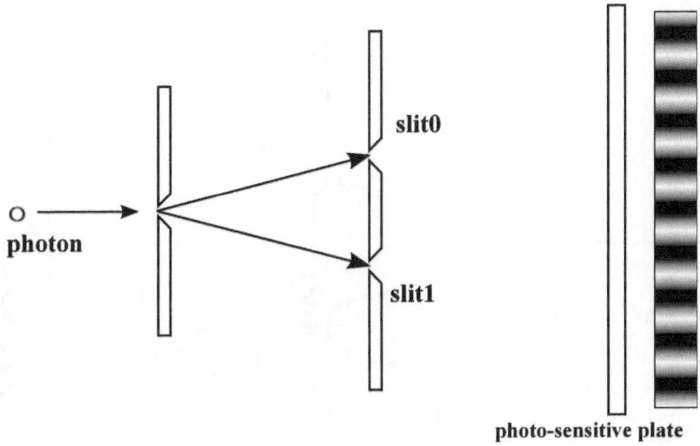

Figure 2.2 Context with both slits open.

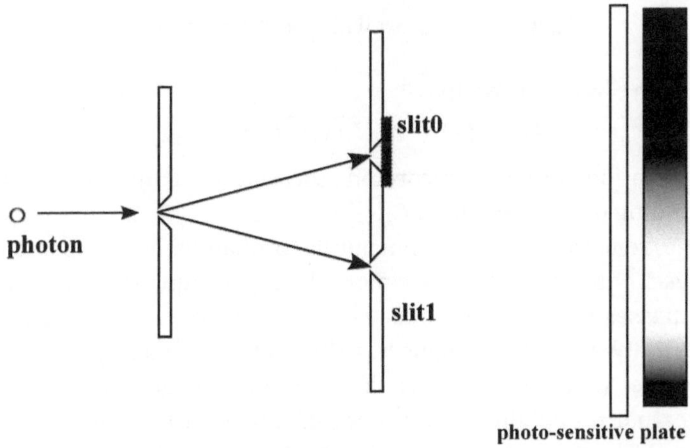

Figure 2.3 Context with one slit is closed, I.

which a photon can pass through either 1 or 2. We represent this fact as

$$C_{1\cup 2} \neq C_1 \cup C_2. \tag{2.43}$$

Let $P_{C_{1\cup 2}}(x)$ be the probability that $b = x$ in context $C_{1\cup 2}$ and let $P_{C_1 \cup C_2}(x)$ be one in context $C_1 \cup C_2$. The probability $P_{C_{1\cup 2}}(x)$ coincides with $P(x)$ in Eq. (2.41), and $P_{C_1 \cup C_2}(x) = 1$ in Eq. (2.42). Then, the

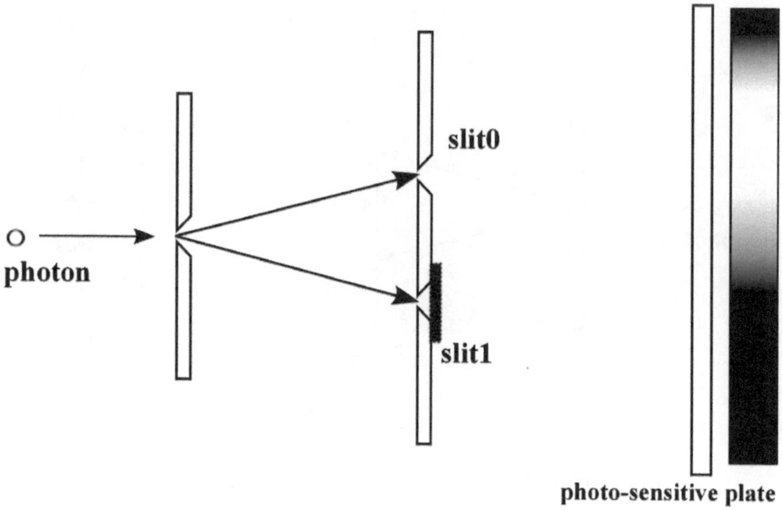

Figure 2.4 Context with one slit is closed, II.

equation (2.41) is rewritten as

$$P_{C_{1\cup 2}}(x) = P_{C_1 \cup C_2}(x) + 2\sqrt{P(x|1)P(1)P(x|2)P(2)}\cos\theta.$$

The violation of FTP is interpreted as a relation of different statistics depending on $C_{1\cup 2}$ and $C_1 \cup C_2$.

We remark that from the purely probabilistic viewpoint this analysis of the two-slit experiment does not imply that each "photon" passes through both slits, cf. Ref. [74]. We just demonstrated that the two-slit experiment has the multicontextual probabilistic structure. (See the monograph [192] on the general theory of contextual probability and a place of quantum probability in this theory.) The physical situation for the two open slits, context $C_{1\cup 2}$, cannot be represented as a straightforward combination, context $C_1 \cup C_2$, of the physical situations for only one slit open, contexts C_1 and C_2. The violation of classical FTP shows that it is impossible to embed all contexts under consideration into a single Kolmogorov probability space. It is impossible to represent these contexts as subsets of some set of random parameters Ω in such a way that $C_{1\cup 2} = C_1 \cup C_2$ and to define a single probability measure $\mathbf{P}$ on some σ-algebra of subsets of Ω in such a way that the corresponding conditional probabilities are defined through the

Bayes formula. (Contexts which cannot be embedded into a single Kolmogorov probability space can be called [160] *incompatible contexts.*) However, the contextuality of physical statistical data can have various physical sources. Hence, the "mystery of statistical data from the two-slit experiment" (in fact, three experiments corresponding to contexts $C_{1\cup 2}$, C_1, C_2 [85]) cannot be resolved through a purely mathematical analysis of the probabilistic structure of this experiment. The resolution of the two-slit mystery can only come from the analysis of the real physical situation, usage of physical models and with the aid of physical intuition.

In principle, even the purely corpuscular model can reproduce contextuality of the two-slit experiment, see, e.g., Refs. [77, 122, 134].

However, one may to ask: *Do such models of corpuscular interference correspond to the real physical situation?*

2.8.2 *Wave Interference*

In our classical field model, PCSFT, violation of the Boolean structure, see (2.43), in the two-slit experiment (the complex of three different experiments) has the most natural physical explanation. A photon is nothing else than a pulse of the classical electromagnetic field. It is spatially distributed and interacts with both slits (when they are both open). One may say a photon goes simultaneously through both slits. Hence, closing one slit, say the first slit, i.e., moving from context $C_{1\cup 2}$ to context C_2, changes the structure of physical interaction and, consequently, the probability distribution of dots on the photosensitive plate. Thus, in prequantum models of the field type contextuality of the statistical data is easily explained by the spatial structure of fields. In such models, the main problem is to explain the absence of coincidence counts (see Section 1.3). Consider another experimental context: both slits are open and a photon detector is placed directly behind each slit. Denote this experimental context by the symbol $C_{1\cup 2,\text{detectors}}$ (Fig. 2.5). In this context only one detector can click. This is the prediction of QM. In accordance with Bohr's principle of complementarity context $C_{1\cup 2,\text{detectors}}$ representing the particle features of photons is incompatible with context $C_{1\cup 2}$ representing

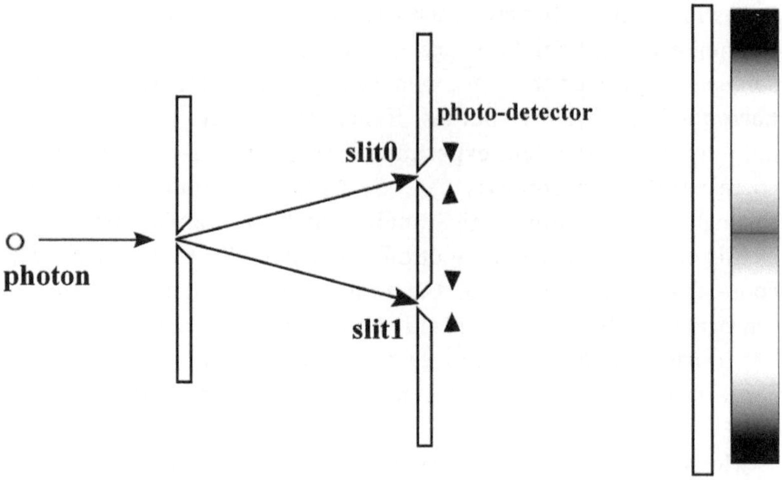

Figure 2.5 Context with two detectors.

the wave features of photons. In some way a photon "knows" that detectors are placed and it does not behave as a wave (as it does in context $C_{1\cup 2}$). Experimentally the absence of coincidence counts in context $C_{1\cup 2,\text{detectors}}$ is confirmed through evaluation of the coefficient of second-order interference (Section 1.3).

PCSFT does not describe clicks of detectors. It describes a prequantum field as it is without interaction with detectors. Nevertheless, the PCSFT averages and correlations coincide with quantum ones. This coincidence motivates searching of a measurement model for PCSFT which would describe transformations of continuous prequantum fields into discrete clicks of detectors. Surprisingly a simplest phenomenological model in which the probability of detection is proportional to the energy of a prequantum field in the corresponding channel solves the measurement problem for PCSFT (Chapter 8). In Chapter 9, this phenomenological model is completed by considering details of interaction of a prequantum random field with a detector. The final model of threshold detection, TSD, explains both the interference picture in context $C_{1\cup 2}$ and the absence of coincidence counts for context $C_{1\cup 2,\text{detectors}}$. (In contrast to QM, TSD describes the real experimental situation and not its operational idealization. In TSD the "absence of coincidence counts" means that the coefficient of second-order interference is small enough.)

The contextual structure of the two-slit experiments was explained already in the monograph by Khrennikov [135] (in the first edition). From purely mathematical viewpoint, this is the end of the story. However, from the physical viewpoint this is just the first step to understanding that any prequantum model combined with the corresponding measurement theory reproducing statistics of individual quantum events, clicks of detectors, has to be contextual. The example PCSFT/TSD shows (see Chapter 9) that at the subquantum level one can proceed without contextuality, it can appear only on the level of a measurement model.

2.9 Corpuscular Interference

2.9.1 *Memory Effects and Violation of Ergodicity*

In Refs. [77, 134] it was demonstrated that the probabilistic structure of data collected in the two-slit experiment can be reproduced by a model describing deterministic particles, each of them passes through only one of the slits. The crucial assumption of this model of corpuscular interference is the presence of *memory effects* in the experimental equipment: the plate with slits, the source and the photosensitive plate. Deterministic particles hitting the photosensitive plate can produce realistic interference patters. To produce the interference patterns, we used p-adic numbers, p-adic configuration space, and dynamical systems acting in this space. However, in this framework usage of p-adics did not have a fundamental value, cf. Refs. [132, 133, 291]. This was just a good mathematical tool to simplify mathematical modeling. As was pointed out, the crucial physical point is the presence of memory effects in the experimental equipment. This "memory assumption" can be checked experimentally. A possibility to perform such an experimental test in the framework of neutron interferometry was discussed with H. Rauch, at that time the director of Atominstitute, Vienna. Unfortunately, this experiment had never been performed.

We remark that the presence of memory effects implies that registered statistical data is mathematically described by non-stationary stochastic processes. In particular, such processes are nonergodic. The hypothesis of *nonergodicity* of QM was discussed in literature [45]. In particular, Summhammer [285] performed a test of ergodicity of QM in the framework of neutron interferometry. This test confirmed the ergodicity of QM.[a] A recent model [122] also explores the hypothesis on the presence of memory effects in the experimental equipment. This model is supported by numerical simulation. Also, its predictions can be tested experimentally in the manner similar to tests for the models presented in Refs. [77, 134].

2.9.2 *Existence of Indivisible Time-Quantum and Discrete Dynamics*

In this section we consider another model of corpuscular interference [146, 153], see also Ref. [192]. It implies a fundamental departure from Newtonian mechanics (opposite to models presented in Section 2.9.1, which did not question the foundations of classical mechanics, but simply emphasized the possibility of memory effects in experimental equipment). On the subquantum level time is discrete, there exists fundamental and indivisible quantum of time. In reality dynamics of quantum particles are the deterministic discrete time dynamics. Such dynamics are described by difference equations, instead of differential equations of classical Newtonian dynamics.[b] The realistic interference pictures and the violation of the classical FTP were simulated numerically. (Thus, before concentrating efforts on the classical field models, I had tested all possibilities to proceed beyond QM.)

Let us start with a postulate about the discreteness of time.

---

[a] Of course, a single experimental test cannot be used to reject the hypothesis of nonergodicity. A series of more careful experimental texts has to be performed, e.g., those proposed in Refs. [77, 134].

[b] Conventionally difference dynamics are considered as approximations of continuous dynamics of the Newtonian type. We proceed another way around: continuity is just the illusion, the result of impossibility (at the present time) to approach the discrete scale of time.

Postulate(D) *There exists the fundamental (indivisible) quantum of time* τ.

"Which features of the conventional quantum formalism could be reproduced on the basis of postulate D?"

We demonstrated [146, 153, 192] that at least diffraction and interference effects can be reproduced by Newtonian particles with discrete time dynamics.

2.9.2.1 Discrete time dynamics

In classical mechanics a dynamical function $A = A(p, q)$ (here p and q are momenta and coordinates of the system) evolves according to the equation

$$D_t A = \{A, H\} \tag{2.44}$$

where $H = H(p, q)$ is the Hamilton function of the system and in the right-hand side is a Poisson bracket, which could be presented as

$$\{A, B\} = \frac{\partial A}{\partial q}\frac{\partial B}{\partial p} - \frac{\partial A}{\partial p}\frac{\partial B}{\partial q}. \tag{2.45}$$

The left-hand side of (2.44) contains a continuous time derivative

$$D_t A = \frac{dA}{dt}.$$

As was mentioned earlier, we are interested in the study of dynamics with discrete time. This is done with the help of a *discrete derivative* which is postulated to be

$$D_t^{(\tau)} A = \frac{1}{\tau}[A(t+\tau) - A(t)],$$

where τ is the discreteness parameter. This parameter is finite and is treated in the same way as Planck constant in quantum mechanical formalism. In particular if τ is small relative to dimensions of the system, then classical approximation with a continuous derivative might work well (although this could not be the case all the time in the same sense as there are examples when quantum formalism is reasonable even for macroscopic systems, for example, in superfluidity).

Summarizing, the discrete time dynamical equation is postulated to be

$$D_t^{(\tau)} A = \{A, H\}, \tag{2.46}$$

where $A(p, q)$ is a real-valued function of real-valued dynamical variables and in the right-hand side there is a classical Poisson bracket (2.45). The equation (2.46) could be solved in the sense that we can write

$$A(t + \tau) = A(t) + \tau\{A, H\} \tag{2.47}$$

thus providing the evolution of any dynamical function $A = A(p, q)$.

Note that in our model the *coordinate space is continuous.*

Consider the well-known Newton's equation

$$\mathbf{F} = m\ddot{\mathbf{r}}. \tag{2.48}$$

Let us rewrite the second-order differential equation (2.48) as a system of first-order differential equations. We have

$$\begin{aligned} \mathbf{F} &= m\dot{\mathbf{v}}, \\ \mathbf{v} &= \dot{\mathbf{r}}. \end{aligned} \tag{2.49}$$

In the system (2.49) the derivatives assume the continuousness of time. Let us now introduce a discreteness parameter τ. We have

$$\begin{aligned} \mathbf{F} &= m\frac{\mathbf{v}(t + \tau) - \mathbf{v}(t)}{\tau}, \\ \mathbf{v}(t + \tau) &= \frac{\mathbf{r}(t + \tau) - \mathbf{r}(t)}{\tau}. \end{aligned} \tag{2.50}$$

In the limit of $\tau \to 0$, (2.50) is equivalent to (2.48) and (2.49). Now let us rewrite the system (2.50) in a directly computable way

$$\begin{aligned} \mathbf{v}(t + \tau) &= \mathbf{v}(t) + \mathbf{F}\tau/m, \\ \mathbf{r}(t + \tau) &= \mathbf{r}(t) + \mathbf{v}(t)\tau. \end{aligned} \tag{2.51}$$

where $\mathbf{F} = \mathbf{F}(\mathbf{r}(t), \mathbf{v}(t), t)$.

Note that in our model the coordinate space is continuous.

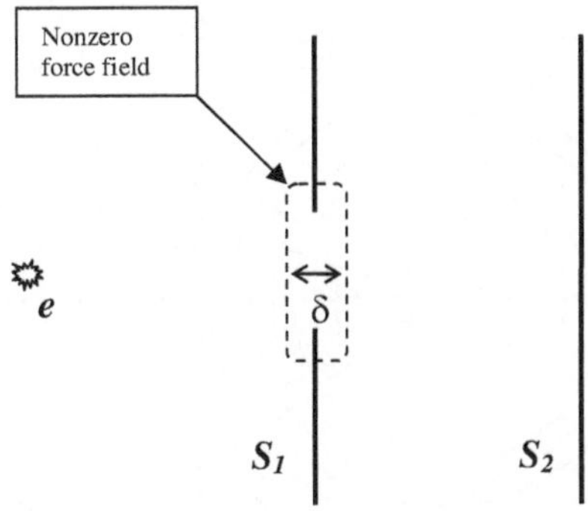

Figure 2.6 Particle scattering on a screen with a slit. Particles are emitted from the source e, pass through a slit in screen S_1 and gather on screen S_2. In the region δ near the slit there is a nonzero force field which affects the particles.

2.9.2.2 Diffraction pattern in a single-slit scattering

Consider the following experimental setup (see Fig. 2.6). A particle source e is located in front of the center of a slit in a screen S_1. Near the slit there is a nonzero force field which affects particles,

$$\mathbf{F}(x, y) \neq 0, \text{ for } (x, y) \in \delta$$

(x and y are coordinates along horizontal and vertical axes respectively, the axes origin being in the center of the slit). Particles pass through the slit and concentrate on a second screen S_2. We study a particle density on the screen S_2.

Let us start from the simplest case when the force field is constant and perpendicular to the screens, $\mathbf{F}(x, y) = \mathbf{e}_x F_0$, for $x \geqslant 0$ and 0 otherwise. In this case the region δ effectively covers the whole space between the screens S_1 and S_2. Although such a case is rather nonphysical, in the general case it allows us to construct a simple model of the particle dynamics. We will come to a more physically natural model below. Let the source e be point-like, emitting particles with constant velocity v_0 under random evenly distributed angles $\alpha \in [0, 2\pi)$. In this case trajectories of particles emitted by e (in discrete time dynamics) in the region $x < 0$ (i.e., before the screen S_1) form concentric circles originating from e with the radii $r_n = v_0 \tau n$, $n = 1, 2, \ldots$ (Fig. 2.7). We get the circles, which are trajectories of many particles emitted under close angles, in the

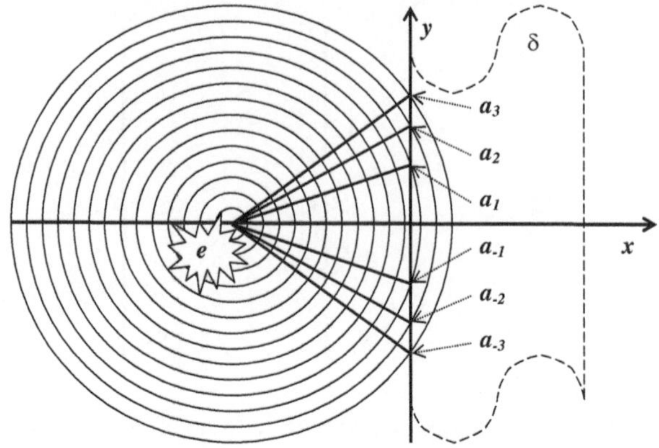

Figure 2.7 Particles emitted from the source e enter the region δ. Points $a_{\pm 1}, a_{\pm 2}, a_{\pm 3}, \ldots$ are origins of deviation which form a diffraction pattern.

region where there is no force field and particles move along straight lines, exactly as in classical dynamics. Let a_i be the points where these circles enter the region δ. The distance between the center of the slit and a_i is given by

$$a_i = \operatorname{sign}(i)\sqrt{v_0^2\tau^2(n_0 + |i|)^2 - d^2}, \quad i = \pm 1, \pm 2, \ldots, \tag{2.52}$$

where d is the distance between e and S_1, and n_0 is the largest integral value not greater than a fraction $d/v_0\tau$,

$$n_0 = \left\lfloor \frac{d}{v_0\tau} \right\rfloor. \tag{2.53}$$

This rather simple setup already produces an interesting nontrivial diffraction pattern (Fig. 2.8).

The points a_i are the origins of deviation from "classical" (i.e., continuous time) trajectories. This deviation forms a diffraction pattern. To argue this let φ_i be an angle between the horizontal axis and a line connecting e and a_i. Particles emitted under angles less than φ_i ($i \geqslant 1$) become affected by the force field in the region δ one step earlier than those emitted under angles greater than φ_i. As a result there appears a "fork"—even with very close trajectories but with the angles above and below φ_i becoming separated. One could get the points of minima of the diffraction pattern by following the trajectory along the line $e \to a_i$ and the parabolic curve (movement

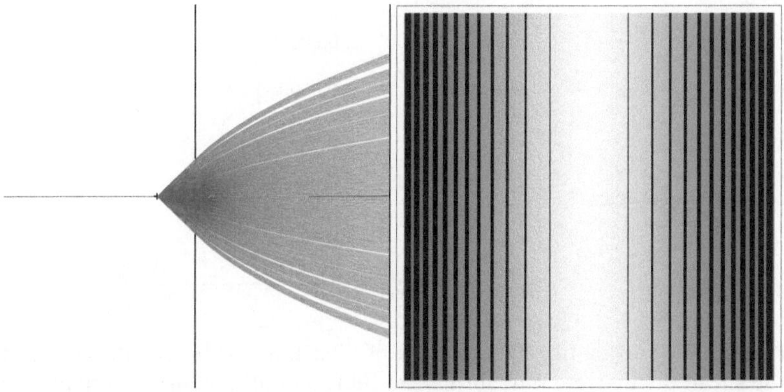

Figure 2.8 Particle trajectories (left) and diffraction pattern (right) for a single-slit scattering computed in the discrete time formalism. Brighter area corresponds to the higher particle density.

in a constant field) in the region $x > 0$ until the trajectory hits the screen S_2.

The case described above is not completely physically reasonable, indeed it requires a force field everywhere in the half-plane $x \geqslant 0$, thus it could be considered only as a simplified model which still demonstrates some interesting properties of discrete time dynamics. An example of such a property is a Fresnel-like phenomenon of a black region behind the center of the slit which is discussed below.

One could see from (2.52) and (2.53) that if

$$d = v_0 \tau k + \varepsilon, \quad \text{where } k = 1, 2, \ldots$$

and $\varepsilon \rightarrow 0$, then $|a_{+1} - a_{-1}| \rightarrow 0$, This means that particle trajectories become separated, forming a region without dots within the center of the screen S_2 just behind the slit (Fig. 2.9).

Now let us consider the case when a field is localized near the slit. We performed numerical simulations taking the force field of the Gaussian form

$$\mathbf{F}(x, y) = \mathbf{e}_x \, F_0 e^{-\sigma x^2}. \tag{2.54}$$

The results of this simulation again show a diffraction-like picture. It is very stimulating that such a setup also produces nontrivial patterns, since it means that one could start thinking of the physical

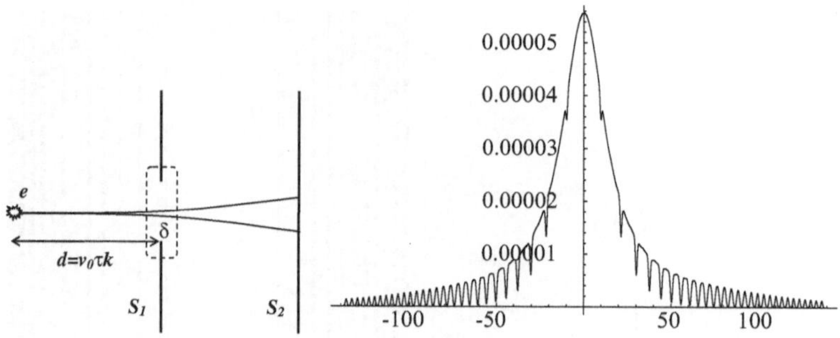

Figure 2.9 Fresnel-like phenomenon of a black region behind the center of the slit in the discrete time formalism (left) and diffraction pattern for the Gaussian force (2.54) (right).

nature of this force field. We have also considered the case of a non-point-like Gaussian source distributed along the vertical line; results of this simulation are presented in Fig. 2.9.

We have shown that a diffraction picture could appear as a statistical effect for deterministic particles, i.e., having trajectories and obeying deterministic equations, if one introduces a discrete time. The nature of the resulting diffraction picture (particle distribution) does not follow from the geometry of a force field alone, but is strongly attached to the discreteness parameter τ.

2.9.2.3 Interference in the two-slit experiment for deterministic particles

We now consider a classical analog of the two-slit experiment (Fig. 2.10). The uniformly charged round particles are emitted at the point e with fixed velocity with the angles evenly distributed in the range $[0, 2\pi)$. Each particle interacts with the uniformly charged flat screen S_1. The charge distributions on the particle and the screen stay unchanged even if the particle comes close to the screen. Physically this is a good approximation when the particle and the screen are both made of dielectric. There are two rectangular slits in the screen (in Fig. 2.10, the slits are perpendicular to the plane of the picture). Particles pass through the slits in screen S_1 and gather on screen S_2.

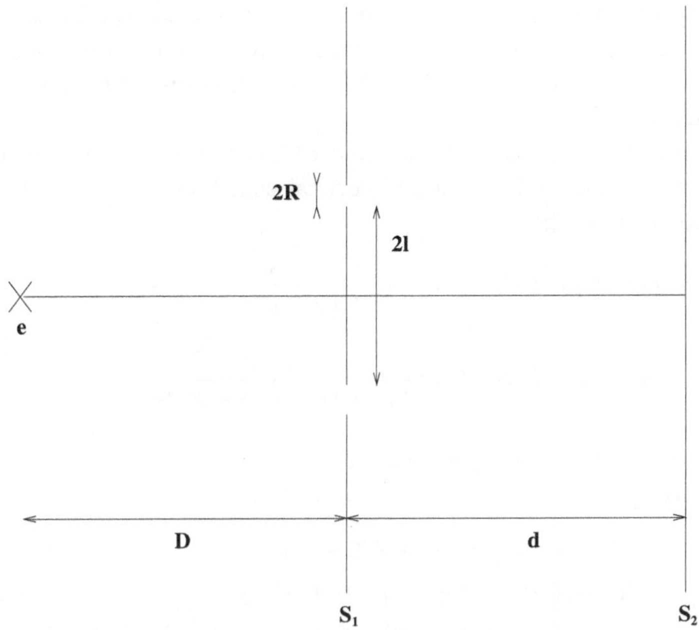

Figure 2.10 Two-slit experiment.

We consider three experiments. In the first one the bottom slit is closed with the shutter, in the second the upper slit is closed, and in the third both slits are left open. The charge distribution on the shutter is the same as on the screen, i.e., in the first two experiments one can consider the uniformly charged screen to have only one slit. In this and several paragraphs below, by the screen we mean screen S_1.

Now let us write the equations of motion in each of three experiments ($i = 1, 2, 3$)

$$m\ddot{\mathbf{r}} = \mathbf{F_i}, \tag{2.55}$$

where **r** determines the place of the particle. Here $\mathbf{F_i}$ is force affecting the particle in each experiment. It is given by Coulomb's law

$$\mathbf{F_i} = \int\limits_{\mathcal{D}_i} \frac{q\sigma}{|\mathbf{r}'|^2} \cdot \frac{\mathbf{r}'}{|\mathbf{r}'|}\, ds, \tag{2.56}$$

where $\mathbf{r}'$ is a vector from an element on the screen to the particle, q is the charge of the particle, σ is charge density on the screen, i.e., charge of a unit square. We integrate over the surface of the screen, the integration region $\mathcal{D}_i$ is the plane of the screen except the slits; as mentioned above it is different in each experiment depending on which slits are opened.

Projecting equations (2.55) and (2.56) to the xy-plane, where x and y denote horizontal and vertical coordinates of the particle respectively, we get

$$m\ddot{x} = q\sigma \int_{\Gamma_i} dy' \int_{\mathbb{R}} dz' \frac{x}{(x^2 + (y - y')^2 + z'^2)^{3/2}},$$
$$m\ddot{y} = q\sigma \int_{\Gamma_i} dy' \int_{\mathbb{R}} dz' \frac{y - y'}{(x^2 + (y - y')^2 + z'^2)^{3/2}}, \tag{2.57}$$

where Γ_i indicates the integration region for the ith experiment. In our previous notation $\mathcal{D}_i = \Gamma_i \times \mathbb{R}$. We have

$$\Gamma_1 = (-\infty, l) \cup (l + 2R, +\infty),$$
$$\Gamma_2 = (-\infty, -l - 2R) \cup (-l, +\infty), \tag{2.58}$$
$$\Gamma_3 = (-\infty, -l - 2R) \cup (-l, l) \cup (l + 2R, +\infty).$$

Here $2l$ is the distance between slits and $2R$ is the height of the slit.

Integrating the right-hand side of (2.57) we get

$$m\ddot{x} = q\sigma \sum_{(a,b)\subset\Gamma_i} 2\left(\arctan\frac{b - y}{x} - \arctan\frac{a - y}{x}\right),$$
$$m\ddot{y} = q\sigma \sum_{(a,b)\subset\Gamma_i} \ln\left(x^2 + (b - y)^2\right) - \ln\left(x^2 + (a - y)^2\right), \tag{2.59}$$

where the notation $(a, b) \subset \Gamma_i$ means that the sum extends over all subranges of Γ_i given in (2.59). For example, for $i = 1$ we have two summands with $(a = -\infty, b = l)$ and $(a = l + 2R, b = +\infty)$, and (2.59) will take the form

$$m\ddot{x} = 2q\sigma\left(\pi + \arctan\frac{l - y}{x} - \arctan\frac{l + 2R - y}{x}\right),$$
$$m\ddot{y} = q\sigma \ln\frac{x^2 + (l - y)^2}{x^2 + (l + 2R - y)^2}. \tag{2.60}$$

Here we took into account that $\arctan(\pm\infty) = \pm\pi/2$ and the sum of the logarithms for $a = -\infty$ and $b = +\infty$ vanishes.

We take the initial values

$$\begin{aligned} x(0) &= -D, \qquad \dot{x}(0) = v_0 \cos\alpha, \\ y(0) &= 0, \qquad \dot{y}(0) = v_0 \sin\alpha, \end{aligned} \tag{2.61}$$

where angle α is a random variable uniformly distributed in $[0, 2\pi)$. The constant parameters v_0 and D are initial velocity and distance between the emitter and the screen.

Particles are emitted at point e (see Fig. 2.10), move obeying (2.59) and (2.61) passing through slit(s) in the screen S_1 and gather on the screen S_2. Having points where particles hit the screen S_2, we compute frequencies with which particles appear on screen S_2 as a function of coordinates on the screen. We interpret these frequencies as probability distributions. We are interested in computing the probability distribution over a vertical line on screen S_2 with $z = 0$. That is why we consider a motion only in the xy-plane and the initial values (2.61) do not contain a z-coordinate.

We solve the equations of motion (2.59) with initial conditions (2.61) numerically. We use Runge–Kutta fourth-order switching to Adams fourth-order method. We used a GNU C++ (g++) compiler to realize the simulation on an Ultra-SPARC computer running Solaris. We had to explore about 10^5 trajectories and we used a four-processor parallel computer located at Växjö University. The computation process was easy to make parallel as moving particles were not interacting, i.e., one could consider them as being emitted with long intervals. The algorithm automatically adjusted the computation precision, making shorter steps when the particle came near to the first screen or the coordinates (x and y) are changed more than minimum precision allowed. The first stage of computation was calibration when the algorithm determined the angle ranges for which the particles passed through the slits and hit the second screen. This reduced the angle range from $[0, 2\pi)$ to a set of ranges, which are different in each experiment. In fact we used symmetry of the first two emissions (when only upper or lower slit is opened) making computations only for the first emission. The second screen was separated with cells of equal size, the diameter of a particle. The number of particles which hit each cell was calculated and interpreted as a probability distribution.

In fact, we present in this picture densities for numbers of particles, i.e., not normalized distributions. In particular, this implies

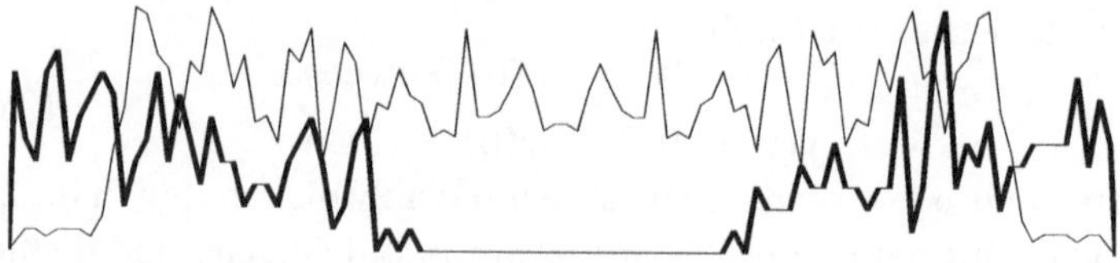

Figure 2.11 Probability distributions. Thin line is $\left(\frac{P_1}{2}+\frac{P_2}{2}\right)$, thick line is P_{12}.

(due to errors of statistical stabilization) asymmetry of densities. This asymmetry would disappear after normalization when the number of particles goes to infinity.

Let us denote the probability distribution in the first experiment (only the upper slit is opened) as $P_1 = P_1(y)$, in the second experiment (only the lower slit is opened) as $P_2 = P_2(y)$, and in the third experiment (both slits are opened) as $P_{12} = P_{12}(y)$. Although the force is different in each experiment, it is quite clear that (Fig. 2.11)

$$P_{12} \neq \frac{P_1}{2} + \frac{P_2}{2}. \tag{2.62}$$

To become an equality the above equation should have an extra term,

$$P_{12} = \frac{P_1}{2} + \frac{P_2}{2} + \sqrt{P_1 P_2}\cos\theta, \tag{2.63}$$

where $\sqrt{P_1 P_2}\cos\theta$ is the interference term (Fig. 2.12), and $\theta = \theta(y)$ is spread along the y-axis.

The function

$$\cos\theta = \frac{2P_{12} - (P_1 + P_2)}{\sqrt{P_1 P_2}} \tag{2.64}$$

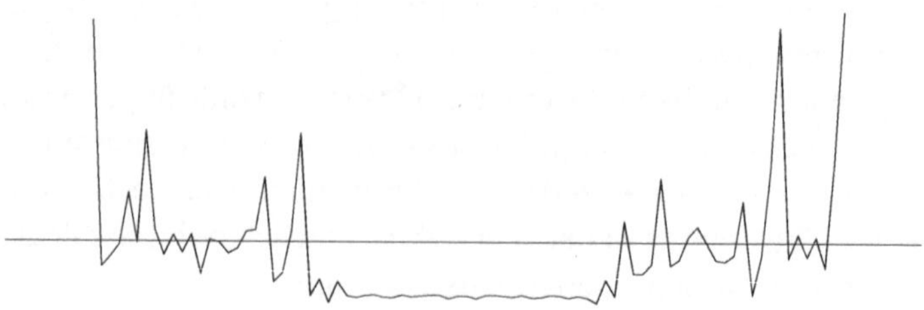

Figure 2.12 Interference term, $\cos\theta = (2P_{12} - (P_1 + P_2))/\sqrt{P_1 P_2}$.

is shown in (Fig. 2.12). Please note that as there are ranges where P_1 or P_2 are equal to zero, i.e., $P_1 P_2 = 0$, the function $\cos\theta$ is not determined and from (2.63) we see that P_{12} does not depend on it.

2.10 Interference of Probabilities in Cognitive Science

As was shown in experiments [54–56] (see also Ref. [206]) the classical FTP can be violated for data collected in cognitive science. In particular, the aforementioned papers were devoted to study of contextuality in recognition of ambiguous figures. In fact, the study of human reactions to ambiguous figures is the important and well-established area of research in cognitive science and psychology. Contextuality (in the form of Gestalt representation) was also discussed a lot, see references in Ref. [206]. In cognitive science contextuality is understood in the same way as in the Växjö interpretation of QM, i.e., cognitive contextuality is not about simultaneous measurement of a few (so to say compatible) observables, but about context dependence of a single mental observable, cf. Bohr [38].

Statistical data collected by Conte and co-workers [54–56] (see also [206]) can be represented in the form of the quantum-like FTP with the additional interference term. One may speak about *cognitive interference*. Later the similar interference effect was found by Busemeyer *et al.* [48–51] in well-known statistical data on the so-called *disjunction effect*, in violation of the *sure thing principle*, see, e.g., Ref. [206]. This principle is one of the basic postulates of modern economic science. It formalizes rationality of behavior of actors of the market. Hence, in cognitive psychology and theory of decision making quantum-like interference is an effect of irrational behavior.

As was already emphasized, in general contextuality can induce violation of the classical FTP and hence generate interference-like terms, see Ref. [157] for cognitive applications. What are neurophysiological sources of mental contextuality? The memory effects definitely play an important role (cf. Section 2.9.1). One may even speculate about a *mental field* or a *field of consciousness* (or an information field), e.g., Ref. [158]. Once some field is introduced in a mental model, the interference-like effects can be explained by the wave mechanics. In principle, one need not to attract different kinds of unphysical fields. It is possible to proceed toward mental interference starting with the classical electromagnetic field in the brain [207, 208]. The author tried to proceed with all possible models, from the field of consciousness [158] and rejection of mental reductionism to deepest mental reductionism of mental processes [207, 208]. The problem of the origin of mental interference is very complex and it is too early to reject any of aforementioned mathematical models.

We can summarize studies on cognitive (mental) interference. Violation of FTP and appearance of the additional interference term are induced by contextuality of mental observables. There are many possible sources of mental contextuality. Only deeper neurophysiological and cognitive studies will determine the real sources of such contextuality.

Chapter 3

Fundamentals of Prequantum Classical Statistical Field Theory

3.1 Noncomposite Systems

Quantum mechanics is a statistical theory (see, e.g., Margenau [239, 240], Ballentine [25, 26], Klyshko [216], Holevo [115, 116], Khrennikov [192]). It cannot tell us anything about an individual quantum system, e.g., an electron or a photon. It predicts only probabilities for results of measurements for large ensembles of quantum systems. *Classical statistical mechanics* (CSM) does the same. Why are QM and CSM based on different probabilistic models?

In CSM averages are given by integrals with respect to probability measures and in QM by operator traces. In CSM we have

$$\langle f \rangle_\mu = \int_M f(\phi) d\mu(\phi), \tag{3.1}$$

where M is the state space. In probabilistic terms: there is given a random vector $\phi(\omega)$ taking values in M. Then

$$\langle f \rangle_\mu = Ef(\phi(\omega)).$$

In QM the average is given by the operator trace formula

$$\langle \widehat{A} \rangle_\rho = \text{Tr} \rho \widehat{A} \tag{3.2}$$

Beyond Quantum
Andrei Khrennikov

ISBN 978-981-4411-73-8 (Hardcover), 978-981-4411-74-5 (eBook)
www.panstanford.com

This formal mathematical difference induces the prejudice that the classical and quantum worlds differ fundamentally. Our aim is to show that, in spite of this common opinion, quantum averages can be easily represented as classical averages and, moreover, even correlations between entangled systems can be expressed as classical correlations with respect to fluctuations of classical random fields.

Einstein's Dreams

It is well known (Chapter 1) that Einstein did not believe in irreducible randomness and completeness of QM. He dreamed of a better, so to say "prequantum," model [79]:

Dream 1. Reduction of quantum randomness to classical.

Dream 2. Renaissance of causal description.

Dream 3. Instead of particles, classical fields will provide a complete description of reality, reality of fields.

3.1.1 *Quantum System as a Classical Random Field*

Einstein's Dreams 1 and 3 (but not Dream 2) came true in PCSFT, a version of CSM in which fields play the role of particles.[a] In particular, composite systems can be described by vector random fields, i.e., by the Cartesian product of state spaces of subsystems, but not the tensor product. The basic postulate of PCSFT can be formulated in the following way [162]:

> *A quantum particle is the symbolic representation of a "prequantum" classical field fluctuating on the space–time scale which is essentially finer than the space–time scale of present measurements.*

The prequantum state space is selected as $M = L_2(\mathbf{R}^3)$. States are fields $\phi : \mathbf{R}^3 \to \mathbf{C}$: the electronic field, the neutronic field, and the photonic field (the classical electromagnetic field of very low

[a] It seems surprising that, although Dream 1 came true, Dream 2 did not. The situation differs essentially from CSM, where dynamics of probability distribution given by the Liouville equation can be reduced to the deterministic Hamiltonian dynamics. The main difference is due to the presence of the background fluctuations.

intensity). An ensemble of quantum particles is represented by an ensemble of classical fields. Mathematically the latter is represented by a probability measure μ on $M = L_2(\mathbf{R}^3)$ or a random field $\phi(x, \omega)$ taking values in $M = L_2(\mathbf{R}^3)$. For each fixed value of the random parameter $\omega = \omega_0$, $x \to \phi(x, \omega_0)$ is a classical field defined on physical space.

Thus the phrase from the canonical QM vocabulary "measurements for an ensemble of quantum particles" is interpreted as "measurements for the classical random field."

3.1.2 *Density Operator Is Covariance Operator*

Each random field has the covariance operator, say D. It describes correlations between various degrees of freedom in the random field.

The map

$$\mu \mapsto D = \rho \tag{3.3}$$

establishes the correspondence between probability distributions of prequantum random fields and density operators on QM.[a]

In PCSFT each quantum state (an element of the QM formalism) is represented by a classical random field. The covariance operator of this field is determined by the density operator. We also postulate that the prequantum random field has *zero mean value.* These two conditions determine uniquely Gaussian random fields. We often restrict our model to such fields; the non-Gaussian framework will be considered in Section 6.1.

3.1.3 *Functional Representation of Quantum Observables*

The map

$$\widehat{A} \mapsto f_A(\phi) = \langle \widehat{A}\phi, \phi \rangle \tag{3.4}$$

establishes a one-to-one correspondence between quantum observables (self-adjoint operators) and classical physical variables (quadratic functionals of the prequantum field).

[a]We still consider the case of noncomposite systems. In the case of composite systems this correspondence is really tricky.

It is easy to prove (see Section 4.4) that the following equality holds:

$$E f_A(\phi(\omega)) = \int f_A(\phi) d\mu(\phi) = \mathrm{Tr}\rho\widehat{A}. \qquad (3.5)$$

In particular, for a pure quantum state Ψ, consider a measure with zero mean value and the covariance operator $\rho_\Psi = \Psi \otimes \Psi$ (the orthogonal projector on the vector Ψ), then $\int f_A(\phi) d\mu(\phi) = \langle \widehat{A}\Psi, \Psi \rangle$. This mathematical formula coupling the integral of a quadratic form and the corresponding trace is well known in measure theory, see, e.g., the classical monograph of Skorohod [279]. Our main contribution is the connection of this mathematical formula with quantum physics.

This is the end of a brief story about quantum noncomposite systems, see Chapter 7 for a detailed presentation.

3.1.4 *Beyond QM*

Main message: *By approaching finer temporal and spatial scales we shall see violation of probabilistic predictions of QM.*

In fact, PCSFT not only reproduces quantum averages, but it also provides a possibility to go beyond QM. Suppose that *not all* prequantum physical variables are given by *quadratic forms.* Consider a more general model, i.e., all smooth functionals $f(\phi)$ of classical fields are considered as "prequantum variables." One may speculate that we only have an illusion of representation of all quantum observables by self-adjoint operators. This illusion is justified by the following considerations.

The map

$$f \mapsto \widehat{A} = f''(0)/2 \qquad (3.6)$$

projects smooth functionals of the prequantum field (physical variables in PCSFT) on self-adjoint operators (quantum observables). Then quantum and classical (prequantum) averages do not coincide, but, nevertheless, they are approximately equal [162, 171, 173, 175]:

$$\frac{1}{\alpha} \int f_A(\phi) d\mu(\phi) = \mathrm{Tr}\rho\widehat{A} + O(\alpha), \qquad (3.7)$$

where

$$\alpha = \int \|\phi\|^2 d\mu(\phi)$$

is the dispersion of the prequantum random field.

In Section 1.2.2 we interpreted the quantity $\pi_2(\phi) = \|\phi\|^2$ as energy of the signal. Thus the prequantum average $\langle f \rangle_\mu = \int f_A(\phi) d\mu(\phi)$ normalized by the average of the signal's energy is approximately equal to the quantum average; the same is valid for probabilities (Chapters 8 and 9) [187, 195, 201].

It is possible to couple the dispersion α with time scales of prequantum fluctuations and measurements, the scales τ and T, respectively [170, 183]:

$$\alpha = \tau / T,$$

(see Sections 3.1.5 and 5.15).

The main problem is that PCSFT does not provide a quantitative estimate of the dispersion of fluctuations of the prequantum random signal, in particular, of the time scale of the prequantum fluctuations. If this scale is too fine, e.g., the Planck scale, then QM is a "too good approximation of PCSFT," i.e., it would be really impossible to distinguish QM from PCSFT experimentally. However, even a possibility to represent QM as the classical wave mechanics can have important theoretical and practical consequences. I optimistically hope that the parameter α is not negligibly small. In particular, the time scale of the prequantum fluctuations is fine compared to the scale approachable in recent experiments, but not extremely fine.

In Section 3.11 we shall present an experimental design which provides a possibility to amplify the "beyond quantum contribution" and hopefully to observe deviations of experimental averages and probabilities from the predictions of QM. Such experiments are complicated from the technical viewpoint. Roughly speaking, in all quantum experiments we observe deviations from the predictions of the quantum formalism. However, such deviations are interpreted as systematic errors (due to improper experimental performance) or contributions of noise. Hence, the main problem is to distinguish such purely technical problems from the additional contribution of the prequantum field, due to nonquadratic effects in detection.

Here we can mention the famous experiment of the group of Gregor Weihs. They tested a model proposed by Sorkin [278], which predicted (as well as PCSFT) a deviation from Born's rule. After the first series of experiments Gregor Weihs reported at the Växjö conference, 2008, that they had found experimental deviations from Born's rule [298]. However, later (in the talk of Gregor Weihs at the Växjö conference, 2010) these deviations have been interpreted as systematic errors.

3.1.5 *Coupling with Classical Signal Theory*

Main message: *QM is a version of classical signal theory. It describes noisy and temporary and spatially singular signals.*

In Section 3.1.1 we used the ensemble approach to random fields based on the Kolmogorov's probability model [218]. However, it is possible to switch from the ensemble representation of random signals (as ensembles of classical fields) to the time representation, as signals fluctuating on fine time scales. This can be done on the basis of the *ergodic hypothesis*, see also Section 9.1.

We make a remark about the debate on quantum randomness: Is it ensemble or individual? In our approach this debate can be considered as a quantum version of the old debate on the interrelation between classical ensemble and time series averages. It is commonly assumed that classical signals are ergodic. Hence, it is always possible to switch freely from ensemble averages to time averages and vice versa [246, 265].

However, in reality the problem of interrelation between various types of classical randomness is extremely complicated [135], and it has not been solved yet. The ensemble approach was formalized by Kolmogorov [218]. As was mentioned, the time-average approach is typically justified on the basis of the ensemble approach via ergodic theory (in particular, in the form of the law of large numbers). However, the real justification can be approached in the von Mises framework [293–295]. Unfortunately, this framework has some mathematical problems and the attempts to solve these problems by using complexity theory, in particular, Kolmogorov's algorithmic complexity [276], cannot be considered as satisfactory. We shall not discuss these purely classical difficulties in this book, see, e.g., Ref. [135] for details.

Thus prequantum random field can be considered as a classical signal fluctuating randomly on some time scale, "the prequantum scale," which is essentially finer than the scale of measurements. Each quantum particle can be considered as such a signal. We cannot observe its wave features directly, because we cannot approach the prequantum scale; however, it is still possible to observe them indirectly, e.g., in the interference experiments. Thus, for example, any electron is a classical random signal; the same is valid for the photon, a pulse of the classical electromagnetic field whose individual fluctuations are unobservable at the moment.

In the PCSFT framework the interrelation between ensemble and time averages makes closer Einstein's ensemble randomness and von Neumann's individual randomness. However, von Neumann would not be happy with such unification, since in PCST even individual randomness is reducible to fluctuations in a signal.

As was mentioned, in signal theory we can switch from the ensemble representation for averages to the time representation. Thus, instead of a random field $\phi(x, \omega)$, which is distributed with some probability distribution $d\mu(\phi)$ on H, we consider a time-dependent signal

$$\phi(s) \equiv \phi(s, x),$$

where $x \in \mathbf{R}^3$, $s \in [0, +\infty)$. Then, for each functional $f(\phi)$ such that

$$\int_H |f(\phi)| d\mu(\phi) < \infty,$$

we have

$$\langle f \rangle_\mu \equiv \int_H f(\phi) d\mu(\phi) = \lim_{T \to \infty} \frac{1}{T} \int_0^T f(\phi(s)) ds \equiv \langle f \rangle_\phi. \quad (3.8)$$

Consider two time scales: τ, a fine scale, and $T \gg \tau$, a rough time scale. In QM the latter is the scale of measurements and Schrödinger's dynamics and τ is the scale of fluctuations of the prequantum field.[a] Thus,

$$\langle f \rangle_\phi \approx \frac{1}{T} \int_0^T f(\phi(s)) ds, \quad (3.9)$$

[a]PCSFT does not predict the magnitude of the scale of prequantum field fluctuations (see Section 5.15 for further discussion).

where s denotes the time variable at the τ-scale. We call the T-scale the *quantum (experimental) time scale.*[a] The τ-scale is called *prequantum time scale.* If a signal is periodic on the prequantum time scale with the period Δ, i.e., $\phi(s+\Delta)=\phi(s)$, $\Delta < T$, then

$$\langle f\rangle_\phi = \frac{1}{\Delta}\int_0^\Delta f(\phi(s))ds, \tag{3.10}$$

For each signal $\phi(s,x)$, $x \in \mathbf{R}^3$, PCSFT determines time averages of physical variables; for example, for the energy variable of the prequantum electromagnetic field, the photon field, we have

$$\langle f\rangle_\mu \approx \frac{1}{T}\int_0^T \left(\int_{\mathbf{R}^3}(E^2(s,x)+B^2(s,x))dx\right)ds. \tag{3.11}$$

In time representation the covariance operator of the signal $\phi(s,x)$ is given by

$$\langle Du, v\rangle = \lim_{T\to\infty}\frac{1}{T}\int_0^T \left(\int_{\mathbf{R}^3} u(x)\overline{\phi(s,x)}dx \int_{\mathbf{R}^3}\phi(s,x)\overline{v(x)}dx\right)ds,$$

$$\approx \frac{1}{T}\int_0^T \left(\int_{\mathbf{R}^3} u(x)\overline{\phi(s,x)}dx \int_{\mathbf{R}^3}\phi(s,x)\overline{v(x)}dx\right)ds \tag{3.12}$$

where $u(x)$, $v(x)$ are two test signals, $u, v \in L_2(\mathbf{R}^3)$.

Thus it is possible to consider PCSFT as a theory of classical prequantum signals. We remark that in classical signal theory some research was done the other way around. The covariance operator of a classical Gaussian signal was represented as a density operator. This representation provided a possibility to apply the quantum information theory to define properly entropy of the classical signal. We recall that there are some mathematical problems with the definition of entropy of classical Gaussian signals for information channels of infinite dimension [252].

3.2 Composite Systems

Main message: *A biparticle quantum system is a pair of correlated classical signals.*

---

[a] This terminology does not match the modern tradition in cosmology and superstring theory to refer to the quantum scale as the Planck scale. However, our terminology matches the Bohr–Heisenberg view on QM as a theory of measurements performed by macroscopic devices.

In CSM a composite system $S = (S_1, S_2)$ is mathematically described by the *Cartesian product* of state spaces of its parts S_1 and S_2. In QM it is described by the *tensor product.* The majority of researchers working in quantum foundations, especially quantum information theory, consider this difference in the mathematical representations as crucial. In particular, the entanglement which is a consequence of the tensor space representation is treated as a totally nonclassical phenomenon. However, Einstein was sure that the EPR states encrypt classical correlations generated by a common preparation. In PCSFT Einstein's dream of the entanglement as a classical correlation will come true.

Let $S = (S_1, S_2)$, where S_i has the state space H_i, which is a complex Hilbert space. Then, according to CSM, the state space of S is $H_1 \times H_2$. By PCSFT an ensemble of composite systems is described by a probability distribution on this Cartesian product or by a vector random field $\phi(\omega) = (\phi_1(\omega), \phi_2(\omega)) \in H_1 \times H_2$.

In our approach each subsystem is described by its own random field: S_i by $\phi_i(\omega)$, $i = 1, 2$. These fields are *correlated* in the completely classical sense. A correlation created at the initial instance of time propagates in time in complete accordance with laws of QM. There is no action at the distance. This is classical dynamics of two stochastic processes which were correlated at the beginning. In fact, the situation is more complex as there is also the common random background, vacuum fluctuations. This common background is a source of additional correlations. Thus the total correlations are even stronger than the correlations in the initial states. We shall come back to this question a little bit later.

3.2.1 *Operator Realization of Wave Function*

Main message: *Mathematics provides an alternative representation of the wave function of a biparticle system: not as a vector belonging to the tensor product, but as an operator acting between state spaces of subsystems.*

Consider now the QM model, take a pure state $\Psi \in H_1 \otimes H_2$. Can one connect the QM and PCSFT formalisms? Yes! But Ψ has to be interpreted in a way completely different from that of the conventional QM. The main mathematical point is that the QM wave function Ψ is not a vector. This is an operator! It is, in fact, the off-diagonal block of the covariance operator of the corresponding prequantum random field: $\phi(\omega) \in H_1 \times H_2$.

The detailed presentation of the operator view on the state of a composite system can be found in Section 7.2. Here we restrict our consideration to the case of the L_2-space. The wave function $\Psi(x, y)$ of a composite system determines the integral operator

$$\widehat{\Psi}\phi(x) = \int \Psi(x, y)\phi(y)dy. \tag{3.13}$$

This operator is of the Hilbert–Schmidt type; moreover, each operator of the Hilbert–Schmidt type in the L_2-space can be represented as such integral operator.

3.2.2 *Basic Equality*

Main message: *Quantum correlations can be written by using the operator representation of the wave function.*[a]

Consider $\Psi \in H_1 \otimes H_2$ which is normalized by 1 and its representation by the operator $\widehat{\Psi} : H_2 \to H_1$. We use Eq. (3.13) in the case of L_2-spaces, see Section 7.2 for the general case and the proofs of formulas which will appear in the PCSFT approach to correlations. In the general case one can define the operator $\widehat{\Psi}$ with the aid of its bilinear form: $\langle \widehat{\Psi}\phi_2, \phi_1 \rangle = \langle \Psi, \phi_1 \otimes \bar{\phi_2} \rangle$, $\phi_k \in H_k$, $k = 1, 2$, and, for $\phi = q+ip$, its complex conjugate is given by $\bar{\phi} = q-ip$.

Take a pair of linear bounded self-adjoint operators $\widehat{A}_j : H_j \to H_j$, $j = 1, 2$. Assume, for the sake of mathematical simplicity, that they are both *real operators*, i.e., the following equality holds:

$$\langle \widehat{A}_i u, v \rangle = \overline{\langle \widehat{A}_i \bar{u}, \bar{v} \rangle}. \tag{3.14}$$

A self-adjoint operator $\widehat{A}$ is real if its matrix elements are real in any real basis of the complex Hilbert space. Let $\{e_j\}$ be an orthonormal basis which is real, i.e., $\bar{e_j} = e_j$. Then, for matrix elements, equality (3.14) implies $a_{jk} = \langle \widehat{A}e_j, e_k \rangle = \overline{\langle \widehat{A}e_j, e_k \rangle} = \overline{a_{jk}}$. The position operator is real, but the momentum operator is not. Any Hamiltonian

[a]This fact is trivial from the purely mathematical viewpoint. However, this step from vectors to operators is very important from the physical viewpoint.

of the form $(\widehat{p}^2/2m+V(q)$, where V is a real-valued potential, is real. In general, one can proceed without this, rather restrictive, condition with a slight modification of equalities used in this chapter (see Chapter 7). At the moment, we do not want to make the presentation too complicated technically.

We have the following operator representation of the quantum average:

$$\mathrm{Tr}\widehat{\Psi}\widehat{A}_2\widehat{\Psi}^*\widehat{A}_1 = \langle \widehat{A}_1 \otimes \widehat{A}_2\rangle_\Psi \equiv \langle \widehat{A}_1 \otimes \widehat{A}_2\Psi, \Psi\rangle. \tag{3.15}$$

This is a mathematical theorem. In Section 7.4 it will be proven in a more general framework. It will play a fundamental role in the further consideration. However, this operator equality is not coupled directly to PCSFT; this is just another mathematical representation of quantum averages; operators are used instead of vectors from the tensor product.

3.2.3 *Coupling of Classical and Quantum Correlations*

Main message: *Quantum correlations can be represented as correlations of quadratic functionals of a prequantum random field.*

In PCSFT, a composite system $S = (S_1, S_2)$ is mathematically represented by a random field $\phi(\omega) = (\phi_1(\omega), \phi_2(\omega)) \in H_1 \times H_2$. Its covariance operator D has the block structure

$$D = \begin{pmatrix} D_{11} & D_{12} \\ D_{21} & D_{22} \end{pmatrix}, \tag{3.16}$$

where $D_{ii} : H_i \to H_i, D_{ij} : H_j \to H_i$. The covariance operator is self-adjoint. Hence $D_{ii}^* = D_{ii}$, and $D_{12}^* = D_{21}$. The diagonal blocks D_{ii} describe internal correlations in subsystems. The block D_{12} describes cross-correlations between subsystems. Here, by definition,

$$\langle D_{ij}u_j, v_i\rangle = E\langle u_j, \phi_j(\omega)\rangle\langle\phi_i(\omega), v_i\rangle, u_i \in H_i, v_j \in H_j.$$

For any Gaussian random vector $\phi(\omega) = (\phi_1(\omega), \phi_2(\omega))$ having zero average and any pair of operators satisfying conditions of Section 3.2.2, $\widehat{A}_i : H_i \to H_i, i = 1, 2$, the following equality takes place:[a]

$$\langle f_{A_1}, f_{A_2}\rangle \equiv Ef_{A_1}(\phi_1(\omega))f_{A_2}(\phi_2(\omega))$$
$$= (\mathrm{Tr}D_{11}\widehat{A}_1)(\mathrm{Tr}D_{22}\widehat{A}_2) + \mathrm{Tr}D_{12}\widehat{A}_2D_{21}\widehat{A}_1. \tag{3.17}$$

[a] In Section 7.3, it will be proven in a more general framework, namely, for arbitrary self-adjoint operators, see equality (7.17).

We remark that

$$\mathrm{Tr} D_{ii} \widehat{A}_i = E f_{A_i}(\phi_i(\omega)),\ i = 1, 2. \tag{3.18}$$

Thus we have

$$\langle f_{A_1}, f_{A_2} \rangle = E f_{A_1} E f_{A_2} + \mathrm{Tr} D_{12} \widehat{A}_2 D_{21} \widehat{A}_1. \tag{3.19}$$

Consider a Gaussian vector random field such that $D_{12} = \widehat{\Psi}$. Then combining (3.19) with the operator wave function representation of quantum correlations (3.15), we obtain

$$E(f_{A_1} - E f_{A_1})(f_{A_2} - E f_{A_2}) = \langle \widehat{A}_1 \otimes \widehat{A}_2 \Psi, \Psi \rangle \equiv \langle \widehat{A}_1 \otimes \widehat{A}_2 \rangle_\Psi. \tag{3.20}$$

or, for covariance of two classical random variables f_{A_1}, f_{A_2}, we have

$$\mathrm{cov}\,(f_{A_1}, f_{A_2}) = \langle \widehat{A}_1 \otimes \widehat{A}_2 \rangle_\Psi. \tag{3.21}$$

As already mentioned, we have the following equality for averages of quadratic forms of coordinates: $E f_{A_i}(\phi_i(\omega)) = \mathrm{Tr} D_{ii} \widehat{A}_i$. We want to construct a random field such that these averages will match those given by QM. For the latter, we have[a]

$$\langle \widehat{A}_1 \rangle_\Psi = \langle \widehat{A}_1 \otimes I \Psi, \Psi \rangle = \mathrm{Tr}(\widehat{\Psi} \widehat{\Psi}^*) \widehat{A}_1; \tag{3.22}$$

$$\langle \widehat{A}_2 \rangle_\Psi = \langle I \otimes \widehat{A}_2 \Psi, \Psi \rangle = \mathrm{Tr}(\widehat{\Psi}^* \widehat{\Psi}) \widehat{A}_2, \tag{3.23}$$

where I denotes the unit operator. Thus it would be natural to take

$$D = \begin{pmatrix} \widehat{\Psi} \widehat{\Psi}^* & \widehat{\Psi} \\ \widehat{\Psi}^* & \widehat{\Psi}^* \Psi \end{pmatrix}.$$

However, *this operator is not positively defined!* It does not determine any probability distribution on the space of classical fields. To obtain a positively defined operator, we modify D. Originally this modification was done by purely mathematical reasons, but there are deep physical grounds for it. The operator

$$D \equiv D_\Psi = \begin{pmatrix} \widehat{\Psi} \widehat{\Psi}^* + \epsilon I & \widehat{\Psi} \\ \widehat{\Psi}^* & \widehat{\Psi}^* \widehat{\Psi} + \epsilon I \end{pmatrix} \tag{3.24}$$

is positively defined if $\epsilon > 0$ is large enough (Section 7.6.1). Hence it determines uniquely the Gaussian measure on the space of classical fields.

[a] In Section 7.3 these equalities will be proven in more general framework.

We point to some additional mathematical difficulties in the infinite-dimensional case. The additional term in the covariance operator is of the white noise type. Therefore the corresponding Gaussian measure is not σ-additive on $H_1 \times H_2$. We also remark that the infinite-dimensional case, $H_i = L_2(\mathbf{R}^3)$, is the most important for applications. We shall study this case in detail in Section 6.2. At the moment we just remark that to obtain the σ-additive Gaussian measure, one has to consider a "good extension" of $L_2(\mathbf{R}^3)$, e.g., the space of Schwartz distributions $\mathcal{S}'(\mathbf{R}^3)$. The Gaussian measure with the covariance operator (3.24) is σ-additive on this larger space. In the L_2-case the covariance operator is an integral operator with the kernel

$$D(x, y) = \begin{pmatrix} D_{11}(x, y) & D_{12}(x, y) \\ D_{21}(x, y) & D_{22}(x, y) \end{pmatrix}, \tag{3.25}$$

where

$$D_{12}(x, y) = \Psi(x, y), \tag{3.26}$$

$$D_{11}(x, y) = \int dz \Psi(x, z)\bar{\Psi}(y, z)dz + \epsilon\delta(x - y), \tag{3.27}$$

$$D_{22}(x, y) = \int dz \bar{\Psi}(x, z)\Psi(y, z) + \epsilon\delta(x - y). \tag{3.28}$$

Suppose, now $\phi(\omega)$ is a random vector with the covariance operator D_Ψ given by (3.24). Then

$$\langle \widehat{A}_1 \rangle_\Psi = E f_{A_1}(\phi_1(\omega)) - \epsilon \mathrm{Tr} \widehat{A}_1, \tag{3.29}$$

$$\langle \widehat{A}_2 \rangle_\Psi = E f_{A_2}(\phi_2(\omega)) - \epsilon \mathrm{Tr} \widehat{A}_2. \tag{3.30}$$

These relations for averages and relation (3.21) for the correlation provide coupling between PCSFT and QM. Quantum statistical quantities can be obtained from corresponding quantities for classical random field. One may say that *"irreducible quantum randomness" is reduced to randomness of classical prequantum fields.* However, the situation is more complicated.

The equalities (3.29) and (3.30) imply that quantum averages are obtained as the shift-type renormalizations of averages with respect to classical random fields. The shift corresponds to elimination of the contribution of the background field. Thus quantum averages are not simply classical averages. Therefore, in some sense von Neumann was right. Quantum averages are not represented in the form

$$\langle \widehat{A}_2 \rangle_\Psi = E g_A(\phi(\omega)), \tag{3.31}$$

where $\phi_\Psi \equiv \phi(\omega)$ is a prequantum random field corresponding to a quantum state Ψ and $g_A(\phi)$ a functional of the prequantum field. By ignoring mathematical difficulties induced by the infinite dimension we can write the equalities (3.29) and (3.30) as

$$\langle \widehat{A}_1 \rangle_\Psi = E f_{A_1}(\phi_1(\omega)) - \epsilon E f_{A_1}(\eta(\omega)), \tag{3.32}$$

$$\langle \widehat{A}_2 \rangle_\Psi = E f_{A_2}(\phi_2(\omega)) - \epsilon E f_{A_2}(\eta(\omega)), \tag{3.33}$$

where $\eta \sim N(0, I)$ is white noise: the Gaussian random variable with zero mean value and the unit covariance matrix. Hence, although we do not have the representation (3.31), the representation (3.32) and (3.33) can be interpreted as a reduction of quantum randomness to classical ensemble randomness. Therefore, in some sense Einstein was right as well. In PCSFT framework the positions of von Neumann and Einstein have been peacefully unified. However, I am not sure that either von Neumann or Einstein would have been happy with such a peaceful agreement between the Copehnagen and classical camps. Von Neumann was definitely sure that the mystery of quantum randomness is not reduced to the shift-type transformation, Eqs. (3.32) and (3.33). And Einstein was not interested in the reduction of quantum effects to a random background. In any event he did not support Planck's attempt to explain spontaneous emission by a random disturbance effect induced by the background field.

At the same time the presentations (3.32) and (3.33) of quantum averages matches quantum experimental science which uses the detectors with thresholds to exclude the contribution of the random background. Subtraction of the contribution of the random background is an important part of the experimental procedure in quantum physics.

3.2.4 *Vacuum Fluctuations*

The additional term given by the unit operator in the diagonal blocks of the covariance operator of the prequantum vector field corresponds to the field of the white noise type. Such a field can be considered as vacuum fluctuations, the background field. PCSFT induces the following picture of reality:

Fluctuations of the background field are combined with random fields representing quantum systems. Since we cannot separate, e.g., the electron from the background field, we cannot separate totally any two quantum systems. Thus all quantum systems are "entangled" via the background field. White noise couples everything in Nature.

For many years Hida has developed and applied a theory of white noise and white noise functionals [109]. He always emphasized the role of white noise in physical processes, especially in quantum. Conversations with prof. Hida played an important role in the formation of my scientific views. *Hida's Dream* of white noise as the basic physical entity came true in PCSFT.

3.2.5 *Superstrong Quantum Correlations*

According to PCSFT, quantum correlations (in particular, violating Bell's inequality) are so strong simply due to the presence of the background field. The off-diagonal term $\widehat{\Psi}$ of the covariance operator (3.24) can be so large only if the diagonal terms are completed by the contribution of the background field. Mathematics tells us this. Thus quantum correlations are so strong because the vacuum field really couples any two systems; they are in the same fluctuating space.

Space is a huge random wave; quantum systems are spikes on this wave; they are correlated via this space-wave. Thus quantum correlations have two contributions:

(1) initial preparation;
(2) coupling via the vacuum field.

The picture is purely classical. In this model the background field is the source of additional correlations. It seems that *this background field is an additional (purely classical) computational resource increasing the power of quantum algorithms.* According to PCSFT, quantum computers really have additional computational resource given by the background field.

3.3 Stochastic Process Corresponding to Schrödinger's Evolution

Main message: *Schrödinger's dynamics can be considered as dynamics of classical random field, a stochastic process in the space of fields.*

In the present section we restrict our consideration to the finite-dimensional case. Set

$$\mathbf{H} = H_1 \times H_2, \tag{3.34}$$

the *Cartesian product* of Hilbert spaces. The proofs presented in this section can be omitted by physically oriented readers. Typically physicists are not so much interested in existence theorems. However, we have to prove a few theorems of existence of classical random processes matching Schrödinger's dynamics for composite systems.

Since at the moment we do not try to go beyond QM, but only reproduce its predictions, we use Schrödinger equation for dynamics of the wave function. But we change the QM interpretation of the Ψ-function. By our model the standard Schrödinger equation describes dynamics of the off-diagonal block of the covariance operator of the prequantum random field. The time variable t is the "quantum time variable," i.e., related to the scale of measurements. We start with Schrödinger equation for a composite system $S = (S_1, S_2)$.

$$i\frac{d\Psi}{dt}(t) = \widehat{H}\Psi(t), \ \ \Psi(0) = \Psi_0, \tag{3.35}$$

where $\widehat{H}$ is Hamiltonian of S.

Hence, at the instance t, the covariance matrix of the prequantum random field $\phi(t, \omega)$ has the form

$$D_{\Psi(t)} \equiv D(t) = \begin{pmatrix} \widehat{\Psi(t)}\widehat{\Psi(t)}^* + \epsilon I & \widehat{\Psi(t)} \\ \widehat{\Psi(t)}^* & \widehat{\Psi(t)}^*\widehat{\Psi(t)} + \epsilon I \end{pmatrix} \tag{3.36}$$

The following fundamental question having both mathematical and physical counterparts immediately arises:

> "Can one construct a stochastic process valued in the Cartesian product $\mathbf{H}$ such that at each $t \in [0, \infty)$ its covariance matrix coincides with $D(t)$?"

3.3.1 *Bernoulli-Type Process*

The formal mathematical answer to above question is yes! It is easy to construct such a stochastic process. Take space $\Omega = \prod_{t \in [0,\infty)} \mathbf{H}$ as the space of random parameters, points of this space $\omega = (\omega_t)$ can be considered as functions $\omega : [0, \infty) \to \mathbf{H}$, trajectories. Consider the family of Gaussian measures $\mu_t \equiv \mu_{D(t)}$ on $\mathbf{H}$ having zero mean value and covariance operators $D(t)$, $t \in [0, \infty)$. Consider the direct product of these measures $P = \prod_{t \in [0,\infty)} \mu_t$ on the space Ω.

Proposition 3. 1. *Let $\phi(t, \omega)$ be a stochastic process having the probability distribution P on Ω. Then, for any pair of vectors $y_1, y_2 \in$ $\mathbf{H}$ and any instance of time $t \geq 0$,*

$$E\langle y_1, \phi(t, \omega)\rangle\langle\phi(t, \omega), y_2\rangle = \langle D(t)y_1, y_2\rangle, \qquad (3.37)$$

where $D(t)$ is given by (3.37).

Proof. First we remark that the measure P is σ-additive by generalization of the Kolmogorov theorem [218]. The product space Ω can be represented as the space of functions $u : [0, \infty) \to \mathbf{H}$. Any fixed instance of time t and any vector $y \in \mathbf{H}$ determine a linear functional $u \to \langle u(t), y\rangle$ on Ω. This is a cylindric function on Ω. By the definition of a product measure we have

$$\begin{aligned} E\langle y_1, \phi(t, \omega)\rangle\langle\phi(t, \omega), y_2\rangle &= \int_\Omega \langle y_1, u(t)\rangle\langle u(t), y_2\rangle dP(u) \\ &= \int_{\mathbf{H}} \langle y_1, \phi\rangle\langle\phi, y_2\rangle d\mu_t(\phi) \\ &= \langle D(t)y_1, y_2\rangle, \end{aligned}$$

by the definition of the covariance operator.

For any composite system prepared initially in a pure state, there exists a classical stochastic process inducing the standard Schrödinger evolution of correlations. This equation describes dynamics of the off-diagonal block of the covariance matrix of this prequantum stochastic process. (Our approach has to be distinguished from other attempts to find classical stochastic processes matching Schrödinger's dynamics. Take Nelson's model [248] as an example. Nelson constructed a classical stochastic process for dynamics of particles. He reduced QM to CSM of particles.

In contrast to him, I reduced QM to CSM of fields, PCSFT.) The PCSFT is a tricky theory: a *prequantum stochastic process is not uniquely determined* by dynamics of the covariance operator $D(t)$! To determine uniquely a Gaussian process (up to natural equivalence), one has to define not only the covariance for each instance of time, i.e., $E\langle y_1, \phi(t,\omega)\rangle\langle\phi(t,\omega), y_2\rangle$, but the so-called *covariance kernel* $D(t_1, t_2)$

$$E\langle y_1, \phi(t_1,\omega)\rangle, \langle\phi(t_2,\omega), y_2\rangle = \langle D(t_1,t_2)y_1, y_2\rangle.$$

3.3.2 *Stochastic (Local) Dynamics in the Absence of Interaction*

We restrict our consideration to dynamics in the absence of interactions between S_1 and S_2 after the preparation procedure. Thus we are interested in propagation of initially correlated random fields. Although this is a rather special dynamics, it plays an important role in quantum foundations. In particular, it describes the evolution of entanglement in the EPR–Bohm type experiments.

We start with the standard quantum description of two noninteracting systems and consider Hamiltonian

$$\widehat{H} = \widehat{H_1} \otimes I + I \otimes \widehat{H_2}, \tag{3.38}$$

where $\widehat{H_j}$ is Hamiltonian of S_j. Here we use the QM terminology. (Hats denote operators.) Now we represent quantum dynamics of a composite system by using the operator representation of the wave function. We state again that this is still the standard QM, but in a new mathematical representation.

Lemma 3.1. *Let Hamiltonian have the form (3.38). Then*

$$\widehat{\Psi(t)} = e^{-i\widehat{H_1}t}\widehat{\Psi_0}e^{-i\widehat{H_2}t}. \tag{3.39}$$

$$\widehat{\Psi(t)}^* = e^{i\widehat{H_2}t}\widehat{\Psi_0}^* e^{-i\widehat{H_1}t}. \tag{3.40}$$

Proof. In this case

$$\Psi(t) = e^{-it(\widehat{H_1}\otimes I + I\otimes\widehat{H_2})}\Psi_0.$$

We expand the initial state $\Psi_0 : \Psi_0 = \sum_{ij}^k \Psi_{ij}e_i \otimes f_j$, $\Psi_{ij} \in \mathbf{C}$, where $\{e_i\}$ and $\{f_i\}$ are two orthonormal bases, in H_1 and H_2, respectively. Then

$$\Psi(t) = \sum_{ij} \Psi_{ij}e^{-i\widehat{H_1}t}e_j \otimes e^{-i\widehat{H_2}t}f_j.$$

Thus, for $v \in H_2$, we get

$$\widehat{\Psi(t)}v = \sum_{ij}^{k} \Psi_{ij} \langle v, e^{i\widehat{H_2}t} \bar{f}_j \rangle e^{-i\widehat{H_1}t} e_j$$

$$= e^{-i\widehat{H_1}t} \left[\sum_{ij}^{k} \Psi_{ij} \langle e^{-i\widehat{H_2}t} v, \bar{f}_j \rangle e_j \right].$$

Lemma 3.2. *Let the condition of Lemma 3.1 hold. Then*

$$\widehat{\Psi(t)}\widehat{\Psi(t)}^* = e^{-i\widehat{H_1}t}\widehat{\Psi_0}\widehat{\Psi_0}^* e^{i\widehat{H_1}t}. \tag{3.41}$$

$$\widehat{\Psi(t)}^*\widehat{\Psi(t)} = e^{i\widehat{H_2}t}\widehat{\Psi_0}^*\widehat{\Psi_0} e^{-i\widehat{H_2}t}. \tag{3.42}$$

Proof. We prove, e.g., (3.41). By (3.39) we have

$$\widehat{\Psi(t)}\widehat{\Psi(t)}^* = e^{-i\widehat{H_2}t}\widehat{\Psi_0}^* e^{i\widehat{H_1}t} e^{-i\widehat{H_2}t}\widehat{\Psi_0}^* e^{i\widehat{H_1}t} = e^{-i\widehat{H_1}t}\widehat{\Psi_0}\widehat{\Psi_0}^* e^{i\widehat{H_1}t}.$$

Now we leave the domain of standard QM and come back to the PCSFT model; we have

Lemma 3.3. *Let the condition of Lemma 3.1 hold. Then the operator $D(t)$ given by Eq. (3.36) can be represented in the form*

$$D(t) = \begin{pmatrix} e^{-i\widehat{H_1}t}\widehat{\Psi_0}\widehat{\Psi_0}^* e^{i\widehat{H_1}t} + \epsilon I & e^{-i\widehat{H_1}t}\widehat{\Psi_0} e^{-i\widehat{H_2}t} \\ e^{i\widehat{H_2}t}\widehat{\Psi_0}^* e^{i\widehat{H_1}t} & e^{i\widehat{H_2}t}\widehat{\Psi_0}^*\widehat{\Psi_0} e^{-i\widehat{H_2}t} + \epsilon I \end{pmatrix} \tag{3.43}$$

Proof. We put (3.41), (3.41), and (3.39) into the matrix (3.36).

By using representation (3.43) it is easy to prove:

Proposition 3.2. *Let the condition of Lemma 3.1 hold. Then the operator $D(t)$ given by (3.36) is the covariance operator of the vector process with coordinates*

$$\phi_1(t, \omega) = e^{-i\widehat{H_1}t}\xi_{01}(\omega),\ \phi_2(t, \omega) = e^{i\widehat{H_2}t}\xi_{02}(\omega), \tag{3.44}$$

where the initial random vector, $\xi_0(\omega) = (\xi_{01}(\omega), \xi_{02}(\omega)) \sim N(0, D(0))$, is Gaussian with zero mean value and the covariance operator $D(0)$.

Proof. We will find not only the covariance operator for a fixed instance of time, but even the covariance kernel. We have, for any pair of vectors $u, w \in H_1$,

$$E \langle u, \phi_1(t_1, \omega)\rangle \langle \phi_1(t_2, \omega), w\rangle = E \langle e^{i\widehat{H_1}t_1} u, \xi_{01}(\omega)\rangle \langle \xi_{01}(\omega)), e^{i\widehat{H_1}t_2} w\rangle$$
$$= \langle e^{-i\widehat{H_1}t_2}(\widehat{\Psi_0}^*\widehat{\Psi_0} + \epsilon I_1) e^{i\widehat{H_1}t_1} u, w\rangle.$$

The same calculations can be done for the second diagonal block. Thus the diagonal blocks of the covariance operator of the stochastic process given by (3.44) coincide with the diagonal blocks of the operator $D(t)$. We now consider the off-diagonal blocks. Let now $u \in H_1$, $v \in H_2$. We have

$$\begin{aligned} E\langle u, \phi_1(t_1, \omega)\rangle\langle \phi_2(t_2, \omega), v\rangle &= E\langle e^{i\widehat{H_1}t_1} u, \xi_{01}(\omega)\rangle\langle \xi_{02}(\omega)), e^{-i\widehat{H_2}t_2} v\rangle \\ &= \langle \widehat{\Psi_0}^* e^{i\widehat{H_1}t_1} u, e^{-i\widehat{H_2}t_2} v\rangle \\ &= \langle e^{i\widehat{H_2}t_2} \widehat{\Psi_0}^* e^{i\widehat{H_1}t_1} u, v\rangle. \end{aligned}$$

The same calculations can be done for the second off-diagonal block. Thus the covariance kernel has the form

$$D(t_1, t_2) = \begin{pmatrix} e^{-i\widehat{H_1}t_2}\widehat{\Psi_0}\widehat{\Psi_0}^* e^{i\widehat{H_1}t_1} + \epsilon e^{i\widehat{H_1}(t_1 - t_2)} I & e^{-i\widehat{H_1}t_2}\widehat{\Psi_0} e^{\widehat{i - H_2}t_1} \\ e^{i\widehat{H_2}t_2}\widehat{\Psi_0}^* e^{i\widehat{H_1}t_1} e^{i\widehat{H_2}t_2}\widehat{\Psi_0}^*\widehat{\Psi_0} & e^{-i\widehat{H_2}t_1} + \epsilon e^{-i\widehat{H_2}(t_1 - t_2)} I \end{pmatrix} \tag{3.45}$$

Hence, for $t_1 = t_2 = t$, $D(t, t) = D(t)$.

Dynamics (3.44) is purely deterministic, stochasticity is generated by initial conditions. A crucial point is that the initial conditions contain the contribution of the random background field.

Propositions 3.1 and 3.2 describe two different stochastic processes. The covariance kernel (3.45) differs from the covariance kernel of the process which has been constructed by considering the product of Gaussian distributions μ_t (Section 3.3.1). The latter has the covariance kernel

$$D(t_1, t_2) = D(t_1)\delta(t_2 - t_2).$$

For this process, its realizations at different instances of time are independent. The process defined by (3.44) contains nontrivial dependence between its realizations at different times. The following interesting problem arises:

For an arbitrary Hamiltonian, to construct a stochastic process such that in the case of the absence of interactions this construction gives the process (3.44).

Remark 3.1. (Nonuniqueness of prequantum dynamics) This problem has its counterpart in the conventional QM. The Schrödinger equation for a composite system $S = (S_1, S_2)$ is written with respect to a single time parameter t. QM cannot describe dynamics of the wave function with respect to two time parameters related to subsystems S_i, $i = 1, 2$, say $\Psi(t_1, t_2, x_1, x_2)$. (This was already discussed by von Neumann [296].) As a consequence, PCSFT cannot borrow from QM the covariance kernel of the prequantum stochastic process. Our analysis demonstrated that it is possible to construct a variety of prequantum stochastic processes reproducing predictions of QM. From the mathematical viewpoint, all such processes are "equally good." Selection of a real "physical process" from a variety of mathematical models is a complicated problem. Some physical reasons have to be taken into account. In the next section we shall see that the situation is even worse. In contrast to the processes constructed in the present section, a process constructed in the next section is nonvocal! Thus we cannot even guarantee locality of prequantum dynamics. Moreover, we cannot exclude the possibility that the unique "real physical prequantum process" does not exist at all. It may be that depending on the experimental context the same quantum dynamics $t \to \Psi(t)$ can be induced by different prequantum processes.

3.3.3 *Stochastic Nonlocal Dynamics*

We now consider another classical stochastic process reproducing dynamics of quantum correlations.

Proposition 3.3. *Let the operator $D(t)$ be defined by (3.36). Then the stochastic process*

$$\xi(t, \omega) = \sqrt{D(t)}\eta_0(\omega), \tag{3.46}$$

where $\eta_0 \sim N(0, I)$, has the covariance operator $D(t)$ for any $t \geq 0$.

Proof. Let $y_1, y_2 \in \mathbf{H}$. Then

$$E\langle y_1, \xi(t, \omega)\rangle\langle\xi(t, \omega), y_2\rangle = E\langle\sqrt{D(t)}y_1, \eta_0(\omega)\rangle\langle\eta_0(\omega), \sqrt{D(t)}y_2\rangle$$
$$= \langle\sqrt{D(t)}y_1, \sqrt{D(t)}y_2\rangle = \langle D(t)y_1, y_2\rangle.$$

The covariance kernel is given by

$$D(t_1, t_2) = \sqrt{D(t_2)D(t_1)}. \tag{3.47}$$

We state again that, for Hamiltonian without interaction, we constructed the stochastic process $\phi(t, \omega)$ given by (3.44). In general, stochastic processes $\xi(t, \omega)$ and $\phi(t, \omega)$ given by (3.46) and (3.44) do not coincide:

We can write the process (3.44) as

$$\phi(t) = V(t)\sqrt{D(0)}\eta_0,$$

where $V(t) = \text{diag}(e^{-it\widehat{H}_1}, e^{it\widehat{H}_2})$. Hence,

$$D_\phi(t, s) = V(t_2)D(0)V(t)^*.$$

On the other hand, the covariance kernel of process (3.46) is given by $D_\xi(t, s) = \sqrt{D(t_2)D(t_1)}$. We remark that

$$D(t) = V(t)D(0)V(t)^*.$$

Hence, $\sqrt{D(t)} = V(t)\sqrt{D(0)}V(t)^*$. Thus,

$$D_\xi(t_1, t_2) = V(t_2)\sqrt{D(0)}V^*(t_2)V(t_1)\sqrt{D(0)}V^*(t_1).$$

We remark that $V^*(t_2)V(t_1) \neq I$, $t_1 \neq t_2$.

The process $\xi(t, \omega)$ is nonvocal in the following sense. Its component $\xi_1(t, \omega)$ is guided not only by the Hamiltonian of S_1, but also of S_2; the same is true for $\xi_2(t, \omega)$. Thus PCSFT cannot give a definite answer to the question on locality of the prequantum world. Quantum correlations can be produced by local as well as nonvocal prequantum stochastic processes. However, a *possibility to proceed with local stochastic dynamics* is an attractive feature of PCSFT.

Proposition 3.4. *In the case of Hamiltonian without interaction, see (3.38), the stochastic process (3.46) can be represented in the form*

$$\xi_1(t) = e^{-it\widehat{H}_1} Q_{11}^0 e^{it\widehat{H}_1}\eta_{01} + e^{-it\widehat{H}_1} Q_{12}^0 e^{-it\widehat{H}_2}\eta_{02}, \tag{3.48}$$

$$\xi_2(t) = e^{it\widehat{H}_2} Q_{21}^0 e^{it\widehat{H}_1}\eta_{01} + e^{it\widehat{H}_2} Q_{22}^0 e^{-it\widehat{H}_2}\eta_{02}, \tag{3.49}$$

where $\sqrt{D(0)} = \begin{pmatrix} Q_{11}^0 & Q_{12}^0 \\ Q_{21}^0 & Q_{22}^0 \end{pmatrix}$, *and* $\eta_0 \sim N(0, I)$.

Proof. For example, take $y_1, y_2 \in H_1$ and consider the average

$$\begin{aligned} &E\langle y_1, \xi(t_1, \omega)\rangle\langle\xi(t_2, \omega), y_2\rangle \\ &= E\langle e^{-it_1\widehat{H}_1} Q_{11}^0 e^{it_1\widehat{H}_1} y_1, \eta_{02}(\omega)\rangle\langle\eta_{01}(\omega), e^{-it_2\widehat{H}_1} Q_{11}^0 e^{it_2\widehat{H}_1} y_2\rangle \\ &\quad + E\langle e^{it_1\widehat{H}_2}(Q_{12}^0)^* e^{it_1\widehat{H}_1} y_1, \eta_{02}(\omega)\rangle\langle\eta_{02}(\omega), e^{it_2\widehat{H}_2}(Q_{12}^0)^* e^{it_2\widehat{H}_1} y_2\rangle \\ &\quad + E\langle e^{-it_1\widehat{H}_1} Q_{11}^0 e^{it_1\widehat{H}_1} y_1, \eta_{10}(\omega)\rangle\langle\eta_{20}(\omega), e^{it_2\widehat{H}_2}(Q_{12}^0)^* e^{it_2\widehat{H}_1} y_2\rangle \\ &\quad + E\langle e^{it_1\widehat{H}_2}(Q_{12}^0)^* e^{it_1\widehat{H}_1} y_1, \eta_{10}(\omega)\rangle\langle\eta_{20}(\omega), e^{-it_2\widehat{H}_1} Q_{11}^0 e^{it_2\widehat{H}_1} y_2\rangle. \end{aligned} \tag{3.50}$$

Two last terms are equal to zero, since $E\langle z_1, \eta_{10}(\omega)\rangle\langle\eta_{20}(\omega), z_2\rangle = 0$ for any pair $z_1 \in H_1$, $z_2 \in H_2$. The first two terms give

$$\langle e^{-it_1\widehat{H}_1} Q_{11}^0 e^{it_1\widehat{H}_1} y_1, e^{-it_2\widehat{H}_1} Q_{11}^0 e^{it_2\widehat{H}_1} y_2\rangle$$
$$+ \langle e^{it_1\widehat{H}_2} (Q_{12}^0)^* e^{it_1\widehat{H}_1} y_1, e^{it_2\widehat{H}_2} (Q_{12}^0)^* e^{it_2\widehat{H}_1} y_2\rangle.$$

Thus,

$$D_{11}(t_1, t_2) = e^{-it_1\widehat{H}_1} Q_{11}^0 e^{i(t_2-t_1)\widehat{H}_1} Q_{11}^0 e^{it_1\widehat{H}_1}$$
$$+ e^{-it_2\widehat{H}_1} Q_{12}^0 e^{i(t_1-t_2)\widehat{H}_2} (Q_{12}^0)^* e^{it_1\widehat{H}_1}.$$

Representations (3.48) and (3.49) imply that even in the absence of interaction between the subsystems S_1 and S_2 dynamics of S_1 depends on Hamiltonian $\widehat{H}_2$ and vice versa. It can be interpreted as a sign of action at the distance. Thus the stochastic process $\xi(t)$ is nonvocal, opposite to the process $\phi(t)$ given by (3.44).

3.4 Correlations of the Components of the Prequantum Field

We remark that PCSFT predicts not only correlations of quadratic functionals of the prequantum field (they correspond to correlations of quantum observables), but correlations of any two functionals $f_1(\phi_1)$ and $f_2(\phi_2)$. Thus PCSFT provides new predictions comparing to QM. The simplest class of field functionals which do not match quantum observables are linear functionals of the form

$$\phi \mapsto f_g(\phi) = \langle\phi, g\rangle = \int_{\mathbf{R}^3} dx\phi(x)\overline{g(x)}, \qquad (3.51)$$

where $g \in L_2$. The effects of such field functionals are not visible in measurements described by QM. To find such effects, finer measurement procedures have to be designed. In Section 3.1.4 we considered the approximation of averages of nonquadratic functionals by averages of quadratic functionals. However, it was the case of noncomposite systems. For composite systems, the PCSFT functionals matching QM observables are fourth-order forms

$$\widehat{A}_1 \otimes \widehat{A}_2 \mapsto f(\phi) = f_{A_1}(\phi_1) f_{A_2}(\phi_2). \qquad (3.52)$$

Thus functionals (3.51) are "invisible in QM."

The correlation of two linear functionals of the prequantum field is given by

$$\langle f_{g_1}, \bar{f}_{g_2}\rangle = \int \langle\phi_1, g_1\rangle\langle\bar{\phi_2}, g_2\rangle d\mu(\phi_1, \phi_2) = \langle D_{21}g_1, \bar{g_2}\rangle$$
$$= \int dxdy\overline{\Psi(x, y)}g_1(x)g_2(y).$$

Choose now $g_1(x) = \delta(x - x_0)$ and $g_2(y) = \delta(y - y_0)$. We proceed formally by operating with functions which do not belong to L_2, Here $f_{g_1}(\phi_1) = \phi_1(x_0)$ and $f_{g_2}(\phi_2) = \phi_2(y_0)$,

$$\langle\phi_1(x_0), \overline{\phi_2(y_0)}\rangle = \int \phi_1(x_0)\overline{\phi_2(y_0)}d\mu(\phi_1, \phi_2) = \overline{\Psi(x_0, y_0)}.$$

It is more convenient to consider the correlation

$$c(x_0, y_0) = \langle\overline{\phi_1(x_0)}, \phi_2(y_0)\rangle = \Psi(x_0, y_0).$$

We have made the following "discovery":

The wave function of the composite system $S = (S_1, S_2)$ is nothing else than the covariance between the corresponding prequantum classical fields.

This implication of PCSFT may have important experimental consequences. However, before we come closer to a fundamental discussion about "beyond quantum" experiments, we would like to apply the mathematical scheme of PCSFT to the classical macroscopic electromagnetic field. The idea behind this is to explore the analogy between PCSFT framework for macroscopic electromagnetic fields and photons.

3.5 PCSFT Formalism for Classical Electromagnetic Field-1

In this section we show that one can mimic QM by using the PCSFT formalism for a macroscopic electromagnetic field, i.e., having sufficiently high intensity.

Consider a classical electromagnetic field $\phi(x) = (E(x), B(x))$, i.e., a field of sufficiently high intensity. Components of this field and not only the energy patterns can be measured. It is convenient to operate in the *Riemann–Silberstein representation*:

$\phi(x) = E(x) + iB(x)$, i.e., with the complex field. Consider now a random electromagnetic signal $\phi(x, \omega) = E(x, \omega) + iB(x, \omega)$, where the parameter ω describes randomness. Randomness is generated by the source of the signal as well as by the noise present in the media of signal propagation. We consider only Gaussian signals with zero mean value: $\langle \phi \rangle = \langle E \rangle + i\langle B \rangle = 0$.

We operate with two random signals $\phi_1(x, \omega) = E_1(x, \omega) + iB_1(x, \omega)$ and $\phi_2(y, \omega) = E_2(y, \omega) + iB_2(y, \omega)$; here $x, y \in \mathbf{R}^3$. Suppose that these signals are correlated with the correlation operator D. These correlations are due to the common preparation procedure. This operator acts in the L_2-space. This is the integral operator with the kernel $D(x, y)$. Take the off-diagonal block of this kernel, $D_{21}(x, y)$, and denote it by $\Psi(x, y)$. By scaling the Gaussian random fields we can always assume that the L_2-norm of this function of two variables is equal to one. We call this function a *wave function of the bi-signal system.* We emphasize that this wave function is a purely classical object.

Suppose that the character of correlations is such that the diagonal blocks are expressed in terms of the off-diagonal block $\Psi(x, y)$ as it is assumed in PCSFT, see (3.26) and (3.27). The crucial point of the present discussion is that the "classical wave function" $\Psi(x, y)$ is measurable! It describes correlations of two random signals: $c(x, y) = \langle \overline{\phi_1(x)}, \phi_2(y) \rangle = \Psi(x, y)$.

The main mathematical problem is the presence of the white noise. Its probability distribution is not concentrated on the L_2-space. To obtain a rigorous mathematical description, we have to use the probability distribution on, e.g., $\mathcal{S}' \equiv \mathcal{S}'(\mathbf{R}^3)$, the space of Schwartz distributions on $\mathbf{R}^3$. The Gaussian probability μ_D with the covariance operator D determined by (3.26)–(3.28) is well defined on $\mathcal{S}'$, in spite of the presence of the white noise. Thus each random signal is described by a random field in $\mathcal{S}'(\mathbf{R}^3)$. A mathematical difficulty arises as a consequence of the fact that the functional $f_{x_0}(\phi) = \phi(x_0)$ is not well defined on $\mathcal{S}'(\mathbf{R}^3)$. Therefore one has to consider field-averaging with respect to the spatial x-variable

$$f_g(\phi) = \langle \phi, g \rangle,$$

where g is a test-function, $g \in \mathcal{S}(\mathbf{R}^3)$. Thus the correlations of the classical fields $\phi_1(x, \omega)$ and $\phi_2(y, \omega)$ are well defined for all possible test functions g_1 and g_2. The presence of such x-averaging is natural from the physical viewpoint. Measurement devices are not able to measure anything at a single point. Any apparatus has spatial dimensions and it performs integration over the corresponding spatial domain, cf. Haag [101] and, especially, Ref. [102]. Thus only x-averaged fields can be measured:

$$E_1(x_0; g) = \int E_1(u) g_1(x_0 - u) du, \; B_1(x_0; g_1) = \int B_1(u) g_1(x_0 - u) du$$

and

$$E_2(y_0; g_2) = \int E_2(v) g_2(y_0 - v) dv, \; B_2(y_0; g_2) = \int B_2(v) g_2(y_0 - v) dv.$$

Since signals are random, these are random fields:

$$\phi_1(x_0, \omega; g_1) = E_1(x_0, \omega; g_1) + i B_1(x_0, \omega; g_1),$$

$$\phi_2(y_0, \omega; g_2) = E_2(y_0, \omega; g_2) + i B_2(y_0; \omega, g_2).$$

Spatial averaging can be interpreted as the use of two antennas described by test functions, smoothing functions. These antennas perform cut-off of signals. Thus, for two classical random signals, we can find correlations

$$\begin{aligned} c(x_0, y_0; g_1, g_2) &= \langle \overline{\phi_1(x_0; g_1)}, \phi_2(y_0; g_2) \rangle \\ &= \int dx dy \Psi(x, y) g_1(x - x_0) g_2(y - y_0). \end{aligned}$$

Take now a δ-sequence: $g^{(\epsilon)}(x) \to \delta(x), \epsilon \to 0$, in the sense of the distribution theory; we select $g_1 = g_2 = g^{(\epsilon)}$. Then

$$\langle \overline{\phi_1(x_0; g^{(\epsilon)})}, \phi_2(y_0; g^{(\epsilon)}) \rangle \to \Psi(x_0, y_0), \epsilon \to 0.$$

Consider now classical energy functionals for the electromagnetic pulses:

$$\mathcal{E}(\phi_1) = \int_{O_1} dx_0 |\phi_1(x_0; g_1)|^2 = \int_{O_1} dx_0 (E_1^2(x_0; g_1) + B_1^2(x_0; g_1)),$$

$$\mathcal{E}(\phi_2) = \int_{O_2} dy_0 |\phi_2(y_0; g_2)|^2 = \int_{O_2} dy_0 (E_2^2(y_0; g_2) + B_2^2(y_0; g_2)),$$

where O_1 and O_2 are two domains in $\mathbf{R}^3$. These are quadratic functionals of the macroscopic electromagnetic field. We can now apply equality (3.21) for these functionals. We obtain a quantum-like formula for the covariance of "smoothed" energy densities in domains O_1 and O_2. The right-hand side of (3.21) contains operators $\widehat{A}_1$ and $\widehat{A}_2$ with the kernels

$$k_1(x_1, x_2) = \int_{O_1} dx_0 g_1(x_0 - x_1) g_1(x_0 - x_2),$$

$$k_2(y_1, y_2) = \int_{O_2} dy_0 g_2(y_0 - y_1) g_2(y_0 - y_2).$$

In principle, to make probabilistic predictions for quadratic functionals of the classical electromagnetic field, one can apply the mathematical formalism of QM, i.e., use only self-adjoint operators and define correlations in accordance with QM. In such a coarse description components of the classical field are not considered at all. However, as we know, the classical electromagnetic field of sufficiently high intensity permits a finer description given by the classical electrodynamics. According to the latter, not only quadratic functionals such as the energy of the field, but even linear functionals such as coordinates of the electric and magnetic fields can be measured.

We also remark that in the signal theory correlations between different components of a signal are a subject of the intensive study. For example, in the case of partially polarized radiation, the Jones vector $\phi = (\phi_x, \phi_y)$ varies in time and space in a way that differs from the constant rate of phase rotation of monochromatic, purely polarized waves. In this case, the wave field is likely to be stochastic, and only statistical information can be gathered about the variations and correlations between components of the electric field. This information is embodied in the *coherency matrix*

$$U_\phi = \begin{pmatrix} \langle \phi_x \phi_x \rangle & \langle \phi_y \phi_x \rangle \\ \langle \phi_x \phi_y \rangle & \langle \phi_y \phi_y \rangle \end{pmatrix},$$

where angular brackets denote averaging over many wave cycles. The PCSFT approach is based on the generalization of the coherency matrix for bi signals. For the classical electromagnetic field, we consider two correlated signals with the Jones vector $\phi_1 = (\phi_{1x}, \phi_{1y})$, $\phi_2 = (\phi_{2x}, \phi_{2y})$. Our model is based on the covariance matrix

$$D = \begin{pmatrix} \langle\phi_{1x}\phi_{1x}\rangle\langle\phi_{1x}\phi_{1y}\rangle & \langle\phi_{2x}\phi_{1x}\rangle\langle\phi_{2x}\phi_{1y}\rangle \\ \langle\phi_{1y}\phi_{1x}\rangle\langle\phi_{1y}\phi_{1y}\rangle & \langle\phi_{2y}\phi_{1x}\rangle\langle\phi_{2y}\phi_{1y}\rangle \\ \langle\phi_{1x}\phi_{2x}\rangle\langle\phi_{1x}\phi_{2y}\rangle & \langle\phi_{2x}\phi_{2x}\rangle\langle\phi_{2x}\phi_{2y}\rangle \\ \langle\phi_{1y}\phi_{2x}\rangle\langle\phi_{1y}\phi_{2y}\rangle & \langle\phi_{2y}\phi_{2x}\rangle\langle\phi_{2y}\phi_{2y}\rangle \end{pmatrix}.$$

The diagonal blocks of D coincide with the coherency matrices U_{ϕ_1}, U_{ϕ_2} of the signals ϕ_1, ϕ_2.

Remark 5.1. (Objectification problem) By comparing PCSFT to classical signal theory we expect that PCSFT can overcome the problems that arise whenever an interpretation of QM is adopted which is not reduced to the minimal (statistical) interpretation: in particular, the objectification problem, which occurs if a very weak "realistic" position is adopted assuming that QM deals with individual objects and their properties [46]. In classical signal theory switching from the ensemble description to the individual description is achieved on the basis of the *ergodic hypothesis.* Ensemble averages can be reduced to time averages. Of course, the objectification problem is important even in classical physics.

In this section the PCSFT like treatment of classical electromagnetic field was performed to illustrate similarities in PCSFT description of classical macroscopic fields and prequantum fields. We shall study these similarities in more detail in Section 5.8, where we consider nonquadratic functionals of the classical electromagnetic field, random fields with nonzero average, and the representation of the system of the Maxwell equations in the form of the Schrödinger equation.

3.6 Discussion of a Possible Experimental Verification of PCSFT

We see how well the classical electromagnetic field mimics quantum behavior. One can guess that this matching is not of a purely mathematical character. In the PCSFT framework it is natural to assume that photons are simply classical electromagnetic pulses. If we were able to perform measurements of electric and magnetic components of these, so to say photonic, fields, then according to PCSFT, we would obtain the covariance of these field components which coincides with the wave function given by QM, $c(x_0, y_0) = \Psi(x_0, y_0)$. The analogy with well studied macroscopic electromagnetic field makes promising the project of measurement of components of the photonic pulses.

The situation is essentially more complicated for massive particles or, in the PCSFT framework, massive prequantum fields. According to PCSFT, e.g., a neutron is a classical field. At the moment we do not have measurement procedures for components of such fields. However, we expect that such procedures will be found.

3.7 Photonic Field

In spite of a rather common opinion that a photon wave function is not well defined or that at least there are serious difficulties, we can mention one consistent approach to this problem based on *Bialynicki–Birula (BB) photon wave function* [33]. It is closely related to the *Riemann–Silbertstein representation* of the classical electromagnetic field. We remark that the BB wave function for a single photon has the form

$$\Psi(x) = \begin{pmatrix} \Psi_+(x) \\ \Psi_-(x) \end{pmatrix},$$

where $\Psi_\pm(x) = \mathbf{E}(x) \pm i\mathbf{B}(x)$ (We use bold $\mathbf{E}$ and $\mathbf{B}$ to distinguish them from coming components of classical electromagnetic field E and B).

Thus it has six coordinates, which are related by the condition $\Psi = \sigma_3 \overline{\Psi}$, where σ_3 is the corresponding Pauli matrix. Thus $\Psi(x, y)$ takes values in $\mathbf{C}^6$.

Let $\Psi_1(x)$ and $\Psi_2(x)$ be two single photon BB wave functions. It is easy to construct the BB wave function of two entangled photons:

$$\Psi(x, y) = \Psi_1(x)\Psi_2(y) + \Psi_2(x)\Psi_1(y).$$

In our approach $\Psi(x, y)$ has a new interpretation, it is the off-diagonal block of the matrix of correlations of two classical electromagnetic pulses. To realize it as an integral operator, we consider fields

$$\phi(x) = \begin{pmatrix} \phi_+(x) \\ \phi_-(x) \end{pmatrix},$$

where $\phi_\pm(x) = E(x) \pm i B(x)$. Thus $\Psi(x, y)$ determines the operator D_{12} acting on such classical fields:

$$D_{12}\phi = \begin{pmatrix} \int \Psi_+(x, y)\phi_+(y)dy \\ \int \Psi_-(x, y)\phi_-(y)dy \end{pmatrix}.$$

In the same way we define operators D_{11}, D_{22} corresponding to the BB wave function, see (3.27) and (3.28).

Technological improvement of photon detectors is an important step toward the experimental verification of predictions of PCSFT for, so to say, the photonic classical field. The majority of photon detectors are based on either scattering of electrons by photons (photomultipliers tubes, PMTs) from a photodiode or creation by photon pairs electron–hole (avalanche photodiodes, APDs, and the visible light counters, VLPCs). Thus, photon-like discreteness of counting might be just an illusion induced by the discreteness of electron emission. New possibilities to test PCSFT may be provided by W-TESs, microcalorimeters. In particular, PCSFT demands essential improvement of the detector efficiency in fundamental experiments to test indivisibility of photons [95, 96]. Grangier-type experiment with W-TESs will be an important step toward testing the PCSFT prediction about divisibility of photons.

3.8 Correlation between Polarization Vectors of Entangled Photons

In this section we work only with polarizations, space dependence is not taken into account. The symbols $|x\rangle$, $|y\rangle$ denote the vectors of the basis corresponding to the x- and y-axes.

According to PCSFT, each photon is a classical wave with well defined polarization given by the Jones vector. We consider a linearly polarized photonic field with the Jones vector

$$\phi = \mathcal{E}_0(\cos\theta|x\rangle + \sin\theta|y\rangle) = \mathcal{E}_0 u,$$

where $u = \cos\theta|x\rangle + \sin\theta|y\rangle$ is the normalized Jones vector. Let this classical signal pass through a PBS with the a-orientation, where $a = \cos\alpha|x\rangle + \sin\alpha|y\rangle$. PBS splits the photonic signal into two orthogonal components

$$\phi_{+,a} = \mathcal{E}_0 \cos(\theta - \alpha)a, \;\; \phi_{-,a} = \mathcal{E}_0 \sin(\theta - \alpha)a^\perp,$$

where

$$a = \cos\alpha|x\rangle + \sin\alpha|y\rangle, \ a^{\perp} = -\sin\alpha|x\rangle + \cos\alpha|y\rangle.$$

We remark that the intensities of these signals are given by

$$I_{+} = \mathcal{E}_0^2 \cos^2(\theta - \alpha), \ I_{-} = \mathcal{E}_0^2 \sin^2(\theta - \alpha).$$

The aforesaid is well known in classical theory. To distinguish PCSFT from QM, we study not correlations of intensities, i.e., squared field amplitudes, but correlations of amplitudes.

The PBS-splitting corresponds to measurements of the linear field functionals

$$f_a(\phi) = \langle \phi, a \rangle = \mathcal{E}_0 \cos(\theta - \alpha),$$

$$f_{a^{\perp}}(\phi) = \langle \phi, a^{\perp} \rangle = \mathcal{E}_0 \sin(\theta - \alpha)$$

We call them *polarization functionals.*

Take a pair of photonic signals ϕ_1, ϕ_2 and do PBS-splitting of each of them. By fixing two orientations of PBSs,

$$a = \cos\alpha|x\rangle + \sin\alpha|y\rangle, \ b = \cos\beta|x\rangle + \sin\beta|y\rangle,$$

we obtain four components. In PCSFT the polarizations of these components are described by the linear functionals

$$f_a(\phi_1), \ f_{a^{\perp}}(\phi_1), \ f_b(\phi_2), \ f_{b^{\perp}}(\phi_2).$$

Suppose the original photonic signals ϕ_1 and ϕ_2 were jointly prepared and the preparation procedure does not provide a possibility to produce constant linear polarizations of these signals. It produces *randomly (linear) polarized radiation.* Such a preparation situation can be described by two random photonic signals $\phi_k(\omega)$, $k = 1, 2$, where ω is a random parameter:

$$\phi_{kx}(\omega) = \mathcal{E}_{k0}(\omega) \cos\theta_k(\omega), \ \phi_{ky}(\omega) = \mathcal{E}_{k0}(\omega) \sin\theta_k(\omega).$$

As a consequence of correlations in the process of preparation, the off-diagonal term of the covariance matrix of these two signals is nonzero. We prepare two photonic random signals with $D_{12} = \widehat{\Psi}$, where the operator $\widehat{\Psi}$ is given by the Bell state:

$$\Psi = (|xx\rangle + |yy\rangle)/\sqrt{2} \mapsto \widehat{\Psi} = I/\sqrt{2},$$

where

$$I = \begin{pmatrix} 1 & 0 \\ 0 & 1 \end{pmatrix}$$

is the unit matrix. Thus the x and y-coordinates of ϕ_1 and ϕ_2 are independent: $\langle \phi_{1x}\phi_{2y} \rangle = \langle \phi_{1y}\phi_{2x} \rangle = 0$. Their x-coordinates are strongly, in fact, maximally, correlated: $\langle \phi_{1x}\phi_{2x} \rangle = 1/\sqrt{2}$. The same is valid for the y-coordinates: $\langle \phi_{1y}\phi_{2y} \rangle = 1/\sqrt{2}$. We remark that in the operator form the covariance operator corresponding to the Bell state Ψ can be written as

$$D_\Psi = \begin{pmatrix} \widehat{\Psi}\widehat{\Psi}^* + \epsilon I & \widehat{\Psi} \\ \widehat{\Psi}^* & \widehat{\Psi}^*\widehat{\Psi} + \epsilon I \end{pmatrix}.$$

We can select $\epsilon = \frac{\sqrt{2}-1}{2}$. This is the minimal ϵ guaranteeing the condition $D_\Psi \geq 0$. Here

$$D_\Psi = \frac{1}{\sqrt{2}} \begin{pmatrix} 1 & 0 & 1 & 0 \\ 0 & 1 & 0 & 1 \\ 1 & 0 & 1 & 0 \\ 0 & 1 & 0 & 1 \end{pmatrix} = \frac{1}{\sqrt{2}} \begin{pmatrix} I & I \\ I & I \end{pmatrix}.$$

We find the correlation coefficient

$$r_{xx} = \frac{\langle \phi_{1x}\phi_{2x} \rangle}{\sqrt{\langle \phi_{1x}^2 \rangle \langle \phi_{2x}^2 \rangle}} = 1$$

(we remark that $\langle \phi_{1x} \rangle = \langle \phi_{2x} \rangle = 0$). Thus the x-coordinates are *maximally correlated.* The same is valid for the y-coordinates.

We analyze the classical probabilistic description of components of the biphotonic field. It is convenient to subtract the background field. After such a renormalization we obtain the random vectors $\tilde{\phi}_1$ and $\tilde{\phi}_2$. Each of them has the covariance matrix

$$\tilde{D}_{11} = \tilde{D}_{22} = \frac{1}{2} \begin{pmatrix} 1 & 0 \\ 0 & 1 \end{pmatrix}.$$

(We remark that $\tilde{D}_{11}$ and $\tilde{D}_{22}$ coincide with the reduced density matrices corresponding to the Bell state Ψ.) For each photon, its polarization vector is the Gaussian random variable. Its probability distribution is given by $p(x, y) = \frac{1}{\pi} e^{-(x^2+y^2)}$. It is rotationally invariant. The angles θ_k of linear polarization are random variables

$\theta_k(\omega) = \arctan \frac{\tilde{\phi}_{kx}(\omega)}{\phi_{ky}(\omega)}$, $k = 1, 2$. This is the classical picture of polarization of each photon of the Bell state.

We now study the geometric structure of the joint probability distribution of two polarization vectors, ϕ_1, ϕ_2. The crucial point is that the *covariance matrix is degenerate.* For any vector $u = (x_1, y_1, x_2, y_2)$, we have

$$\langle Du, u \rangle = (x_1 + x_2)^2 + (y_1 + y_2)^2.$$

Thus $\mathrm{Ker}D = \{u : [x_1 = -x_2, y_1 = -y_2]\}$ is a plane in the four-dimensional space. The Gaussian distribution of the biphotonic random field (corresponding to the Bell state Ψ) is concentrated on the orthogonal complement of this plane, i.e., on the plane $L_\Psi = \{u : [x_1 = x_2, y_1 = y_2]\}$. This is the classical probabilistic picture of the exact correlations between the respective coordinates of photonic fields ϕ_1 and ϕ_2. We have

$$x_1(\omega) = x_2(\omega), y_1(\omega) = y_2(\omega)$$

for almost all ω. Thus $\phi_1(\omega) = \phi_2(\omega)$ for almost all ω.

We now find correlations between the polarization functionals

$$\langle f_a f_b \rangle = \frac{1}{\sqrt{2}} \langle a, b \rangle = \frac{1}{\sqrt{2}} \cos(\alpha - \beta).$$

These correlations are not predicted by QM. In our approach they can be measured after the elaboration of a proper procedure of measurement of polarization of the photonic pulses.

3.9 Functionals of Prequantum Fields Corresponding to Operators of Photon Polarization

We remark that

$$f_a^2(\phi) = |\langle \phi, a \rangle|^2 = \langle \widehat{P}_a \phi, \phi \rangle,$$

where $\widehat{P}_a$ is the projector on the vector a. In QM the quadratic functional $f_a^2(\phi)$ (the object of PCSFT) is represented by the operator $\widehat{P}_a$, the quantum observable. Hence the classical correlations of these quadratic functionals coincide with the quantum correlations. We form the linear combination of two quadratic functionals

$$F_a(\phi) = f_a^2(\phi) - f_{a^\perp}^2(\phi) = \langle \widehat{P}_a \phi, \phi \rangle - \langle \widehat{P}_{a^\perp} \phi, \phi \rangle = \langle \widehat{P}_a - \widehat{P}_{a^\perp} \phi, \phi \rangle.$$

Let us set $\widehat{\sigma}_a = \widehat{P}_a - \widehat{P}_{a^\perp}$. This is a photon polarization operator corresponding to the a-oriented PBS. Thus, quantum observables of photon polarization do not correspond to projections of polarization of the prequantum photonic field. These are linear combinations of squared polarization functionals.

We remark that the classical average $\langle F_a \rangle = 0$. By (3.20),

$$\langle F_a F_b \rangle = \langle \widehat{\sigma}_a \otimes \widehat{\sigma}_b \Psi, \Psi \rangle = \cos 2(\alpha - \beta).$$

Thus PCSFT reproduces correlations of quantum polarizations. However, PCSFT also provides a possibility of going beyond QM. It predicts not only correlations of squared polarization functionals, but even of polarization functionals.

Conclusion. *By exploring a purely wave model we found correlations between components of prequantum fields which are symbolically represented by quantum particles in QM. In particular, we predict correlations between components of electromagnetic classical fields corresponding to entangled photons.*

3.10 Classical Representation of Heisenberg's Uncertainty Relation

The Heisenberg's uncertainty relation is the very heart of QM. In fact, this relation was the main source of the Copenhagen interpretation of QM. Bohr proposed his principle of complementarity after a number of conversations with Heisenberg who was advocating his uncertainty principle. Roughly speaking, the fathers of Copenhagen QM (Bohr, Heisenberg, Pauli,...) were not interested in various no-go theorems (such as, e.g., von Neumann theorem).[a] For them, it was completely clear that the Heisenberg's uncertainty relation is the first and final no-go theorem, see Stanford Encyclopedia of Philosophy (article: Uncertainty Principle):

> One striking aspect of the difference between classical and quantum physics is that whereas classical mechanics presupposes that exact simultaneous values can be assigned to all physical

[a]I am practically sure that they would not demonstrate any interest even in the Bell theorem.

> quantities, quantum mechanics denies this possibility, the prime example being the position and momentum of a particle. According to quantum mechanics, the more precisely the position (momentum) of a particle is given, the less precisely can one say what its momentum (position) is.

Margenau [239, 240] was the first who presented an argument against the Copenhagen interpretation of the Heisenberg's uncertainty relation. He emphasized that the Heisenberg's uncertainty relation is about statistical quantities, namely, dispersions of two random variables. These statistical quantities can be measured separately in series of two different experiments. Therefore the Heisenberg's uncertainty relation has nothing to do with joint determination of values of these random variables. This is a trivial argument. It is surprising that the majority of physicists still totally ignore it.

3.10.1 *The Role of the Random Background*

In PCSFT approach to the Heisenberg's uncertainty relation, the presence of the random background (vacuum fluctuations) will play a crucial role. Roughly speaking, all quantum mysteries arise from the ignorance to the presence of these fluctuations. To be more precise, from the ignorance of the fact that QM is a special mathematical formalism elaborated to cancel the contribution of vacuum fluctuations. Although this is a fruitful strategy for dealing with experimental statistical data (because otherwise we would operate with noisy data) one cannot simply forget about these fluctuations. For example, the right interpretation of the Heisenberg's uncertainty relation can be obtained only by taking vacuum fluctuations into account. The crucial point is that the "quantum dispersion" is not really the statistical dispersion, but it gets rise from ignoring the dispersion of the random background.

In Section 3.2.3 we have seen that, to construct a proper probability measure for a composite system, one should change the correspondence between density operators of QM and covariance operators of PCSFT even for a single system, and not simply identify a density operator, an element of the quantum formalism, with the covariance operator (see 3.1.2), but take into account the background of the white noise type, see (3.24). The prequantum random field corresponding to the density operator ρ should be taken as Gaussian with zero mean value and the covariance operator given by the correspondence

$$\rho \to D_\rho = \rho + \epsilon I, \tag{3.53}$$

see (3.24). To simplify the presentation in this section, we set $\epsilon = 1$. For a pure state Ψ, we set

$$D_{\rho_\Psi} \equiv D_\Psi,$$

where, as always, $\rho_\Psi = \Psi \otimes \Psi$ or in Dirac's notation $|\Psi\rangle\langle\Psi|$.

From this viewpoint, QM is a special mathematical formalism designed to eliminate effects of vacuum fluctuations of the white noise type. It is a natural formalism to describe observations performed on the random background. The contribution of this background should be subtracted. For a trace-class operator $\widehat{A}$, we have

$$E_{\mu_{D_\rho}} f_A = \mathrm{Tr}\rho\widehat{A} + \mathrm{Tr}\widehat{A},$$

where $\mu_{D_\rho} \sim N(0, D_\rho)$ is the Gaussian distribution with zero mean value and the covariance operator D_ρ and $E_{\mu_{D_\rho}}$ denotes the average with respect to this probability distribution. Hence,

$$\langle\widehat{A}\rangle_\rho = E_{\mu_{D_\rho}} f_A - \mathrm{Tr}\widehat{A} \tag{3.54}$$

or at least formally (there are some mathematical difficulties in the case of the infinite-dimensional Hilbert space):

$$\langle\widehat{A}\rangle_\rho = E_{\mu_{D_\rho}} f_A - E_{\mu_I} f_A,$$

where $\mu_I \sim N(0, I)$ is the standard Gaussian distribution, white noise. QM formalism can be interpreted as a rather special *regularization procedure.* If $\mathrm{Tr}\widehat{A} = \infty$, then PCSFT average with respect to the Gaussian distribution μ_{D_ρ} is not defined: the Gaussian integral diverges. Of course, this effect is a consequence of the infinite dimension of the state space. However, the QM formalism provides its regularization.

3.10.2 *Coupling of Dispersions*

We proceed with finite-dimensional Hilbert spaces and restrict our consideration to pure states. Let $\widehat{A}$ be self-adjoint. Its dispersion in a pure state Ψ is defined as

$$\sigma_\Psi^2(\widehat{A}) = \langle(\widehat{A} - \langle\widehat{A}\rangle_\Psi I)^2\rangle_\Psi = \langle\widehat{A}^2\rangle_\Psi - \langle\widehat{A}\rangle_\Psi^2.$$

Let ξ be a classical random variable. Its dispersion is given by

$$\sigma_P^2(\xi) = E_P(\xi - E_P\xi)^2 = E_P\xi^2 - (E_P\xi)^2,$$

where P is a probability measure.

Lemma 10. 1. *Let the operator $\widehat{A}$ be self-adjoint and let Ψ be a pure state. Then*

$$Tr(D_\Psi\widehat{A})^2 = Tr\widehat{A}^2 + 2\langle\widehat{A}^2\rangle_\Psi + \langle\widehat{A}\rangle_\Psi^2 \tag{3.55}$$

where the operator D_Ψ is defined by (3.53).

Proof. For an orthonormal basis, we have

$$\begin{aligned}\sum_k\langle D_\Psi\widehat{A}e_k, \widehat{A}D_\Psi e_k\rangle &= \sum_k\langle(I+\rho_\Psi)\widehat{A}e_k, \widehat{A}(I+\rho_\Psi)e_k\rangle \\ &= \sum_k\langle\widehat{A}e_k, \widehat{A}e_k\rangle + \sum_k\langle\widehat{A}e_k, \widehat{A}\rho_\Psi e_k\rangle \\ &\quad+\sum_k\langle\rho_\Psi\widehat{A}e_k, \widehat{A}e_k\rangle + \sum_k\langle\rho_\Psi\widehat{A}e_k, \widehat{A}\rho_\Psi e_k\rangle \\ &= \mathrm{Tr}\widehat{A}^2 + \sum_k\langle\widehat{A}e_k, \langle e_k, \Psi\rangle\widehat{A}\Psi\rangle \\ &\quad+\sum_k\langle\langle\widehat{A}e_k, \Psi\rangle\Psi, \widehat{A}e_k\rangle \\ &\quad+\sum_k\langle\langle\widehat{A}e_k, \Psi\rangle\Psi, \langle e_k, \Psi\rangle\widehat{A}\Psi\rangle \\ &= \mathrm{Tr}\widehat{A}^2 + \sum_k\langle\Psi, e_k\rangle\langle e_k, \widehat{A}^2\Psi\rangle \\ &\quad+\sum_k\langle\widehat{A}\Psi, e_k\rangle\langle e_k, \widehat{A}\Psi\rangle \\ &\quad+\sum_k\langle\Psi, \widehat{A}\Psi\rangle\langle e_k, \widehat{A}\Psi\rangle\langle\Psi, e_k\rangle \\ &= \mathrm{Tr}\widehat{A}^2 + 2\langle\widehat{A}^2\Psi, \Psi\rangle + \langle\widehat{A}\Psi, \Psi\rangle^2.\end{aligned}$$

We recall once again that each pure quantum state Ψ determines the density operator ρ_Ψ. The latter determines the covariance operator D_Ψ of the Gaussian measure μ_{D_Ψ} by (3.53). To simplify the notation, we will use the symbol μ_Ψ for this measure. Thus μ_Ψ is the prequantum Gaussian distribution corresponding to the pure quantum state Ψ. It describes prequantum random field of "quantum system coupled to vacuum thermostat." We remark that, in particular, our activity concerns the translation of the operator language of the traditional quantum formalism into the language of traditional probability theory.

We shall use formula (3.17) for the Gaussian integral of the product of two quadratic forms on the complex Hilbert space: $E_\mu f_{A_1} f_{A_2} \equiv \int_H f_{A_1}(\phi) f_{A_2}(\phi) d\mu_D(\phi) = \mathrm{Tr} D\widehat{A}_1 \mathrm{Tr} D\widehat{A}_2 + \mathrm{Tr} D\widehat{A}_2 D\widehat{A}_1$. By setting $\widehat{A}_1 = \widehat{A}_2 = \widehat{A}$ and $D = D_\Psi$ in this equality and using (3.55), we obtain the following result.

Theorem 10.1. *Let conditions of Lemma 10.1 hold. Then*

$$E_{\mu_\Psi} f_A^2 = (E_{\mu_\Psi} f_A)^2 + \mathrm{Tr}\widehat{A}^2 + 2\langle \widehat{A}^2 \rangle_\Psi + \langle \widehat{A} \rangle_\Psi^2. \tag{3.56}$$

By (3.56) we have

$$\sigma_{\mu_\Psi}^2 (f_A) = \mathrm{Tr}\widehat{A}^2 + 2\sigma_\Psi^2(\widehat{A}) + 3\langle A \rangle_\Psi^2. \tag{3.57}$$

The classical dispersion $\sigma_{\mu_\Psi}^2(f_A)$ is always larger than the quantum dispersion $\sigma_\Psi^2(\widehat{A})$ since in the last one we ignore the dispersion of vacuum fluctuations. Moreover, the classical dispersion is larger than the dispersion produced by vacuum fluctuations which is given by $\mathrm{Tr}\widehat{A}^2$. Indeed, we have

$$E_{\mu_I} f_A^2 = (\mathrm{Tr}\ \widehat{A})^2 + \mathrm{Tr}\ \widehat{A}^2.$$

Taking into account that $E_{\mu_I} f_A = \mathrm{Tr}\ \widehat{A}$, we get $\sigma_{\mu_I}^2(f_A) = \mathrm{Tr}\ \widehat{A}^2$.

Let now $\langle A \rangle_\Psi = 0$. Then

$$\sigma_\Psi^2(\widehat{A}) = \frac{1}{2}[\sigma_{\mu_\Psi}^2(f_A) - \mathrm{Tr}\widehat{A}^2] = \frac{1}{2}[\sigma_{\mu_\Psi}^2(f_A) - \sigma_{\mu_I}^2(f_A)] \equiv \Gamma_{\mu_\Psi}(f_A).$$

By ignoring the contribution of vacuum fluctuations the QM dispersion can be obtained as regularization, shift by $\mathrm{Tr}\widehat{A}^2$, of the classical dispersion up to the factor $1/2$. We remark that *the equality of the quantum dispersion to zero is nothing else than the reduction of the classical dispersion to the dispersion of vacuum fluctuations, i.e., the dispersion of the irreducible background noise.*

Consider two quantum observables, symmetric operators $\widehat{A}_1$ and $\widehat{A}_2$; in general, noncommutative: $[\widehat{A}_1, \widehat{A}_2] \neq 0$. Take a pure quantum state Ψ. We can always assume that these observables have zero averages in this state:

$$\langle \widehat{A}_i \rangle_\Psi = \langle \widehat{A}_i \Psi, \Psi \rangle = 0, i = 1, 2. \tag{3.58}$$

If $\langle \widehat{A}_i \rangle_\Psi \neq 0$, we just consider shifted observables

$$\widehat{C}_i = \widehat{A}_i - \langle \widehat{A}_i \rangle_\Psi I. \tag{3.59}$$

Since the QM dispersions satisfy the Schrödinger inequality

$$\sigma_\Psi^2(\widehat{A}_1)\sigma_\Psi^2(\widehat{A}_2) \geq \frac{1}{4}|\langle[\widehat{A}_1, \widehat{A}_2]\rangle_\Psi|^2, \tag{3.60}$$

the dispersions of the corresponding PCSFT variables shifted by the background white noise satisfy its classical counterpart

$$\Gamma_{\mu_\Psi}(f_{A_1})\Gamma_{\mu_\Psi}(f_{A_2}) \geq \frac{1}{4}|\langle[\widehat{A}_1, \widehat{A}_2]\rangle_\Psi|^2. \tag{3.61}$$

We now represent even the right-hand side as classical (PCSFT) average. Set

$$\widehat{K} = [\widehat{A}_1, \widehat{A}_2].$$

It is a skew-symmetric operator. By applying (3.5) and taking into account that this formula is valid not only for quantum observables, i.e., symmetric operators, but even arbitrary linear operators (see Chapter 4, the proof of equality (4.29) is valid for any bounded operator), we obtain:

$$E_{\mu_\Psi} f_K = \text{Tr } \rho_\Psi \widehat{K} + \text{Tr } \widehat{K}.$$

Thus,

$$\langle \widehat{K} \rangle_\Psi = E_{\mu_\Psi} f_K - \text{Tr } \widehat{K}.$$

Inequality (3.61) can be written in purely classical terms:

$$\Gamma_{\mu_\Psi}(f_{A_1})\Gamma_{\mu_\Psi}(f_{A_2}) \geq \frac{1}{4}|E_{\mu_\Psi} f_K - \text{Tr } \widehat{K}|^2. \tag{3.62}$$

Thus, noncommutativity has its trace even in PCSFT. In contrast to QM, Schrödinger's PCSFT inequality does not have drastic consequences. The main difference is the presence of the shift of the dispersion. This shift, which is by the way produced by the contribution of fluctuations of vacuum, makes the restriction on the product of the dispersions not so rigid as it is in QM. However, we state again that even the conventional Schrödinger inequality has a statistical interpretation which has nothing to do with incompatibility of observables in the sense of impossibility of joint measurement. This statistical interpretation is due to Margenau [239, 240] and Ballentine [25–28].

Conclusion. *Prequantum field theory contains a counterpart of the Heisenberg's uncertainty relation, the Schrödinger-like inequality (3.62). This classical analog of the Schrödinger inequality provides an estimate from below not for the dispersions, but for dispersions shifted by contribution of vacuum fluctuations. Such estimate is not so rigid as that provided by the Schrödinger inequality for the quantum dispersions. Paradoxical consequences of the Heisenberg's uncertainty principle were induced by neglecting the aforementioned shift.*

3.11 Toward Violation of Born's Rule: Description of a Simple Experiment

As mentioned above (see Section 3.1.4), in PCSFT Born's rule is violated due to the presence of nonquadratic nonlinearities in the process of detection (QM describes only quadratic terms), see Chapter 8 for the detailed presentation. The simplest nonquadratic nonlinearity, which is taken into account by PCSFT, is of the fourth-order (third-order nonlinearities do not produce any statistical effect, if the prequantum random field is of the Gaussian type; and we proceed with Gaussian fields).

Take a quantum wave function for one-dimensional system $\Psi(x)$. Then by Born's rule the probability to find a system in an interval I of the real line is given by

$$\mathbf{p}(x \in I) = \int_I |\Psi(x)\|^2 dx. \qquad (3.63)$$

PCSFT predicts the appearance of an additional term which contains the contribution of the type

$$\int_I |\Psi(x)|^4 dx. \qquad (3.64)$$

In fact, the situation is a bit more complicated and the precise form of the deviation from Born's rule will be presented later. Now we would like to discuss another important issue of the model. The fourth-order term (3.64) contributes to the deviation with some coefficient $\alpha > 0$, the dispersion of prequantum fluctuations, see Chapter 8 for details.[a] PCSFT does not provide a numerical value of this parameter of the model or its magnitude. Therefore the experiment has to be performed for quantum states such that the contribution of the term (3.64) will be large enough.

If 't Hooft [287–289] is right and prequantum model works only on the Planck scale, then the scale α of fluctuations of the prequantum field has to be very small (of the magnitude of the Planck time). (See Elze [81–83] for the mathematical justification of 't Hooft's approach.) However, I am not so pessimistic as 't Hooft and I hope that the scale of prequantum fluctuations is not so fine. Therefore the contribution of fourth-order nonlinearities may be strong enough to compensate smallness of α.

Recently a preliminary experimental evidence of violation of Born's rule has been found in the *triple slit interference* experiment [278, 282]. However, the experiment proposed in this section on the basis of PCSFT is essentially simpler than the triple-slit experiment performed by the team of Gregor Weihs. Moreover, our model points to a source of violation of Born's rule: the contribution of nonquadratic nonlinearities to the process of detection. Weihs' experiment was based on Sorkin's model, see, e.g., Ref. [282], which is formal non-Kolmogorovian model. It cannot say anything about the process of detection, see Chapter 8 for discussion.

3.11.1 *Deviation from Born's Rule for Fourth-Order Nonlinearities in Detection*

According to PCSFT, QM describes the contribution of quadratic nonlinearities in the process of detection. This "quadratic contribution" is the main term in detection probability. According to PCSFT, detectors can also take into account field nonlinearities of

[a] In this section the presentation is very schematic and it may induce the feeling of dissatisfaction. It would be better to present this topic in Chapter 8 devoted to measurement theory based on PCSFT. However, I decided to discuss an experiment which may demonstrate violation of Born's rule in this chapter. I am not sure that experimenters will be able to proceed through mathematics of the chapters preceding Chapter 8.

higher orders. In the case of Gaussian prequantum random field, the simplest one is of the fourth order. The latter contributes with a small parameter α. In modern experiments such contribution is not visible. That is why the quantum formalism matches so well experimental statistical data.

We now present the results of calculation of the effect of the fourth-order nonlinearity in detection. The calculations (see Chapter 8) are not so tricky, but they are based on integration over the L_2-space. For a pure state given by the wave function $\Psi(x)$, the deviation from the basic quantum probabilistic law, Born's rule, is approximately given by

$$\Delta(I, \Psi, \alpha) = \alpha\Big[\int_I |\Psi(x)|^4 dx - \int_I |\Psi(x)|^2 dx \int_{\mathbf{R}^3} |\Psi(x)|^4 dx\Big]. \tag{3.65}$$

Thus "generalized Born's rule," which takes into account nonlinear fourth-order effects in detection, can be written as

$$\mathbf{p}(x \in I) \approx \int_I |\Psi(x)\|^2 dx + \Delta(I, \Psi, \alpha). \tag{3.66}$$

The main difficulty is the presence of the small parameter α, the dispersion of fluctuations of prequantum random fields. It is quite small, otherwise Born's rule would be violated long ago.

Suppose that supp $\Psi \subset I$, so the wave function is zero outside the set I.[a] Then $\Delta \equiv 0$. In particular,

$$\mathbf{p}(x \in \mathbf{R}) = \int_{\mathbf{R}} |\Psi(x)\|^2 dx = 1. \tag{3.67}$$

Let now $\Psi(x) = H,\ L/2 \leq x \leq L/2$. Thus $H^2 L = 1$, so $L = 1/H^2$. We choose $I = [0, L/2]$:

$$\int_I |\Psi(x)|^2 dx = 1/2, \int_{\mathbf{R}^3} |\Psi(x)|^4 dx = H^4 L = H^2,$$

and

$$\int_I |\Psi(x)|^4 dx = \frac{H^4 L}{2} = \frac{H^2}{2}, \ \Delta = \alpha(\frac{H^2}{2} - \frac{H^2}{2}) = 0.$$

This calculation gave a hint that an *asymmetric probability distribution* may induce nontrivial Δ.

---

[a] To be mathematically rigorous, we consider $\Psi \in L_{2,4}(\mathbf{R}^3)$: both integrals $\int |\phi(x)|^2 dx$ and $\int |\phi(x)|^4 dx$ are finite.

We choose

$$\Psi(x) = \begin{cases} H, & -L/2 \le x \le 0 \\ kH, & 0 < x \le L/2 \end{cases}$$

Hence, $1 = ||\Psi||^2 = LH^2(k^2+1)/2$, so $L = 2/(H^2(k^2+1))$. Here $I = [L/2, 0]$, $\int_I |\Psi(x)|^2 dx = H^2 L/2 = 1/(k^2+1)$;

$$\int_{\mathbf{R}^3} |\Psi(x)|^4 dx = \left(\frac{1+k^4}{1+k^2}\right) H^2, \quad \int_I |\Psi(x)|^4 dx = \frac{H^2}{k^2+1}.$$

$$\Delta = \frac{\alpha H^2 k^2 (1-k^2)}{(1+k^2)^2}.$$

If $k > 1$, then $\Delta(I, \Psi, \alpha) < 0$. Suppose, H increases (and k is fixed), then the deviation from Born's rule will be always negative and this deviation will be increasing. For large H, the probability to find a system in I will be essentially less than it is predicted by QM. For example, choose $k = 2$, then

$$\Delta = 0, 48\alpha H^2.$$

On the other hand, by choosing $k < 1$, we shall get the positive deviation. For $k = 1$, we have $\Delta = 0$ and there will be no deviation from Born's rule.

The concrete form of the wave function inducing nontrivial violation of Born's rule is not important. There are many other possibilities to make Δ large enough by taking into account the behavior of $|\Psi(x)|^4$ on the segment I.

3.11.2 *Violation of Born's Rule for Gaussian States*

Consider a Gaussian state

$$\Psi(x) = \frac{1}{(2\pi b)^{\frac{1}{4}}} e^{-\frac{x^2}{4b} + ikx}. \tag{3.68}$$

We select the interval $I = [-L/2, L/2]$ for some $L > 0$ and consider the following integrals:

$$c_1 = \int_{-L/2}^{L/2} |\Psi(x)|^2 dx = \frac{1}{\sqrt{2\pi b}} \int_{-L/2}^{L/2} e^{-\frac{x^2}{2b}} dx = \frac{1}{\sqrt{\pi b}} \int_{-L/2\sqrt{2}}^{L/2\sqrt{2}} e^{-\frac{x^2}{b}} dx; \tag{3.69}$$

$$c_2 = \int_{-L/2}^{L/2} |\Psi(x)|^4 dx = \frac{1}{2\pi b} \int_{-L/2}^{L/2} e^{-\frac{x^2}{b}} dx; \tag{3.70}$$

$$c_3 = \int_{-\infty}^{+\infty} |\Psi(x)|^4 dx = \frac{1}{2\pi b}\int_{-\infty}^{+\infty} e^{-\frac{x^2}{b}} dx = \frac{1}{2\sqrt{\pi b}}. \quad (3.71)$$

Born's rule gives the probability to find a particle in the interval I :

$$\mathbf{p}(x \in I) = c_1. \quad (3.72)$$

Our prequantum model predicts the deviation from this probability; this deviation is approximately equal to

$$\Delta(I, \Psi, \alpha) \equiv \Delta(L, b, \alpha) = \alpha[c_2 - c_1 c_3]$$
$$= \frac{\alpha}{2\pi b}\left[\int_{-L/2}^{L/2} e^{-\frac{x^2}{b}} dx - \int_{-L/2\sqrt{2}}^{L/2\sqrt{2}} e^{-\frac{x^2}{b}} dx\right].$$

Thus

$$\Delta(L, b, \alpha) = \frac{\alpha}{\pi b}\int_{-L/2\sqrt{2}}^{L/2} e^{-\frac{x^2}{b}} dx. \quad (3.73)$$

For a fixed state Ψ, we are interested in approaching the maximal deviation from Born's rule. We shall see that deviation is maximal for some special L depending on the dispersion of the Gaussian state. We have

$$\frac{\partial\Delta(L, b, \alpha)}{\partial L} = \frac{\alpha}{2\sqrt{2\pi} b}\left[\sqrt{2}e^{-\frac{L^2}{4b}} - e^{-\frac{L^2}{8b}}\right] = 0.$$

Then

$$L_{\max} = 2\sqrt{b\ln 2}. \quad (3.74)$$

We can easily check that this is the point of maximum and that

$$\Delta(L_{\max}, b, \alpha) = \frac{\alpha}{\pi\sqrt{b}}\int_{\sqrt{\frac{\ln 2}{2}}}^{\sqrt{\ln 2}} e^{-x^2} dx. \quad (3.75)$$

Set

$$\gamma = \frac{1}{\pi}\int_{\sqrt{\frac{\ln 2}{2}}}^{\sqrt{\ln 2}} e^{-x^2} dx.$$

Then

$$\Delta_{\max} = \gamma\frac{\alpha}{\sqrt{b}}. \quad (3.76)$$

Let $\alpha \sim 10^{-m}$, where m is sufficiently large. Then, to get the deviation of the magnitude $\sim 10^{-s}$, we should be able to prepare a Gaussian state with the dispersion $b \sim 10^{-2m+2s}$.

3.11.3 *Coupling with Measurement Theory for Prequantum Random Signals*

According to PCSFT, "quantum particles" are symbols used to denote classical random signals fluctuating on a fine time scale. A position detector performs spatial integration of such a signal. The main contribution is given by the quadratic term; this is Born's rule. However, a detector might integrate not only quadratic nonlinearity, but even nonlinearities of higher orders. (This is our conjecture.) In the simplest model a detector integrates the following functional of the prequantum field

$$\pi_{2,4}(\phi) = |\phi(x)|^2 + \alpha|\phi(x)|^4, \tag{3.77}$$

where $\alpha > 0$ is a small parameter (see Chapter 8).

Conclusion. *We presented an experimental design, which can be used to test violation of Born's rule, as a consequence of nonlinear (fourth-order) effects in detection. To perform such experiment, Gaussian states with very small dispersion have to be prepared.*

3.12 Why Gaussian?

Although we try to proceed as far as possible with prequantum random fields having arbitrary probability distributions, the assumption that these fields are Gaussian essentially simplifies the theory. In particular, only for Gaussian fields we were able to establish the correspondence between classical (prequantum) and quantum correlations. Only for Gaussian fields the correspondence between quantum states (density operators) and prequantum random fields is one-to-one. The latter is very important for objectification of quantum states. In the absence of the one-to-one correspondence quantum states are merely labels for huge classes of prequantum random fields. In this case the preparation procedures of the modern quantum physics are not good enough to generate the same type of random signals in all runs of the experiment.

What can be the physical reasons for usage of Gaussian probability distributions for prequantum random fields?

We shall explore the analogy between PCSFT and the classical signal theory (see also Chapter 8).

3.12.1 *Pulse Processes*

An important class of random signals is given by pulse processes, see, e.g., Ref. [265],

$$\xi(s) = \sum_m a_m F(s - s_m), \tag{3.78}$$

where s_m is the instance of generation of the mth pulse, a_m is its amplitude and $F(s)$ describes the form of a pulse. As always, it is assumed that $|F(s)| \to 0$, $|s| \to \infty$, sufficiently quickly. For example, $\xi(s) = \sum_m a_m \delta(s - s_m)$, the sum of δ-pulses.

The random parameters a_m and s_m of the random process $\xi(s)$ satisfy the following natural assumptions:

(a) All amplitudes $\{a_m\}$ and instances $\{s_m\}$ of generation of pulses are independent; amplitudes are equally distributed and instances of time are equally distributed as well

$$w(a_1 \ldots, a_m, \ldots; s_1, \ldots, s_m, \ldots)da_1 \ldots da_m \ldots ds_1 \ldots ds_m \ldots$$

$$= \Pi_m w_a(a_m) da_m w_s(s_m) ds_m.$$

(b) The probability of appearance of a pulse in the interval $[s, s+ds]$ does not depend on s and it is proportional to ds, i.e.,

$$w_s(s)ds = n_1 ds,\ n_1 = \text{const.} \tag{3.79}$$

The latter implies that the probability of appearance of n pulses in the interval T is given by the Poisson distribution

$$P(n) = \frac{\bar{n}^n e^{-\bar{n}}}{n!},\ \bar{n} = n_1 T. \tag{3.80}$$

Set

$$\bar{a} = \bar{a}_m = \int_{-\infty}^{+\infty} a_m w_a(a_m) da_m,$$

the amplitude mean value (we assumed that amplitudes of all pulses are equally distributed, so the average does not depend on m) and set

$$\overline{a^2} = \overline{a_m^2} = \int_{-\infty}^{+\infty} a_m^2 w_a(a_m) da_m.$$

Then the dispersion of $\xi(s)$ is given by

$$\sigma_\xi^2 \equiv \sigma_{\xi(s)}^2 = n_1\overline{a^2}\int_{-\infty}^{+\infty} F^2(\theta)d\theta.$$

It does not depend on s (a feature of stationarity of $\xi(s)$). Here we consider the case of symmetric fluctuations of amplitudes, i.e., $\bar{a} = 0$. In this case $\bar{\xi}(s) = 0$ and $\sigma_{\xi(s)}^2 = E\xi^2(s)$.

It can be proved that the density of the probability distribution $w(\xi)$ of the process $\xi(s)$ can be approximated (under some conditions) by the density of the Gaussian distribution with zero mean value and the dispersion which is equal to the dispersion of $\xi(s)$:

$$w(\xi) \approx \frac{1}{\sqrt{2\pi\sigma_\xi^2}} e^{-\frac{\xi^2}{2\sigma_\xi^2}} + \ldots$$

(As a consequence of stationarity of $\xi(s)$, its probability distribution does not depend on s.) If $\int_{-\infty}^{+\infty} F(\theta)d\theta = 1$ and the pulse duration is of the magnitude $\delta > 0$, then the approximation is good under the condition

$$n_1\delta \gg 1 \tag{3.81}$$

This means that the number of pulses during the interval δ must be very large. In other words, we can say, at the moment s the process $\xi(s)$ is composed of pulses which appeared in the interval $[s - \delta, s]$, because the earlier pulses died by the moment s. Thus $n_1\delta$ is the mean value of the summands $x_m = a_m F(s - s_m)$ composing $\xi(s)$ at each fixed s. Condition (3.81) can be interpreted in the following way: the distribution $w(\xi)$ of the pulse process $\xi(s)$ is close to the Gaussian distribution if a huge number of pulses are mixed in $\xi(s)$ at each moment s. If the frequency of generation of pulses is low, then the distribution of $\xi(s)$ would depend essentially on the form of an individual pulse and the Gaussian approximation would not be useful.

Thus the Gaussian law can be used to describe (of course, approximately) random pulse processes for sources of pulses which produce pulses with very high frequency comparing to the average duration of pulses.

3.12.2 *Wave Pulses*

Consider now a source of wave pulses, in (3.78) the coefficients belong not to the field of complex numbers **C**, but to the space of square-integrable fields $L_2(\mathbf{R}^3)$

$$\phi(s) = \sum_m \phi_m F(s - s_m), \tag{3.82}$$

where $\phi_m \equiv \phi_m(x, \omega)$ are random fields and $s_m \equiv s_m(\omega)$ are random variables. At each instance of prequantum time s_m, a spatial wave $\phi_m(x)$ is produced. In the abstract formalism the coefficients ϕ_m of randomly generated pulses belong to the complex Hilbert space H. We can repeat the previous reasoning for such random fields. We again assume that all random fields ϕ_m have the same probability distribution on H, say μ. The only restriction is that its mean value equals zero. The covariation operator of ν is denoted by D_ν. Consider now the probability distribution, say w_ϕ, of the wave pulses process $\phi(s)$. It is possible to prove that $\phi(s)$ is a stationary process and its probability distribution can be approximated by the Gaussian probability distribution ν_{D_ϕ}, where D_ϕ is the covariance operator of $\phi(s)$.

$$D_\phi = \left(n_1 \int_{-\infty}^{+\infty} F^2(\theta)d\theta\right) D_\mu \tag{3.83}$$

For the pulse function $F(s) = 1/\delta$, if $s \in [0, \delta]$, and 0, if $s \notin [0, \delta]$, we obtain $\int_{-\infty}^{+\infty} F(\theta)d\theta = 1$ and $\int_{-\infty}^{+\infty} F^2(\theta)d\theta = 1/\delta$. Hence,

$$D_\phi = \frac{n_1}{\delta} D_\mu. \tag{3.84}$$

If condition (3.81) holds, i.e., if the wave pulses are produced densely on the duration interval δ, then

$$\sigma^2_{\omega_\phi} = \sigma^2_{\mu_{D_\phi}} = \frac{n_1^2 \mathrm{Tr} D_\mu}{n_1 \delta} \ll 1 \tag{3.85}$$

for

$$\mathrm{Tr} D_\mu \sim \frac{1}{n_1^2}. \tag{3.86}$$

Thus if

$$\sigma_\mu \sim \frac{1}{n_1}, \tag{3.87}$$

then, instead of the wave pulses process $\phi(s)$ we can consider a Gaussian random field $\sim N(0, D_\phi)$, where D_ϕ is given by (3.84). If the prequantum random field is really produced at the prequantum time scale as a stationary process of the wave pulses type, then it is *mathematically* justified to operate everywhere with Gaussian random field. Of course, the same Gaussian random field approximates a huge variety of wave pulses processes. From this viewpoint, the QM approximation of the prequantum randomness can be represented by the following scheme:

[prequantum RF] $\to$ [Gaussian RF] $\to$ [covariance operator] $\to$ [density operator],

where RF is the abbreviation for random field. The first step is important from the mathematical viewpoint, it simplifies essentially the description of coupling of the prequantum and quantum structures, e.g., for composite systems. However, from the physical viewpoint this step can be omitted. We state again that we proceed under the assumption that the prequantum random field is a wave pulses process.

This assumption is also a question for the debate (see Section 3.62). Under this assumption the dream of one-to-one correspondence between prequantum random fields and quantum states does not come true! The quantum language was elaborated to introduce a kind of proper order in prequantum randomness. Instead of describing a huge class of random fields fluctuating on very fine prequantum time scale and composed of a huge number of wave pulses, QM invented the notion of a quantum state. The latter is mathematically presented by a density operator ρ, which is nothing else than the normalization by dispersion of the covariance operator of the prequantum stationary random field $\phi(s)$.

3.12.3 *Where Are the Sources of Prequantum Pulses?*

The picture of prequantum random fields as a stochastic processes which is stationary at the prequantum time scale and composed of wave pulses is very attractive. In particular, it provides a coupling between classical statistical radiophysics and electro-engineering. The main physical problem of justification of this viewpoint for prequantum random fields is the identification of physical sources of wave pulses. Two fundamentally different approaches to this problem can be discussed.

(A) **Operational approach on the prequantum time scale.** The operational approach [46, 59, 61, 70, 115, 116, 216, 232, 255, 256] teaches us that the QM formalism describes preparation and measurement procedures. By extending this approach beyond QM we might identify sources of wave pulses as ordinary sources of quantum signals (quantum particles). We just consider a fine time scale (the prequantum scale) and speculate that each quantum particle is, in fact, a huge train of wave pulses produced on this time scale. The duration of a prequantum pulse is so short that it cannot be observed at the time scale of measurement, the quantum scale, i.e., the scale described by QM.[a]

Thus the individual summands in (3.82) are unobservable; only the integral result of summation of a huge number of such wave pulses is observable in the form of a quantum particle. This approach works well in standard experiments, where we can identify sources of ensembles of quantum particles and hence of trains of wave pulses. Such a source produces a stationary process on the prequantum time scale and a detector, which is always of a threshold type, divides this process into portions. These portions are called quantum particles. This model works especially well for experiments of quantum optics, where electromagnetic signals, including entangled bi signals, are identified (in the quantum formalism) with ensembles of photons. The question whether the electromagnetic field is quantized in the empty space or only in the process of interaction with matter is in the very heart of QM. It is still the subject of intensive debates; in particular, during the series of conferences "The Nature of Light: What are Photons?" (San-Diego, 2005, 2007, 2009 [219, 220, 220, 268, 269]).

(B) **Background field as a source of everything.** The background field can play the role of a source of stationary prequantum fields, which are considered as quantum systems in QM. By

[a] For example, the Schrödinger dynamics is performed on this time scale.

this interpretation, each isolated quantum system is nothing else than a source of a special stationary random field. The background field in a special domain of space produces a train of wave pulses which are dissipated on this field. The balance between generation and dissipation produces a stationary random field. The background field has to be considered as a fundamental field, which is able to produce random fields corresponding to all possible sorts of quantum systems. In the future we can hope to see a fantastic picture of radiation of wave pulses of, e.g., electron random field and the corresponding stationary signal. Different types of quantum particles correspond to different structures of time series (3.82): combination of wave amplitudes, intensity of pulses n_1, and the duration of δ.

3.13 On Correspondence between Quantum Observables and Classical Variables

Although equality (3.21) establishes the coupling between classical correlations of random signals and quantum correlations, it is not completely satisfactory from the purely probabilistic viewpoint. On the left-hand side of Eq. (3.21) we have the classical *covariation*, cov (f_{A_1}, f_{A_2}), but on the right-hand side we have just the quantum average of the correlation observable $\widehat{A}_1 \otimes \widehat{A}_2$. We want to modify the correspondence between quantum and classical models to obtain a symmetric relation between classical and quantum covariances. We recall that the latter is given by

$$\text{cov}\,(\widehat{A}_1, \widehat{A}_2) = \langle A_1 \otimes A_2 \rangle - \langle A_1 \rangle \langle A_2 \rangle = \langle (\widehat{A}_1 - \langle \widehat{A}_1 \rangle I) \otimes (\widehat{A}_2 - \langle \widehat{A}_2 \rangle I) \rangle. \tag{3.88}$$

We set

$$\widehat{A}_{0i} = \widehat{A}_i - \langle A \rangle I, \; i = 1, 2, \tag{3.89}$$

Then $\langle \widehat{A}_{0i} \rangle = 0$ and

$$\text{cov}\,(\widehat{A}_{01}, \widehat{A}_{02}) = \langle \widehat{A}_{01} \otimes \widehat{A}_{02} \rangle = \text{cov}\,(\widehat{A}_1, \widehat{A}_2). \tag{3.90}$$

Let us modify the correspondence between classical and quantum variables (see section 3.1.3). Instead of the formerly used correspondence

$$\widehat{A} \to f_A(\phi) = \langle \widehat{A}\phi, \phi \rangle, \tag{3.91}$$

we introduce a new map from the quantum model to the classical prequantum model

$$\widehat{A}_0 \to f_{A_0}(\phi) = \langle \widehat{A}_0\phi, \phi \rangle, \ \widehat{A}_0 = \widehat{A} - \langle A \rangle I. \tag{3.92}$$

The crucial difference between these two correspondence maps is that the former is state independent and the latter depends on a state Ψ, the random field corresponding to this state. By using (3.21) for $\widehat{A}_{0i}$ instead of $\widehat{A}_i$, we obtain

$$\text{cov}\,(f_{A_{01}}, f_{A_{02}}) = \langle \widehat{A}_{01} \otimes \widehat{A}_{02} \rangle. \tag{3.93}$$

Thus

$$\text{cov}\,(f_{A_{01}}, f_{A_{02}}) = \text{cov}\,(\widehat{A}_{01}, \widehat{A}_{02}). \tag{3.94}$$

We have

$$\langle f_{A_{0i}} \rangle \equiv E f_{A_{0i}} = \text{Tr}\,(\rho_\Psi^{(i)} + \epsilon I)\widehat{A}_{0i} = \epsilon \text{Tr}\,\widehat{A}_{0i}.$$

This is the average with respect to vacuum fluctuations. The invention of

$$\langle f_{A_{01}} f_{A_{02}} \rangle - \langle f_{A_{01}} \rangle \langle f_{A_{02}} \rangle$$

can be interpreted as renormalization with respect to vacuum fluctuations. As we have already stated, QM can be interpreted as a formalism for renormalization of averages through ignoring the contribution of vacuum fluctuations.

Thus by considering classical variables $f_{A_{0i}}(\phi_i)$, $i = 1, 2$, instead of $f_{A_i}(\phi_i)$, we treat both formalisms (quantum and classical) in the same way, namely, as theories of signals on the noisy background whose contribution must be discarded. This explains the symmetric form of coupling (3.94) (in contrast to (3.21), where the right-hand side, the quantum one, has already been "calibrated," but the left-hand side, the classical one, has not been).

We also remark that (3.94) and (3.90) imply $\text{cov}\,(f_{A_{01}}, f_{A_{02}}) = \text{cov}(\widehat{A}_1, \widehat{A}_2)$, i.e., $\text{cov}\,(\widehat{A}_1, \widehat{A}_2) \not\to \text{cov}\,(f_{A_1}, f_{A_2})$. Hence, the formerly used correspondence between the quantum and classical models (3.91) does map the quantum covariation to the classical one.[a] This is a consequence of the following property of the quantum $\rightarrow$ classical correspondence

$$\widehat{A} - \langle A \rangle I \not\rightarrow f_A - \langle f_A \rangle.$$

One can say that this correspondence does not preserve algebraic structures, cf. the Kochen–Specker no-go theorem.

3.13.1 *Independence of Components of Prequantum Random Fields Corresponding to Factorizable Quantum States*

Consider now a factorizable quantum state $\Psi = \Psi_1 \otimes \Psi_2$, where $\Psi_i \in H_i, i = 1, 2$. Then $\langle \widehat{A}_{01} \otimes \widehat{A}_{02} \rangle = \langle \widehat{A}_{01} \Psi_1, \Psi_1 \rangle \langle \widehat{A}_{02} \Psi_2, \Psi_2 \rangle = 0$. Hence,

$$\text{cov}\,(\widehat{A}_1, \widehat{A}_2) = \text{cov}\,(\widehat{A}_{01}, \widehat{A}_{02}) = 0. \tag{3.95}$$

Thus by (3.94),

$$\text{cov}\,(f_{A_{01}}, f_{A_{02}}) = 0.$$

Factorization of a pure quantum state Ψ implies that, for any two quantum observables $\widehat{A}_i$ on the subsystems $S_i, i = 1, 2$, of a composite system $S = (S_1, S_2)$, the corresponding prequantum variables, $f_{A_{0i}}(\phi_i)$, where $\widehat{A}_{0i} = \widehat{A}_i - \langle A_i \rangle I$, are uncorrelated.

Are these prequantum variables independent in the sense of classical probability theory, i.e., as two quadratic forms of Gaussian random fields? Surprisingly enough the answer is negative!

We recall that quadratic forms of Gaussian random variables are used in mathematical statistics, in particular, in connection with the usage of the χ^2-distribution, and we can borrow the corresponding result from this domain of classical probability. We shall apply *Craig's theorem.*

Theorem 13.1. (Craig) *Let* $\phi \sim N(m, D)$ *and let* $\widehat{A}$ *and* $\widehat{B}$ *be two bounded self-adjoint operators. Then the corresponding quadratic forms* $f_A(x)$ *and* $f_B(x)$ *are independent iff*

$$\widehat{A} D \widehat{B} = 0. \tag{3.96}$$

[a] However, its modification (3.92) which is based on centering with respect to the average transforms the quantum covariation into the classical covariation.

In our situation $\mathbf{H} = H_1 \times H_2$, $\widehat{A}_{0i}$ is bounded and self-adjoint in H_i, and $\phi = (\phi_1, \phi_2) \sim N(0, D)$, where D has the block structure (3.16). A little bit later we shall also use representation (3.24). But for a moment we are fine with (3.16). First we lift operators $\widehat{A}_{0i}$ to $\mathbf{H}$. Consider two projectors $\pi_i : \mathbf{H} \to H_i$, $i = 1, 2$. Then $\pi_i^* : H_i \to \mathbf{H}$. We have $f_{A_{0i}}(\phi_i) = \langle \widehat{A}_{0i}\phi_i, \phi_i\rangle = \langle \widehat{A}_{0i}\pi_i\phi, \pi_i\phi\rangle = \langle \pi_i^* \widehat{A}_{0i}\pi_i\phi, \phi\rangle$.

Thus we can work with quadratic forms on $\mathbf{H}$ corresponding to self-adjoint operators $\widehat{C}_i = \pi_i^* \widehat{A}_{0i}\pi_i : \mathbf{H} \to \mathbf{H}$. By Craig's theorem the random variables $f_{C_i}(\phi)$, $i = 1, 2$, are independent iff

$$\widehat{C}_1 D \widehat{C}_2 = 0 \tag{3.97}$$

for D given by (3.16). We remark that $\widehat{C}_i$ also have the block structure

$$\widehat{C}_1 = \begin{pmatrix} \widehat{A}_{01} & 0 \\ 0 & 0 \end{pmatrix}, \widehat{C}_2 = \begin{pmatrix} 0 & 0 \\ 0 & \widehat{A}_{02} \end{pmatrix}.$$

Hence (3.97) can be written as

$$\widehat{A}_{01} D_{12} \widehat{A}_{02} = 0 \tag{3.98}$$

or by using (3.24),

$$\widehat{A}_{01} \widehat{\Psi} \widehat{A}_{02} = 0. \tag{3.99}$$

Thus, for any $\phi_2 \in H_2$, $\widehat{A}_{01}\widehat{\Psi}\widehat{A}_{02}\phi_2 = 0$. Let us now use $\Psi = \Psi_1 \otimes \Psi_2$, $\Psi_i \in H_i$. Then $\widehat{\Psi}\phi_2 = \langle \phi_2, \Psi_2\rangle \Psi_1$. Hence, $\langle A_{02}\phi_2, \Psi_2\rangle \widehat{A}_{01}\Psi_1 = 0$. Therefore either

$$\widehat{A}_{01}\Psi_1 = 0 \tag{3.100}$$

or

$$\widehat{A}_{02}\Psi_2 = 0. \tag{3.101}$$

These equations can be written in the form

$$\widehat{A}_i \Psi_i = \lambda_i \Psi_i, \; i = 1, 2, \tag{3.102}$$

Thus Ψ_i is an eigenvector of $\widehat{A}_i$.

Let $\Psi = \Psi_1 \otimes \Psi_2$ be a factorizable pure state of a composite system $S = (S_1, S_2)$. Then the prequantum variables $f_{A_{01}}(\phi_i)$ are independent iff at least one of Ψ_i is an eigenvector of $\widehat{A}_i$. As usual, $\widehat{A}_i$ is a quantum observable on the subsystem S_i or S.

Physical interpretation of this mathematical fact is unclear. The most disappointing point is the asymmetry of the condition of independence: the equality (3.102) has to hold either for $i = 1$ or for $i = 2$. It is sufficient that, e.g., only Ψ_1 is an eigenvector of $\widehat{A}_1$ and there is not any constraint between $\widehat{A}_2$ and Ψ_2.

We first discuss (3.102) by using the language of QM. For a factorizable state $\Psi = \Psi_1 \otimes \Psi_2$, if the value of at least one physical observable $\widehat{A}_i$ on S_i is definite, i.e., obtained with the probability one, then these observables are independent on the state Ψ. If both values are indefinite, then these observables are not independent in spite of factorizability of Ψ.

By using the PCSFT language we can say that $f_{A_{01}}$ and $f_{A_{02}}$ are independent if at least one component $\phi_i(\omega) \equiv \phi_{\Psi_i}(\omega)$ of the prequantum random field $\phi_\Psi(0) = (\phi_{\Psi_1}(\omega), \phi_{\Psi_2}(\omega))$ is concentrated (up to the background field) on one-dimensional eigenspace of $\widehat{A}_i$.

In this case we know that with the probability one the prequantum random signal will be detected in this subspace, after discarding the background field. If both signals $\phi_{\Psi_i}(\omega)$ are distributed between various eigenspaces of $\widehat{A}_i$, then $f_{A_{01}}(\phi_{\Psi_i}(\omega))$ are dependent, in spite of factorizability of Ψ.

The most surprising physical consequence is that, in general, prequantum variables are not independent for a factorisable state, although they are uncorrelated. The quantum intuition tells us that it is impossible. However, one has to be careful when projecting the quantum intuition (which is related to the properties of measured quantities) on prequantum models. We state again that from the PCSFT viewpoint, QM is a very special theory of measurements. It is based on usage of the threshold in detectors, i.e., discarding the background fluctuations. These fluctuations are taken into account in PCSFT. We can speculate that the presence of the common background field can induce dependence of prequantum observables, even for factorizable states.[a]

The quantum situation corresponds to a "trivial independence": if one quantity takes a fixed value with the probability one, then it is clear that randomly realized values of another quantity cannot

---

[a]Another explanation of this rather paradoxical result of our study on independence of prequantum variables for random fields corresponding to factorizable quantum states is that in this special situation we confront with an old problem of the foundations of classical probability theory, which is well known for experts: the standard notion of independence, Kolmogorov [218], may imply counterintuitive conclusions, see Mires [294], see also Ref. [135].

depend statistically on the first quantity. The PCSFT situation is more complicated: if one signal belongs to a fixed eigenspace, then the behavior of any quadratic form of another signal does not depend on the first quadratic form. In fact, the QM and PCSFT interpretations match with each other through a theory of measurement for prequantum random signals (see Chapter 8).

Since centering (3.89) can always be done afterward, we shall proceed without this centering, i.e., we shall use correspondence (3.91) everywhere (see also Section 3.1.3) instead of correspondence (3.92) for centered operators.

Chapter 4

Prequantum Dynamics from Hamiltonian Equations on the Infinite-Dimensional Phase Space

In Section 1.2.8, we postulated that the prequantum field dynamics is given by the Schrödinger equation. In this chapter we present the classical mechanical motivation of this approach. It is based on the more or less well-known coupling between classical Hamiltonian dynamics and Schrödinger dynamics.

We start with the remark that the presence of the *complex structure* on the state space of QM is one of the mysteries of quantum foundations.[a] We shall show that the complex structure is coupled with the symplectic and *Riemann space* structures on the phase space, in general, infinite dimensional. The canonical symplectic form corresponding to the scalar product (Riemannian metric) on the real Hilbert phase space $H = Q \times P$ plays the fundamental role. One can introduce the *Hermitian metric* on the complex realization $H_c = Q \oplus iP$ of H. Thus, we work not on an arbitrary *symplectic manifold*, but on the *Kähler manifold*.

[a]Real quantities, e.g., energy, coordinate, momentum, and the probabilities for these real results also belong to **R**. What is the source of appearance of complex numbers?

Beyond Quantum
Andrei Khrennikov

ISBN 978-981-4411-73-8 (Hardcover), 978-981-4411-74-5 (eBook)
www.panstanford.com

It is well known that the Schrödinger equation can be written in the form of the system of Hamiltonian equations for a quadratic Hamilton function [284]. We show that under a special restriction on the Hamiltonian function the classical Hamiltonian dynamics on the phase space can be written in the form of the Schrödinger equation.[a] As was mentioned in Section 1.2.8, the ordinary Schrödinger equation, but with the random initial condition, describes dynamics of the prequantum random field. By rewriting the Schrödinger equation in the form of Hamiltonian dynamics we show that the prequantum dynamics (see Section 1.2.8) can be considered as Hamiltonian dynamics on the *infinite-dimensional phase space.*

The system of Hamiltonian equations is reduced to the Schrödinger equation under the special condition: the classical Hamilton function $\mathcal{H}(q, p) \equiv \mathcal{H}(\phi), \phi = (q, p)$, has to be quadratic and J *-invariant*

$$\mathcal{H}(J\phi) = \mathcal{H}(\phi). \tag{4.1}$$

Here the prequantum field $\phi = (q, p) \in H = Q \times P$, where $Q = P$ are two copies of the real Hilbert space and $J : Q \times P \to Q \times P$ is the symplectic operator (it will be defined later). In fact, we have to select $P = Q'$, the dual space to Q, but it is convenient to identify Q and Q'.

The main internal reason to consider classical dynamics with J-invariant Hamilton functions is that such dynamics *preserves the magnitude of classical random fluctuations*, i.e., the dispersion of the probability distribution of a prequantum random field. In our approach linear (complex) quantum dynamics is the image of classical dynamics for quadratic Hamilton functions satisfying condition (4.1). Thus any quantum dynamics is dynamics of a classical infinite-dimensional harmonic oscillator. Since all models under consideration are statistical, dynamics of a quantum state (including a pure state) is dynamics of an ensemble of *infinite-dimensional harmonic oscillators.*

For the nonquadratic Hamilton function, classical dynamics on H can be represented as a *nonlinear Schrödinger equation.* Thus by

[a] In particular, we remark that the system of Maxwell equations can be written as the Hamiltonian system on the infinite-dimensional phase space. Hence, the system of Maxwell equations can be written as the Schrödinger equation. This is the Riemann–Silberstein representation of the classical electromagnetic field.

representing QM as the image of PCSFT, we see that the nonlinear Schrödinger equation is not less natural than the conventional linear equation (see Section 6.1 for the detailed presentation, see also, e.g., Refs. [32, 75, 76, 93, 94, 299]).[a]

Any pointwise classical dynamics (in particular, Hamiltonian) can be lifted to spaces of functions and measures. Consider quadratic J-invariant Hamilton function. In this case by mapping these liftings to QM we obtain, respectively, Heisenberg's dynamics for quantum observables and von Neumann's dynamics for density operators.

Let H be a Hilbert space. The space of *all bounded operators* in H is denoted by the symbol $\mathcal{L}(H)$. The space of bounded *self-adjoint operators* is denoted by the symbol $\mathcal{L}_s(H)$.

4.1 Hamiltonian Mechanics

4.1.1 *Quadratic Hamilton Function*

We consider the conventional classical phase space

$$H = Q \times P, \quad Q = P = \mathbf{R}^n$$

Here states are represented by points $\phi = (q, p) \in H$; the evolution of a state is described by the Hamiltonian equations

$$\dot{q} = \frac{\partial \mathcal{H}}{\partial p}, \quad \dot{p} = -\frac{\partial \mathcal{H}}{\partial q}, \tag{4.2}$$

where $\mathcal{H}(q, p)$ is the Hamilton function, a real-valued function on the phase space H.

We consider the scalar product on $\mathbf{R}^n : (x, y) = \sum_{j=1}^n x_j y_j$ and define the scalar product on $H : (\phi_1, \phi_2) = (q_1, q_2) + (p_1, p_2)$. We shall be interested in quadratic Hamilton function

$$\mathcal{H}(q, p) = \frac{1}{2}(\widehat{H}\phi, \phi), \tag{4.3}$$

where $\widehat{H} : H \to H$ is a self-adjoint operator. We remark that any linear operator $\widehat{A} : \mathbf{R}^{2n} \to \mathbf{R}^{2n}$ can be represented in the form

[a] We recall that Mie's nonlinear generalization of Maxwell equation (1912) can be considered as the first attempt to invent a nonlinear Schrödinger's type equation [243, 245].

$$\widehat{A} = \begin{pmatrix} \widehat{A}_{11} & \widehat{A}_{12} \\ \widehat{A}_{21} & \widehat{A}_{22} \end{pmatrix},$$

where $\widehat{A}_{11} : Q \to Q$, $\widehat{A}_{12} : P \to Q$, $\widehat{A}_{21} : Q \to Q$, $\widehat{A}_{22} : P \to P$. A linear operator $\widehat{A} : \mathbf{R}^{2n} \to \mathbf{R}^{2n}$ is self-adjoint if $\widehat{A}_{11}^\star = \widehat{A}_{11}$, $\widehat{A}_{22}^\star = \widehat{A}_{22}$, $\widehat{A}_{12}^\star = \widehat{A}_{21}$, $\widehat{A}_{21}^\star = \widehat{A}_{12}$. Thus the Hamilton function (4.3) can be written as

$$\mathcal{H}(q, p) = \frac{1}{2}[(\widehat{H}_{11}q, q) + 2(\widehat{H}_{12}p, q) + (\widehat{H}_{22}p, p)], \quad (4.4)$$

The system of Hamiltonian equation is linear

$$\dot{q} = \widehat{H}_{21}q + \widehat{H}_{22}p, \quad \dot{p} = -(\widehat{H}_{11}q + \widehat{H}_{12}p). \quad (4.5)$$

As always, we define the canonical symplectic structure on the phase space $H = Q \times P$ starting with the *symplectic operator*

$$J = \begin{pmatrix} 0 & 1 \\ -1 & 0 \end{pmatrix}$$

(here the blocks ± 1 denote $n \times n$ matrices with ± 1 on the diagonal). By using the symplectic operator J we can write these Hamiltonian equations in the operator form:

$$\dot{\phi} = \begin{pmatrix} \dot{q} \\ \dot{p} \end{pmatrix} = J\,\widehat{H}\phi \quad (4.6)$$

Thus,

$$\phi(t) = U_t\phi, \quad \text{where } U_t = e^{J\widehat{H}t}. \quad (4.7)$$

The map $U_t\phi$ is a linear Hamiltonian flow on the phase space H.

We still proceed in the purely symplectic framework, i.e., without appealing to the complex structure.

4.1.2 *Symplectically Invariant Quadratic Forms*

We shall mainly consider J-invariant quadratic forms. It is easy to see that J-invariance of the quadratic form $f_A(\phi) = (\widehat{A}\phi, \phi)$, where $\widehat{A} : H \to H$ is a self-adjoint operator, is equivalent to commuting the operator $\widehat{A}$ with the symplectic operator J. Let us consider the class $\mathcal{L}_{\mathrm{sp}}(H)$ of linear operators $\widehat{A} : H \to H$ which commute with the symplectic operator

$$\widehat{A}J = J\widehat{A}. \quad (4.8)$$

This is a subalgebra of the algebra of all **R**-linear operators $\mathcal{L}(H)$. We call such operators J-commuting. Thus,

$$\mathcal{L}_{\rm sp}(H) = \{\widehat{A} \in \mathcal{L}(H) : [\widehat{A}, J] = 0\}.$$

Proposition 1.1. $\widehat{A} \in \mathcal{L}_{\rm sp}(H)$ *iff* $\widehat{A}_{11} = \widehat{A}_{22} = D$, $\widehat{A}_{12} = -\widehat{A}_{21} = S$:

$$\widehat{A} = \begin{pmatrix} \widehat{D} & \widehat{S} \\ -\widehat{S} & \widehat{D} \end{pmatrix}$$

We remark that an operator $\widehat{A} \in \mathcal{L}_{\rm sp}(H)$ is self-adjoint iff $\widehat{D}^\star = \widehat{D}$ and $\widehat{S}^\star = -\widehat{S}$. Hence any self-adjoint J-commuting operator in the phase space is determined by a pair of operators $(\widehat{D}, \widehat{S})$, where $\widehat{D}$ is self-adjoint and $\widehat{S}$ is skew self-adjoint. Such operator induces the quadratic form

$$f_A(\phi) = (\widehat{A}\phi, \phi) = (\widehat{D}q, q) + 2(\widehat{S}p, q) + (\widehat{D}p, p). \qquad (4.9)$$

4.1.3 *Dynamics for Symplectically Invariant Quadratic Hamilton Functions*

Let us consider an operator $\widehat{H} \in \mathcal{L}_{\rm sp}(H)$

$$\widehat{H} = \begin{pmatrix} \widehat{R} & \widehat{T} \\ -\widehat{T} & \widehat{R} \end{pmatrix}. \qquad (4.10)$$

This operator defines the quadratic Hamiltonian function $\mathcal{H}(q, p) = \frac{1}{2}(\widehat{H}\phi, \phi)$, which can be written as

$$\mathcal{H}(q, p) = \frac{1}{2}[(\widehat{R}p, p) + 2(\widehat{T}p, q) + (\widehat{R}q, q)] \qquad (4.11)$$

where $\widehat{R}^\star = \widehat{R}$, $\widehat{T}^\star = -\widehat{T}$. The corresponding Hamiltonian equations have the form

$$\dot{q} = \widehat{R}p - \widehat{T}q, \quad \dot{p} = -(\widehat{R}q + \widehat{T}p) \qquad (4.12)$$

Proposition 1.2. *For a J-invariant Hamilton function, the Hamiltonian flow U_t, see (4.7), consists of J-commuting operators: $U_t J = J U_t$.*

Example 1.1. (One-dimensional J-invariant harmonic oscillator) Let $\mathcal{H}(q, p) = \frac{1}{2}[\frac{p^2}{m} + mk^2q^2]$. We use the symbol k to denote frequency, since ϕ is already used for the point of the phase space. To get a Hamiltonian function of the form (4.11) and (4.10), we consider the case $\frac{1}{m} = mk^2$. Thus $m = \frac{1}{k}$ and

$$\mathcal{H}(q, p) = \frac{k}{2}\left[p^2 + q^2\right]; \tag{4.13}$$

the Hamiltonian equations are given by

$$\dot{q} = kp, \quad \dot{p} = -kq. \tag{4.14}$$

Here the symmetric J-commuting matrix $\widehat{H} = \begin{pmatrix} k & 0 \\ 0 & k \end{pmatrix}$.

Let us define the *symplectic form* on the phase space:

$$w(\phi_1, \phi_2) = (\phi_1, J\phi_2). \tag{4.15}$$

Thus,

$$w(\phi_1, \phi_2) = (p_2, q_1) - (p_1, q_2)$$

for $\phi_j = \{q_j, p_j\}$, $j = 1, 2$. This is a *skew-symmetric bilinear form.*

Proposition 1.3. *Let $\widehat{A}$ be a self-adjoint operator. Then $\widehat{A} \in \mathcal{L}_{\mathrm{sp}}(H)$ iff it is symmetric with respect to the symplectic form*

$$w(\widehat{A}\phi_1, \phi_2) = w(\phi_1, \widehat{A}\phi_2) \tag{4.16}$$

4.1.4 *Complex Representation of Hamiltonian Dynamics*

Let us introduce the complex structure on the phase space H:

$$H_{\mathrm{c}} = Q \oplus iP.$$

Sometimes we shall proceed only in the complex framework. Then the index c will be omitted and the symbol H will be used. However, in the process of parallel working with real and complex spaces we shall distinguish them by using symbols H and H_{c}, respectively.

We have $i\phi = -p + iq = -J\phi$. A **R**-linear operator $\widehat{A} : H \to H$ is **C**-linear as an operator in H_{c} iff $\widehat{A}(i\phi) = iA\phi$, which is equivalent to $\widehat{A} \in \mathcal{L}_{\mathrm{sp}}(H)$.

Proposition 1.4. *The class of* **C***-linear operators $\mathcal{L}(H_{\mathrm{c}})$ coincides with the class of J-commuting operators $\mathcal{L}_{\mathrm{sp}}(H)$.*

We introduce on H_{c} a complex scalar product (Hermitian metric) based on the **C**-extension of the real scalar product:

$$\begin{aligned}\langle \phi_1, \phi_2 \rangle &= \langle q_1 + ip_1, q_2 + ip_2 \rangle \\ &= (q_1, q_2) + (p_1, p_2) + i((p_1, q_2) - (p_2, q_1)).\end{aligned}$$

Thus,

$$\langle \phi_1, \phi_2 \rangle = (\phi_1, \phi_2) - iw(\phi_1, \phi_2),$$

where w is the symplectic form. This is the canonical Hermitian metric on the Kähler manifold H.

A **C**-linear operator $\widehat{A}$ is symmetric with respect to the complex scalar product $\langle \ldots \rangle$ iff it is symmetric with respect to both real bilinear forms: $(\cdot, \cdot)$ and $w(\cdot, \cdot)$. Since for $\widehat{A} \in \mathcal{L}_{\text{sp}}(H)$ the former implies the latter, we get that a **C**-linear operator in H_{c} is self-adjoint iff it is self-adjoint in the real space H.

Proposition 1.5. *The class of* **C**-*linear self-adjoint operators* $\mathcal{L}_s(H_{\text{c}})$ *coincides with the class of* J*-commuting self-adjoint operators* $\mathcal{L}_{\text{sp,s}}(H)$.

We also remark that for a J-commuting operator $\widehat{A}$, its real and complex adjoint operators (denoted by $\widehat{A}^{\star}$ and $\widehat{A}^{*}$, respectively) coincide. Thus **C**-linear self-adjoint operators appear naturally as complex representations of J-commuting self-adjoint **R**-linear operators.

Proposition 1.6. *For a quadratic* J*-invariant Hamilton function* $\mathcal{H}(\phi)$*, its complexification does not change dynamics.*

Proof. To prove this, we remark that $w(\widehat{H}\phi, \phi) = 0$ and hence,

$$\mathcal{H}(\phi) = \frac{1}{2}\langle \widehat{H}\phi, \phi \rangle = \frac{1}{2}[(\mathbf{H}\phi, \phi) - iw(\widehat{H}\phi, \phi)] = \frac{1}{2}(\widehat{H}\phi, \phi),\ \phi \in H.$$

I consider the introduction of a complex structure on the phase space merely as using a new language: instead of symplectic invariance, we speak about **C**-linearity. By Proposition 1.6 the Hamilton function (4.11) can be written as $\mathcal{H}(\phi) = \frac{1}{2}\langle \widehat{H}\phi, \phi \rangle$, $\widehat{H} \in \mathcal{L}_s(\mathbf{C}^n)$, and the Hamiltonian equation (4.6) can be written in the complex form:

$$i\frac{d\phi}{dt} = \widehat{H}\phi \tag{4.17}$$

Any solution has the following complex representation:

$$\phi(t) = U_t\phi, \quad U_t = e^{-i\widehat{H}t/h}. \tag{4.18}$$

This is the complex representation of flows corresponding to quadratic J-invariant Hamilton functions. It is nothing else than the Schrödinger dynamics.

4.2 Symplectic Representation of Schrödinger Dynamics

Let H_c be a complex Hilbert space (in general, infinite dimensional and separable) and let $\langle\cdot,\cdot\rangle$ be the complex scalar product on H_c. The symbol $\mathcal{L}_s \equiv \mathcal{L}_s(H_c)$ denotes the space of continuous **C**-linear self-adjoint operators. We use the Planck system of units: $h = 1$. The Schrödinger dynamics in H_c is given by the linear equation

$$i\frac{d\phi}{dt} = \widehat{H}\phi \tag{4.19}$$

and hence,

$$\phi(t) = U_t\phi, \quad U_t = e^{-i\widehat{H}t}. \tag{4.20}$$

We see that these are simply infinite-dimensional versions of equations (4.17) and (4.18) obtained from the Hamiltonian equations for a quadratic J-invariant Hamilton function in the process of complexification of the classical mechanics. Therefore we can reverse our previous consideration (with the only remark that now the phase space is infinite-dimensional) and represent the Schrödinger dynamics (4.19) in the complex Hilbert space as the Hamiltonian dynamics in the infinite-dimensional phase space.[a] We emphasize that this Hamiltonian dynamics (4.6) is a dynamics in the phase space H but not in the unit sphere of this Hilbert space! The Hamiltonian flow $\phi(t,\phi) = U_t\phi$ is a flow on the whole phase space H.

We consider in H the **R**-linear operator J corresponding to the multiplication by $-i$; we represent the complex Hilbert space in the form $H_c = Q \oplus iP$, where Q and P are copies of the real Hilbert space. Here $\phi = q + ip$. We emphasize that q and p are not ordinary position and momentum of particles. These are their field analogues (if we choose $Q = P = L_2(\mathbf{R}^3)$): these are functions of $x \in \mathbf{R}^3$. We consider now a real phase space $H = Q \times P$. As in the finite-dimensional case, we have:

Proposition 2.1. *The class of continuous* **C***-linear self-adjoint operators* $\mathcal{L}_s(H_c)$ *coincides with the class of continuous* J*-commuting self-adjoint operators* $\mathcal{L}_{\mathrm{sp,s}}(H)$.

[a] Infinite dimension induces merely mathematical difficulties. The physical interpretation of the formalism is the same as in the finite-dimensional case.

Let us consider a quantum Hamiltonian $\widehat{H} \in \mathcal{L}_s(H_c)$.[a] It determines the classical Hamiltonian function $\mathcal{H}(\phi) = \frac{1}{2}\langle \widehat{H}\phi, \phi\rangle = \frac{1}{2}[(\widehat{R}p, p) + 2(\widehat{T}p, q) + (\widehat{R}q, q)]$. The corresponding system of Hamiltonian equations is given by

$$\frac{dq}{dt} = \widehat{R}p - \widehat{T}q, \quad \frac{dp}{dt} = -(\widehat{R}q + \widehat{T}p) \tag{4.21}$$

If we apply the complexification procedure to this system of Hamiltonian equations, we, of course, obtain the Schrödinger equation (4.19).

Example 2.1. Let us consider an important class of Hamilton functions

$$\mathcal{H}(q, p) = \frac{1}{2}\left[(\widehat{R}p, p) + (\widehat{R}q, q)\right], \tag{4.22}$$

where $\widehat{R}$ is a self-adjoint operator. The corresponding Hamiltonian equations have the form

$$\dot{q} = \widehat{R}p, \quad \dot{p} = -\widehat{R}q. \tag{4.23}$$

We now choose $H = L_2(\mathbf{R}^3)$, so $q(x)$ and $p(x)$ are components of the vector-field $\phi(x) = (q(x), p(x))$. We can call the fields $q(x)$ and $p(x)$ *mutually inducing.* The presence of the field $p(x)$ induces dynamics of the field $q(x)$ and vice versa, cf. electric and magnetic components, $q(x) = E(x)$ and $p(x) = B(x)$, of the classical electromagnetic field, cf. Einstein and Infeld [79], p. 148: "Every change of an electric field produces a magnetic field; every change of this magnetic field produces an electric field; every change of ..., and so on." We can write the form (4.22) as

$$\mathcal{H}(q, p) = \frac{1}{2}\int_{\mathbf{R}^6} R(x, y)[q(x)q(y) + p(x)p(y)]dxdy \tag{4.24}$$

or

$$\mathcal{H}(\phi) = \frac{1}{2}\int_{\mathbf{R}^6} R(x, y)\phi(x)\bar{\phi}(y)dxdy, \tag{4.25}$$

where $R(x, y) = R(y, x)$ is, in general, a distribution on $\mathbf{R}^6$.

We call such kernel $R(x, y)$ a *self-interaction potential* for the field $\phi(x) = (q(x), p(x))$. We remark that $R(x, y)$ induces a self-interaction of each component of $\phi(x)$, but there is no cross-interaction between components $q(x)$ and $p(x)$ of the vector field $\phi(x)$.

---

[a] We may consider operator $\widehat{H} \geq 0$, but for the present consideration this is not important.

4.3 Classical and Quantum Statistical Models

For further considerations, it is useful to formalize the notions of classical and quantum statistical models by establishing a correspondence between the mathematical and physical structures, cf. Section 3.1.

4.3.1 *Classical Statistical Model*

This model is described in the following way

(a) *Physical states* are represented by points of some set Ω (state space).
(b) *Physical variables* are represented by the functions $f : \Omega \to \mathbf{R}$ belonging to some functional space $V(\Omega)$.[a]
(c) *Statistical states* are represented by probability measures on Ω belonging to some class $S(\Omega)$.[b]
(d) The average of a physical variable (which is represented by a function $f \in V(\Omega)$) with respect to a statistical state (which is represented by a probability measure $\mu \in S(\Omega)$) is given by $\langle f \rangle_\mu \equiv \int_\Omega f(\omega) d\mu(\omega)$.
A *classical statistical model* is a pair $M = (S(\Omega), V(\Omega))$.

We recall that CSM on the phase space $\Omega = \mathbf{R}^{2n}$ gives an example of a classical statistical model. But we shall not be interested in this example in our further considerations. We shall develop a classical statistical model with an *infinite-dimensional phase space.*

Remark 3.1. We emphasize that the space of variables $V(\Omega)$ need not coincide with the space of all random variables $RV(\Omega)$, measurable functions $\xi : \Omega \to \mathbf{R}$. For example, if Ω is a differentiable manifold, it is natural to choose $V(\Omega)$ consisting of smooth functions; if Ω is an analytic manifold, it is natural to choose $V(\Omega)$ consisting of analytic functions, and so on. Denote the space of all probability measures by the symbol $PM(\Omega)$. The space of statistical

---

[a] The choice of a concrete functional space $V(\Omega)$ depends on various physical and mathematical factors.

[b] The choice of a concrete space of probabilities $S(\Omega)$ depends on various physical and mathematical factors.

states $S(\Omega)$ need not coincide with $PM(\Omega)$. For example, for some statistical model, $S(\Omega)$ may consist of Gaussian measures.

4.3.2 *Quantum Statistical Model*

In the Dirac–von Neumann formalism [74, 296] with the complex Hilbert space H_c this model is described in the following way:

(a) *Physical observables* are represented by bounded self-adjoint operators[a] $\widehat{A} : H_c \to H_c$, i.e., $\widehat{A} \in \mathcal{L}_s(H_c)$.
(b) *Statistical states* are represented by density operators. The class of such operators is denoted by $\mathcal{D} \equiv \mathcal{D}(H_c)$.
(c) The average of a physical observable (which is represented by the operator $\widehat{A} \in \mathcal{L}_s(H_c)$) with respect to a statistical state (which is represented by the density operator $\rho \in \mathcal{D}(H_c)$) is given by von Neumann's formula:

$$\langle \widehat{A} \rangle_\rho \equiv \mathrm{Tr}\, \rho \widehat{A}.$$

The *quantum statistical model* is the pair $M_q = (\mathcal{D}(H_c), \mathcal{L}_s(H_c))$.

We do not stress the role of the pure quantum states given by normalized vectors $\Psi \in H_c$. These are just special quantum statistical states, cf. Margenau [239, 240] and Ballentine [25–28].

4.4 Measures on Hilbert Spaces

Let μ be a σ-additive measure on the σ-algebra $\mathcal{F}$ of Borel subsets of H. We suppose that μ has the finite second-order momentum

$$\int_H ||\phi||^2 d\mu(\phi) < \infty.$$

In what follows we restrict our consideration to measures with zero mean value $m = 0$. The mean value $m \equiv m_\mu \in H$ is defined by the equality

$$(m, y) = \int_H (y, \phi) d\mu(\phi)$$

[a]For simplicity we shall consider only quantum observables represented by bounded operators. To obtain the general quantum model with observables represented by unbounded operators, we should consider a prequantum classical statistical model based on the rigged Hilbert space (Section 6.2).

for any $y \in H$. We recall that the covariance operator $B \equiv B_\mu$ is defined as

$$(By_1, y_2) = \int (y_1, \phi)(y_2, \phi) d\mu(\phi), \; y_1, y_2 \in H, \tag{4.26}$$

and has the following properties

(a) $B \geq 0$, i.e., $(By, y) \geq 0$, $y \in H$;
(b) B is a self-adjoint operator, $B \in \mathcal{L}_s(H)$;
(c) B is a trace-class operator and

$$\text{Tr } B = \int_H ||\phi||^2 d\mu(\phi) \tag{4.27}$$

The right-hand side of (4.27) defines the *dispersion* of measure μ. Thus for any measure μ, we have

$$\sigma^2(\mu) = \text{Tr } B. \tag{4.28}$$

The list of properties of the covariance operator of a measure differs from the list of properties of a density operator only by one condition: Tr $\rho = 1$, for a density operator ρ.

We can easily find the integral of the quadratic form $f_A(\phi)$ of a bounded (in general, not self-adjoint) operator $\widehat{A}$:

$$\begin{aligned}\int_H f_A(\phi) d\mu(\phi) &= \int_H (\widehat{A}\phi, \phi) d\mu(\phi) \\ &= \sum_{i,j=1}^{\infty} (\widehat{A}e_i, e_j) \int_H (e_i, \phi)(e_j, \phi) d\mu(\phi) \\ &= \sum_{i,j=1}^{\infty} (\widehat{A}e_i, e_j)(Be_i, e_j) \\ &= \sum_i^{\infty} (\widehat{A}e_i, Be_i) = \text{Tr } B\widehat{A},\end{aligned}$$

where $\{e_i\}$ is some orthonormal basis in H. Thus,

$$\int_H f_A(\phi) d\mu(\phi) = \text{Tr } B\widehat{A}. \tag{4.29}$$

An arbitrary measure is not determined by its covariance operator and mean value. However, any Gaussian measure is determined by its covariance operator $B : H \to H$ and mean value

$m \in H$. For example, B and m determine the Fourier transform of μ :

$$\tilde{\mu}(y) = \int_H e^{i(y,\phi)} d\mu(\phi) = e^{\frac{1}{2}(By,y)+i(m,y)}, \; y \in H.$$

Sometimes the symbol μ_B will be used to denote the Gaussian measure with the covariance operator B and $m = 0$.

4.5 Lifting of Pointwise Dynamics to Spaces of Variables and Measures

Let $(X, \mathcal{F})$ be an arbitrary *measurable space.* Here X is a set and $\mathcal{F}$ is the σ-algebra of its subsets. Denote the space of random variables (measurable maps $f : X \to \mathbf{R}$) by the symbol $RV(X)$ and the space of probability measures on (X, F) by the symbol $PM(X)$. Consider a measurable map $g : X \to X$. It induces the maps

$$g^* : RV(X) \to RV(X), \quad g^* f(x) = f(g(x))$$

$$g^* : MP(X) \to MP(X), \quad \int_X f(x) dg^* \mu(x) = \int_X g^* f(x) d\mu(x).$$

We now consider a dynamical system in X :

$$x_t = g_t(x), \tag{4.30}$$

where $g_t : X \to X$ is a one-parametric family of measurable maps. The parameter t is real and plays the role of time. By using lifting of g_t we obtain dynamics in spaces of functions $RV(X)$ and measures $MP(X)$, respectively:

$$f_t = g_t^* f \tag{4.31}$$

$$\mu_t = g_t^* \mu. \tag{4.32}$$

We shall see that for $X = H$ (an infinite-dimensional phase space), quantum images of dynamical systems (4.30), (4.31), and (4.32) are respectively the Schrödinger dynamics for the quantum state, the Heisenberg's dynamics for the operators-observables, and the von Neumann dynamics for the density operator. To obtain quantum mechanics, we have to select adequate spaces of physical variables and measures.

4.6 Dispersion Preserving Dynamics

Everywhere in this section we consider only quadratic Hamilton functions on an infinite-dimensional phase space H (see Section 6.1 for general dynamics). We start with an arbitrary quadratic Hamiltonian function $\mathcal{H}(\phi) = \frac{1}{2}(\widehat{H}\phi, \phi)$ (the operator $\widehat{H}$ need not be J-commuting). Let us consider the Hamiltonian flow $U_t : H \to H$ induced by the Hamiltonian system (4.6). This map is given by (4.7). It is important to remark that the map U_t is invertible; in particular,

$$U_t(H) = H. \tag{4.33}$$

We are interested in a Hamiltonian flow U_t such that the corresponding dynamics in the space of probabilities (4.32) preserves the magnitude of statistical fluctuations:

$$\sigma^2(U_t^* \mu) = \sigma^2(\mu) : \int_H \|\phi\|^2 dU_t^* \mu(\phi) = \int_H \|\phi\|^2 d\mu(\phi) \tag{4.34}$$

or

$$\int_H \|U_t\phi\|^2 d\mu(\phi) = \int_H \|\phi\|^2 d\mu(\phi). \tag{4.35}$$

We present a sufficient condition for preserving the magnitude of statistical fluctuations: the Hamiltonian flow $U_t\phi$ consists of isometric maps:

$$\|U_t\phi\|^2 = \|\phi\|^2, \phi \in H. \tag{4.36}$$

Proposition 6.1. *A Hamiltonian flow $U_t\phi$ is isometric iff the function $\mathcal{H}(\phi)$ is J-invariant.*

Proof. (a) Let $\widehat{H}$ be J-commuting. Then we have

$$\frac{d}{dt}\|U_t\phi\|^2 = 2(\dot{U}_t\phi, U_t\phi) = 2(J\,\widehat{H}U_t\phi, U_t\phi) = 0$$

Here we used the simple fact that the operator $J\widehat{H}$ is skew symmetric: $(J\,\widehat{H})^\star = -\widehat{H}J = -J\,\widehat{H}$. Thus (4.36) holds.

(b) Let (4.36) hold. Then $\frac{d}{dt}\|U_t\phi\|^2 = 0$. By using previous computations and (4.33) we get

$$(J\,\widehat{H}\phi, \phi) = 0, \quad \phi \in H. \tag{4.37}$$

Hence the operator $J\widehat{H}$ is skew symmetric. This implies that $\widehat{H}$ commutes with J.

For our further consideration it is useful to rewrite (4.37) in the form

$$(J\mathcal{H}'(\phi),\ \phi) = 0, \phi \in H. \tag{4.38}$$

Corollary 6.1. *The flow corresponding to a J-invariant Hamilton function preserves the fluctuations of the fixed α-magnitude. For any measure μ (having zero mean value and finite dispersion), if $\sigma^2(\mu) = \alpha$, then $\sigma^2(U_t^* \mu) = \alpha$ for any $t \geq 0$.*

This is our explanation of the exceptional role of J-invariant physical variables on the infinite-dimensional classical phase space. If a Hamilton function is not J-invariant, then the corresponding Hamiltonian flow can increase the magnitude of fluctuations.

Interpretation. According to PCSFT, a quantum system is simply a classical random field. To provide stability of a quantum system, the dispersion of fluctuations of the corresponding prequantum random field has to be preserved. Otherwise the field will be smashed over the whole space. Thus stability of the electron or neutron is a consequence of dispersion conservation or, in mathematical terms, J-invariance. In the absence of interactions with the environment, e.g., the "electron random field" does not spread in the space of fields.

4.7 Dynamics in the Space of Physical Variables

4.7.1 *Arbitrary Quadratic Variables*

Let us consider the Hamiltonian flow $U_t : H \to H$ induced by an arbitrary quadratic Hamilton function. Let $\widehat{A} : H \to H$ be a continuous self-adjoint operator and $f_A(\phi) = (\widehat{A}\phi, \phi)$. We have $U_t^* f_A(\phi) = f_A(U_t\phi) = f_{U_t^*\widehat{A}U_t}(\phi)$. This dynamics can be represented as a dynamics in $\mathcal{L}_s(H)$:

$$\widehat{A}_t = U_t^\star \widehat{A} U_t \tag{4.39}$$

We remark that $U_t = e^{J\widehat{H}t}$, so $U_t^\star = e^{-\widehat{H}Jt}$. Thus,

$$\widehat{A}_t = e^{-\widehat{H}Jt}\widehat{A}e^{J\widehat{H}t}. \tag{4.40}$$

Thus $\frac{d\widehat{A}_t}{dt} = \left(\widehat{A}_t J\, \widehat{H} - \widehat{H} J\, \widehat{A}_t\right)$, or

$$\frac{d\widehat{A}_t}{dt} = [\widehat{A}_t, \widehat{H} J\,] + \widehat{A}_t [J, \widehat{H}\,] \tag{4.41}$$

4.7.2 *J-invariant Variables*

We consider the space of physical variables

$$V_{\text{quad,sp}}(H) = \{f : H \to \mathbf{R} : f \equiv f_A(\phi) = (\widehat{A}\phi, \phi), \widehat{A} \in \mathcal{L}_{\text{sp},s}(H)\}$$

consisting of J-invariant quadratic forms. Let us consider the lifting of the flow corresponding to a J-invariant quadratic Hamilton function to the space $V_{\text{quad,sp}}(H)$. In this case both operators $\widehat{H}$ and $\widehat{A}$ are J-commuting. Therefore the flow (4.40) can be written as

$$\widehat{A}_t = U_t^{\star} \widehat{A} U_t = e^{-J\widehat{H}t} \widehat{A} e^{J\widehat{H}t} \tag{4.42}$$

The evolution equation (4.41) is simplified as

$$\frac{d\widehat{A}_t}{dt} = -J\,[\widehat{H}, \widehat{A}_t] \tag{4.43}$$

4.7.3 *Complex Structure*

Suppose that $[\widehat{H}, J\,] = 0$ and $[\widehat{A}, J\,] = 0$. By considering the complex structure on the phase space and representing the symplectic operator J by $-i$, we write (4.40) in the form of the Heisenberg's dynamics

$$\widehat{A}_t = U_t^{*} \widehat{A} U_t = e^{it\widehat{H}} \widehat{A} e^{-it\widehat{H}}. \tag{4.44}$$

(here U_t^* is the complex adjoint operator to U_t) and the evolution equation (4.41) in the form of the Heisenberg's equation

$$\frac{d\widehat{A}_t}{dt} = i[\widehat{H}, \widehat{A}_t]. \tag{4.45}$$

Thus this equation is just the image of lifting of the classical quadratic Hamiltonian dynamics in the case of J-invariant quadratic variables.

4.8 Probabilistic Dynamics

4.8.1 *Arbitrary Measure*

Let us consider the flow $U_t : H \to H$ induced by an arbitrary quadratic Hamilton function $\mathcal{H}(\phi)$. Let μ be an arbitrary measure with zero mean value and the finite second momentum, i.e., its dispersion and covariance operator are well defined. We find dynamics of the covariance operator of $U_t^*(\mu)$. We have

$$\begin{aligned}(B_{U_t^*\mu}y_1, y_2) &= \int_H (y_1, \phi)(y_2, \phi)d(U_t^*\mu)(\phi)\\ &= \int_H (y_1, U_t\phi)(y_2, U_t\phi)d\mu(\phi)\\ &= (B_\mu U_t^\star y_1, U_t^\star y_2).\end{aligned}$$

Thus, for the covariance operator $B_t = B_{U_t^*\mu}$, we have

$$B_t = U_t B_\mu U_t^\star \equiv e^{J\widehat{H}t} B_\mu e^{-\mathbf{H}Jt} \tag{4.46}$$

Thus,

$$\frac{dB_t}{dt} = \left(J\widehat{H}B_t - B_t\widehat{H}J\right),$$

or

$$\frac{dB_t}{dt} = [J\widehat{H}, B_t] + B_t[J, \widehat{H}] \tag{4.47}$$

4.8.2 *J-invariant Measures*

We now consider the lifting (to the space of measures) of the flow $U_t : H \to H$ induced by a J-invariant quadratic Hamilton function $\mathcal{H}(\phi)$. We start with the following mathematical result:

Proposition 8.1. *Let μ be an arbitrary J-invariant measure. Then its covariance operator is J-invariant.*

Proof. Let $J^*\mu = \mu$. It is sufficient to prove that $B_\mu J$ is skew symmetric. We have

$$
\begin{aligned}
(B_\mu J y_1, y_2) &= \int_H (J y_1, \phi)(y_2, \phi) d\mu(\phi) \\
&= -\int_H (y_1, J\phi)(y_2, J^\star J\phi) d\mu(\phi) \\
&= -\int_H (y_1, J\phi)(J y_2, J\phi) d\mu(\phi) \\
&= -\int_H (y_1, \phi)(J y_2, \phi) dJ^* \mu(\phi) \\
&= -\int_H (J y_2, \phi)(y_1, \phi) d\mu(\phi) \\
&= -(B_\mu J y_2, y_1) = -(y_1, B_\mu J y_2).
\end{aligned}
$$

Proposition 8.2. *A Gaussian measure μ (with zero mean value) is J-invariant iff its covariance operator is J-invariant.*

Proof. Take B_μ belonging to $\mathcal{L}_{\mathrm{sp},s}(H)$. We find the Fourier transform of the Gaussian measure $J^*\mu$:

$$\widetilde{J^*\mu}(y) = \int_H e^{i(y, J\phi)} d\mu(\phi) = \tilde{\mu}(J^\star y) = e^{-\frac{1}{2}(B_\mu J^\star y, J^\star y)} = \tilde{\mu}(y).$$

Since the flow for a J-invariant quadratic Hamilton function consists of J-commuting linear operators, $J U_t = U_t J$, we have that the space of J-invariant measures (with zero mean value and finite second momentum) is invariant with respect to the map U_t^*.

Take μ such that $J^*\mu = \mu$. Then $J^* U_t^* \mu = U_t^* J^* \mu = U_t^* \mu$. For an arbitrary J-invariant measure, we have

$$B_t = U_t B_\mu U_t^* \equiv e^{J\widehat{H}t} B_\mu e^{-J\mathbf{H}t} \tag{4.48}$$

or

$$\frac{dB_t}{dt} = -J[B_t, \widehat{H}] \tag{4.49}$$

4.8.3 *Complex Structure*

Suppose that $[\widehat{H}, J] = 0$ and $[B_\mu, J] = 0$. By considering the complex structure on the phase space and representing the symplectic operator J by $-i$, we write (4.48) in the form

$$B_t = U_t B_\mu U_t^* = e^{-i\widehat{H}t} B_\mu e^{i\widehat{H}t} \tag{4.50}$$

or

$$\frac{dB_t}{dt} = i[B_t, \widehat{H}] \tag{4.51}$$

This is nothing else than the von Neumann equation for the density operator. The only difference is that the covariance operator B_μ is not normalized.

4.8.4 *Dynamics in the Space of Statistical States*

First we consider the space of all Gaussian measures having zero mean value and dispersion α, where $\alpha > 0$ is a fixed parameter of the model (the dispersion of prequantum fluctuations).[a] Denote this space of such measures by the symbol $S_{\mathrm{G}}^{\alpha}(H)$. These are Gaussian measures such that

$$(y, m_\mu) = \int_H (y, \phi) d\mu(\phi) = 0, \, y \in H,$$

and

$$\sigma^2(\mu) = \int_H \|\phi\|^2 d\mu(\phi) = \alpha.$$

Thus,

$$S_{\mathrm{G}}^{\alpha}(H) = \{\mu : \text{ Gaussian, } m_\mu = 0, \sigma^2(\mu) = \alpha\}.$$

For the flow U_t corresponding to a J-invariant quadratic Hamilton function, we have

$$U_t^* : S_{\mathrm{G}}^{\alpha}(H) \to S_{\mathrm{G}}^{\alpha}(H)$$

Denote the subspace of $S_{\mathrm{G}}^{\alpha}(H)$ consisting of J-invariant measures by the symbol $S_{G,\mathrm{sp}}^{\alpha}(H)$. Thus,

$$S_{G,\mathrm{sp}}^{\alpha}(H) = \{\mu \in S_{\mathrm{G}}^{\alpha}(H) : J^*\mu = \mu\}.$$

We also have

$$U_t^* : S_{G,\mathrm{sp}}^{\alpha}(H) \to S_{G,\mathrm{sp}}^{\alpha}(H).$$

---

[a] In this chapter we proceed with neglecting the background field, fluctuations of vacuum. Prequantum fluctuations are "calibrated fluctuations," i.e., the contribution of the background field is neglected.

4.8.5 *Complex Covariance Operator*

In this section we consider arbitrary measures (i.e., in general, non-Gaussian) with finite dispersions. Let us introduce *complex average* m_μ^c *and covariance operator* $D_\mu \equiv B_\mu^c$ by setting

$$\langle m_\mu^c, y\rangle = \int_H \langle y, \phi\rangle d\mu(\phi), \tag{4.52}$$

$$\langle D_\mu y_1, y_2\rangle = \int_H \langle y_1, \phi\rangle\langle\phi, y_2\rangle d\mu(\phi). \tag{4.53}$$

Proposition 8.3. *Let μ be a J-invariant measure. Then*

$$m_\mu^c = 0 \text{ iff } m_\mu = 0. \tag{4.54}$$

Proof. Since μ is J-invariant, for any Borel function $f : H \to \mathbf{R}$, we have

$$\int_H f(\phi_q, \phi_p) d\mu(\phi_q, \phi_p) = \int_H f(\phi_p, -\phi_q) d\mu(\phi_q, \phi_p) \tag{4.55}$$

Let $m_\mu = 0$. Then

$$0 = \int_H (y, \phi) d\mu(\phi) = \int_H [(y_q, \phi_q) + (y_p, \phi_p)] d\mu(\phi)$$
$$= \int_H [(y_q, \phi_p) - (y_p, \phi_q)] d\mu(\phi) = \int_H w(y, \phi) d\mu(\phi),$$

where w is the symplectic form H. Hence the last integral is also equal to zero. On the other hand, for the complex average we have

$$\langle y, m_\mu^c\rangle = \int_H (y, \phi) d\mu(\phi) - i\int_H w(y, \phi) d\mu(\phi). \tag{4.56}$$

Proposition 8.4. *Let μ be an arbitrary J-invariant measure with zero mean value. Then*

$$D_\mu = 2B_\mu \tag{4.57}$$

Proof. We have

$$\langle D_\mu y, y\rangle = \int_H |\langle y, \phi\rangle|^2 d\mu(\phi) = \int_H |(y, \phi) - iw(y, \phi)|^2 d\mu(\phi)$$
$$= \int_H [(y, \phi)^2 + (y, J\phi)^2] d\mu(\phi).$$

By using symplectic invariance of the measure μ we get

$$\int_H (y, J\phi)^2 d\mu(\phi) = \int_H (y, \phi)^2 d\mu(\phi).$$

Thus,

$$\langle D_\mu y, y\rangle = 2\int_H (y, \phi)^2 d\mu(\phi) = 2(B_\mu y, y).$$

Theorem 8.1. *For any measure μ with zero mean value and any J-commuting operator $\widehat{A}$, we have*

$$\int_H \langle \widehat{A}\phi, \phi\rangle d\mu(\phi) = Tr\ D_\mu\ \widehat{A}; \tag{4.58}$$

in particular,

$$\sigma^2(\mu) = \text{Tr}\ D_\mu. \tag{4.59}$$

Proof. Let $\{e_j\}$ be an orthonormal basis in H_c. (We emphasize that orthogonality and normalization are with respect to the complex and not real scalar product.) Then

$$\text{Tr}\ D_\mu\ \widehat{A} = \int_H \sum_j \langle Ae_j, \phi\rangle\langle\phi, e_j\rangle d\mu(\phi) = \int_H \langle\widehat{A}\phi, \phi\rangle d\mu(\phi).$$

We recall that

$$\sigma^2(\mu) = \text{Tr}\ B_\mu. \tag{4.60}$$

It seems that there is a contradiction between equalities (4.60), (4.59), and (4.57). In fact, there is no contradiction, because in (4.60) and (4.59) we use two different traces: with respect to the real and complex scalar products, respectively. This is an important point; even normalization by trace one for the von Neumann density operator is the normalization with respect to the complex scalar product. By using indices **R** and **C** to denote real and complex traces, respectively, we can write

$$\sigma^2(\mu) = \text{Tr}_{\mathbf{R}}\ B_\mu = \text{Tr}_{\mathbf{C}}\ D_\mu.$$

We remark that *the complex average m_μ^c and the covariance operator D_μ are* **C***-linear even if a measure is not J-invariant.* However, in general, real and complex averages do not coincide, and real and complex covariance operators are not coupled by (4.57).

Let us find the relation between B_μ and D_μ in the general case. It is easy to show that for

$$B_\mu = \begin{pmatrix} B_{11} & B_{12} \\ B_{21} & B_{22} \end{pmatrix},\ B_{11}^\star = B_{11},\ B_{22}^\star = B_{22},\ B_{12}^\star = B_{21}$$

and

$$D_\mu = \begin{pmatrix} D & S \\ -S & D \end{pmatrix}$$

we have the following proposition.

Proposition 8.5. *The blocks in real and complex covariance operators are connected by the following equalities:*

$$D = B_{11} + B_{22}, \; S = B_{12} - B_{21}. \tag{4.61}$$

Thus *in the general case the complex covariance operator D_μ does not determine the Gaussian measure μ_B uniquely.*

Let now μ_B be J-invariant. Then

$$B = \begin{pmatrix} B_{11} & B_{12} \\ -B_{12} & B_{11} \end{pmatrix}. \tag{4.62}$$

Thus,

$$D = 2B_{11}, \; S = 2B_{12}, \tag{4.63}$$

so we obtain (4.57) and, hence, we obtain the following corollary.

Corollary 8.1. *There is a one-to-one correspondence between J-invariant Gaussian measures with the zero mean value and complex covariance operators.*[a]

4.8.6 *Dynamics of Complex Covariance Operator: The Case of an Arbitrary Measure*

Consider dynamics with the initial measure μ which is induced by the Hamiltonian dynamics in $H = Q \times P$ with a quadratic Hamilton function $\mathcal{H}(q, p)$. We obtain a one-parameter family of measures $\mu_t = U_t^* \mu$. We remark that representation (4.46) can be obtained not only for the covariance operator of the Gaussian measure μ, but for any measure μ (with the finite dispersion and zero mean value). Thus we obtain the evolution equation (4.47).

Now we consider the Hamiltonian dynamics in $H = Q \times P$ with a quadratic J-invariant Hamilton function $\mathcal{H}(q, p)$. As remarked above, even for a Gaussian measure μ which is not J-invariant,

[a] These are **C**-linear self-adjoint positively defined operators $D_\mu : H_c \to H_c$ belonging to the trace-class.

its complex covariance operator D_μ does not define μ uniquely. Nevertheless, let us represent an arbitrary measure μ (with zero mean value and finite dispersion) by its complex covariance operator D_μ. We "project" measures on their complex covariance operators. It is easy to see that

$$D_\mu(t) = U_t D_\mu U_t^*. \tag{4.64}$$

Since $[D_\mu, J] = 0$ (for any μ), the complex covariance operator $D_t \equiv D_\mu(t)$ satisfies the von Neumann equation (4.51):

$$\frac{dD_t}{dt} = i[D_t, \widehat{H}]. \tag{4.65}$$

As mentioned above, for J-invariant Gaussian measures, this dynamics of the complex covariance operator determines uniquely dynamics of the measure.

In the previous consideration we emphasized the role of the J-invariance of prequantum probability distributions. We remark that in the Gaussian case it is equivalent to the invariance with respect to the orthogonal group acting in the Hilbert phase space (see Chapter 5). This is an important restriction on prequantum random fields. It must have some physical explanation. At the moment we can only point out that the invariance with respect to the orthogonal group is, in fact, the invariance with respect to $U(1)$-gauge group for classical prequantum fields.

4.9 Detailed Analysis of Dispersion Preserving Dynamics

This section is a supplement of Section 4.6.

In the phase space $H = Q \times P$ (considered as a real Hilbert space) we consider dynamics generated by a bounded operator $\widehat{L}$:

$$\frac{dU_t}{dt} = \widehat{L}U_t, \; U_0 = I, \tag{4.66}$$

hence, we have

$$U_t = e^{t\widehat{L}}. \tag{4.67}$$

Proposition 9.1. *Let an operator $\widehat{L}$ be skew symmetric. Then equality (4.36) holds (i.e., U_t is a one parametric group of isometric operators).*

Proof. We have

$$\frac{d\|U_t\phi\|^2}{dt} = ((\widehat{L} + \widehat{L}^\star)U_t\phi, U_t\phi). \tag{4.68}$$

(We state again that the real adjoint operator is denoted by the symbol $\widehat{L}^\star$). Thus $\frac{d\|U_t\phi\|^2}{dt} = 0$.

Proposition 9.2. *Let an operator $\widehat{L}$ generate the one parametric group of isometric surjective operators U_t, i.e., $U_t(H) = H$ for every t. Then the generator $\widehat{L}$ is skew-symmetric.*

Proof. By using (4.68) we obtain that, for any $\phi \in H$, $((\widehat{L} + \widehat{L}^\star)U_t\phi, U_t\phi) = 0$. As a consequence of surjectivity, $((\widehat{L}+\widehat{L}^\star)y, y) = 0$ for any $y \in H$. Since the operator $(\widehat{L} + \widehat{L}^\star)$ is self-adjoint, it can be reconstructed from its quadratic form. Hence, $\widehat{L} + \widehat{L}^\star = 0$.

If operators U_t are not surjective, then, in principle, the norm might be preserved for groups whose generators need not be skew-symmetric.

Preservation of the norm, of course, implies preservation of dispersion for any measure μ on the phase space. Thus we found an essentially more general sufficient condition for preservation of the dispersion than in Section 4.6. We remark that a skew symmetric generator $\widehat{L}$ need not be of the form $\widehat{L} = J\widehat{H}$, where $\widehat{H}$ is a self-adjoint operator.

Example 9.1. Take a four-dimensional phase space H, where $Q = P = \mathbf{R}^2$, and take the generator

$$\widehat{L} = \begin{pmatrix} 0 & 1 & 0 & 0 \\ -1 & 0 & 0 & 0 \\ 0 & 0 & 0 & -1 \\ 0 & 0 & 1 & 0 \end{pmatrix}. \tag{4.69}$$

It is skew-symmetric, however, $\widehat{L} \notin \mathcal{L}_{\text{sp}}(H)$, i.e., $[\widehat{L}, J] \neq 0$. We find the operators

$$\widehat{L}J = \begin{pmatrix} 0 & 0 & 0 & 1 \\ 0 & 0 & -1 & 0 \\ 0 & 1 & 0 & 0 \\ -1 & 0 & 0 & 0 \end{pmatrix}, \quad \widehat{H} \equiv J\widehat{L} = \begin{pmatrix} 0 & 0 & 0 & -1 \\ 0 & 0 & 1 & 0 \\ 0 & -1 & 0 & 0 \\ 1 & 0 & 0 & 0 \end{pmatrix}. \tag{4.70}$$

We see that $\widehat{H}^\star \neq \widehat{H}$. Thus, the corresponding dynamics is norm preserving (and, hence, dispersion preserving), but it is not unitary.

4.9.1 *Physical Discussion on Dispersion Preserving Prequantum Dynamics*

The PCSFT image of a quantum particle is given by a random wave with small dispersion (and zero average), see Chapter 5. (The dispersion, say α, is a small parameter of the model; QM is obtained from PCSFT by setting $\alpha = 0$.) Although occasionally (at some instances of prequantum time) random fluctuations can produce wave components of sufficiently high magnitude, in average their magnitudes cannot be very high, since the average is zero and the dispersion is small. In the absence of interactions with the environment dispersion should be preserved; otherwise a system will exhibit fluctuations of higher magnitude. This increasing of magnitude cannot be done without energy supply. Since it is assumed that the system is isolated, there are no sources of this supply.

In Section 4.6 we saw that the unitary dynamics used in the conventional QM preserves the dispersion of a prequantum random field (e.g., representing the electron). However, we have seen that dispersion preserving dynamics are not restricted to unitary maps. Whether such dynamics are artifacts of our mathematical model or they can be physically realized, i.e., a quantum system, e.g., an electron can evolve in nonunitary way but with preservation of dispersion of the prequantum fluctuations, is an open question.

We state again that in this chapter we consider dynamics of a single system, e.g., an electron, and we ignore the contribution of vacuum fluctuations (cf. Section 3.3). The presence of vacuum fluctuations makes the physical and mathematical pictures more complicated. The prequantum process is not concentrated in the L_2-space, but in the space of distributions. Hence the dispersion is infinite. Thus the problem of preservation of dispersion of prequantum fluctuations can be treated only after subtraction of the contribution of vacuum fluctuations.

Chapter 5

Quantum Mechanics as Approximation of Statistical Mechanics of Classical Fields

In Section 3.1, we showed that QM can be represented as a classical field model in which physical quantities are given by quadratic functionals of fields. In this chapter we shall show that in complete accordance with Einstein's views QM is an approximation of CSM of fields, i.e., classical statistical mechanics with the infinite-dimensional phase space $H = L_2(\mathbf{R}^3; \mathbf{R}) \times L_2(\mathbf{R}^3; \mathbf{R})$, where $L_2(\mathbf{R}^3; \mathbf{R})$ is the space of real-valued square-integrable functions. (In the previous considerations the symbol $L_2(\mathbf{R}^3)$ was used to denote the space of complex-valued square-integrable functions.) QM can be represented as the quadratic Taylor approximation of CSM of fields. By expanding classical quantities, functionals of the prequantum field, $\phi \mapsto f(\phi)$, into the Taylor series up to the second term, we obtain quantum quantities. By averaging these expansions we establish a coupling between classical average (so to say, prequantum) and quantum average. The latter appears as the main term in the expansion of the classical phase space average with respect to a small parameter α, the dispersion of the prequantum fluctuations, $\alpha = \sigma^2(\mu)$. Here μ is the probability distribution

Beyond Quantum
Andrei Khrennikov

ISBN 978-981-4411-73-8 (Hardcover), 978-981-4411-74-5 (eBook)
www.panstanford.com

of a prequantum random field. In the limit $\alpha \to 0$ the classical prequantum and quantum averages coincide. Thus QM can be interpreted as the limit (as $\alpha \to 0$) of the theory of classical random fields, cf. Section 3.11 for an experiment which may show a deviation from the prediction of QM, namely, violation of Born's rule.

The real physical model is essentially infinite dimensional: the field phase space is infinite dimensional (even in the classical signal theory [246]). The analysis and theory of stochastic processes on infinite-dimensional spaces does not differ essentially from the finite-dimensional one, e.g., Refs. [124–130, 138]. However, operating with differentiable and analytic functions of the infinite number of variables needs some experience. I do not want to make physical considerations too complicated because of the aforementioned difficulties arising in infinite-dimensional mathematics. Therefore I start with the presentation of QM as an approximative model in the finite-dimensional case. Here $H = \mathbf{C}^n$. The real physical model with the infinite-dimensional phase space will be presented in Section 5.4.

However, before going to the interplay between classical and quantum averages, we recall the well-known procedure of approximative calculation of averages, which is widely used in various applications. We follow chapter 11 of the book [290] of Elena Ventzel. It was written for Soviet artillery officers. This chapter was recommended me by Alexander Choustov, a naval officer. He pointed out that if I was looking for approximative procedures for calculation of averages, then I should definitely try a procedure widely used by artillery officers. And it works! This was the origin of PCSFT. Therefore the presentation (in this chapter) of the quantum formalism as an approximative procedure for calculation of (classical) averages can be called *quantum mechanics for naval officers.* Surprisingly, the trace formula for calculation of averages was used by officers and engineers; of course, without any coupling to QM.

As we have seen in Section 3.2, to proceed to the classical field description of composite systems, we have to restrict the class of possible prequantum distributions to Gaussian, see Chapter 7 for details; see also Section 3.12 for motivation of the use of Gaussian random fields on the prequantum level. Therefore (to have consistency between the presentations in this chapter and Chapter 7), the main presentation of QM as an approximation of classical (prequantum) field theory is done for Gaussian random fields.

However, it may happen that this restriction, i.e., the usage of Gaussian distributions for composite quantum systems, was of a purely mathematical origin, cf. Section 3.12. Therefore in the introductory section (quantum mechanics for naval officers) we consider arbitrary probability distributions.

5.1 The Taylor Approximation of Averages for Functions of Random Variables

Elena Ventzel wrote her book in the form of precise instructions, which students should follow to solve a problem.

> In practice we have very often situations in that, although investigated function of random arguments is not strictly linear, but it differs practically so negligibly from a linear function that it can be approximately considered as linear. This is a consequence of the fact that in many problems fluctuations of random variables play the role of small deviations from the basic law. Since such deviations are relatively small, functions which are not linear in the whole range of variation of their arguments are *almost linear* in a restricted range of their random changes. (Ref. [290], p. 238.)

Let $y = f(x)$. Here, in general, f is not linear, but it does not differ so much from a linear function on some interval $[m_x - \delta, m_x + \delta]$, where $x = x(\omega)$ is a random variable and

$$m_x \equiv E\, x = \int x(\omega)\, d\mathbf{P}(\omega)$$

is its average. Here $\delta > 0$ is sufficiently small. A student of a military college should approximate f by using the first-order Taylor expansion at the point m_x :

$$y(\omega) \approx f(m_x) + f'(m_x)(x(\omega) - m_x). \tag{5.1}$$

By taking the average of both sides he obtains:

$$m_y \approx f(m_x). \tag{5.2}$$

The crucial point is that the linear term $f'(m_x)(x(\omega) - m_x)$ does not give any contribution! Further Ventzel pointed out

> For some problems the above linearization procedure may be unjustified, because the method of linearization may not produce a sufficiently good approximation. In such cases to test the applicability of the linearization method and to improve results there can be applied the method which is based on preserving not only the linear term in the expansion of function, but also some terms of higher orders. (Ref. [290], p. 245.)

Let $y = f(x)$. The student now should preserve the first three terms in the expansion of f into the Taylor series at the point m_x :

$$y(\omega) \approx f(m_x) + f'(m_x)(x(\omega) - m_x) + \frac{1}{2} f''(m_x)(x(\omega) - m_x)^2. \quad (5.3)$$

Hence,

$$m_y \approx f(m_x) + \frac{\sigma_x^2}{2} f''(m_x), \quad (5.4)$$

where

$$\sigma_x^2 = E\ (x - m_x)^2 = \int (x(\omega) - m_x)^2\, d\mathbf{P}(\omega)$$

is the dispersion of the random variable x.

Let us now consider the special case of symmetric fluctuations, $m_x = 0$, and let us restrict our consideration to functions f such that $f(0) = 0$. Then we obtain the following special form of (5.4):

$$m_y \approx \frac{\sigma_x^2}{2} f''(0). \quad (5.5)$$

We emphasize again that the first derivative does not give any contribution to the average.

We can calculate averages approximately not by using the Lebesgue integral (as we do in classical probability theory), but by finding the second derivative. Such "calculus of probability" matches well with experiment. I hope that the reader has already found analogy with the quantum calculus of probabilities. To express better this analogy, we shall consider the multidimensional case. Let now $x = (x_1, \ldots, x_n)$, so we consider a vector of n random variables.

We use the vector average $m_x = (m_{x_1}, \ldots, m_{x_n})$ and the covariance matrix

$$B = (B_{ij}), \;\; B_{ij} = E\,(x_i - m_{x_i})\,(x_j - m_{x_j}).$$

We now consider the random variable

$$y(\omega) = f(x_1(\omega), \ldots, x_n(\omega)).$$

By using the Taylor expansion of $f(x)$ we obtain an algorithm for approximation of the average m_y. We start directly from the second-order Taylor expansion

$$y(\omega) \approx f(m_{x_1}, \ldots, m_{x_n}) + \sum_{i=1}^{n} \frac{\partial f}{\partial x_i}(m_{x_1}, \ldots, m_{x_n})(x_i(\omega) - m_{x_i})$$

$$+\frac{1}{2}\sum_{i,j=1}^{n} \frac{\partial^2 f}{\partial x_i \partial x_j}(m_{x_1}, \ldots, m_{x_n})(x_i(\omega) - m_{x_i})(x_j(\omega) - m_{x_j}), \quad (5.6)$$

and hence,

$$m_y \approx f(m_{x_1}, \ldots, m_{x_1}) + \frac{1}{2}\sum_{i,j=1}^{n} \frac{\partial^2 f}{\partial x_i \partial x_j}(m_{x_1}, \ldots, m_{x_1}) B_x^{ij}. \quad (5.7)$$

By using the vector notation, we can rewrite the previous formulas as

$$y(\omega) \approx f(m_x) + (f'(m_x), x(\omega) - m_x) + \frac{1}{2}(f''(m_x)(x(\omega) - m_x), x(\omega) - m_x). \quad (5.8)$$

and

$$m_y \approx f(m_x) + \frac{1}{2}\mathrm{Tr}\; Bf''(m_x). \quad (5.9)$$

Let us again consider the special case: $m_x = 0$ and $f(0) = 0$. We have

$$m_y \approx \frac{1}{2}\mathrm{Tr}\; Bf''(0). \quad (5.10)$$

We now remark that the Hessian $f''(0)$ is *always a self-adjoint operator.* Let us now represent f by its second derivative at zero:

$$f \to \widehat{A} = f''(0)/2.$$

Then we see that at some level of approximation, instead of operating with Lebesgue integrals, one can use linear algebra:

$$m_y \approx \mathrm{Tr}\; B\widehat{A} \quad (5.11)$$

In Section 5.3 we shall proceed in the mathematically rigorous way, namely, we shall estimate the reminder, which was neglected in the approximative formula for average. We shall also formalize the correspondence between classical and quantum statistical models.

5.2 Quantum Model: Finite-Dimensional Case

We now specify the notions of quantum model (see Section 4.3) in the finite-dimensional case.

5.2.1 *Real Quantum Mechanics*

We shall start with a toy model of quantum mechanics which is based on the real space. Statistical features of the correspondence between a prequantum classical statistical model and quantum mechanics are more evident for this toy model.

Denote the algebra of all $(m \times m)$ real matrices by the symbol $M^{(r)}(m)$. We denote by $\mathcal{D}^{(r)}(m)$ the class of nonnegative self-adjoint trace-one matrices $\rho \in M^{(r)}(m)$. We call them real density matrices. We denote by $L_{\rm s}^{(r)}(m)$ the class of all symmetric matrices.

In the quantum model (for the m-dimensional real space) *statistical states* (describing ensembles of systems prepared for measurement) are represented by density matrices and *quantum observables* by matrices belonging to $L_{\rm s}^{(r)}(m)$.

The *quantum average* of an observable $A \in L_{\rm s}^{(r)}(m)$ with respect to a statistical state $\rho \in \mathcal{D}^{(r)}(m)$ is given by the von Neumann trace formula $\langle A \rangle_\rho = \mathrm{Tr}\, \rho A$.

In the operator representation observables and density matrices are the corresponding classes of **R**-linear operators. Denote the quantum model by

$$M_{\rm q}^{(r)} = (\mathcal{D}^{(r)}(m), L_{\rm s}^{(r)}(m)).$$

If $m = 1$, then quantum observables are given by real numbers (operators of multiplication by real numbers on the real line) and there is only one statistical state $\rho = 1$. Here $\langle A \rangle_\rho = \rho A = A$.

5.2.2 *Complex Quantum Mechanics*

Denote the algebra of all $(m \times m)$ complex matrices by the symbol $M^{(c)}(m)$. We denote by $\mathcal{D}^{(c)}(m)$ the class of nonnegative symmetric trace-one matrices $\rho \in M^{(c)}(m)$, density matrices. We denote by $L_{\rm s}^{(c)}(m)$ the class of all Hermitian matrices.

In the quantum model (for the m-dimensional complex space) statistical states (describing ensembles of systems prepared for measurement) are represented by density matrices and quantum observables by matrices belonging to $L_s^{(c)}(m)$.

The *quantum average* is given by the same trace formula as in the real case (by using the complex trace).

In the operator representation observables and density matrices are given by the corresponding classes of **C**-linear operators. Denote the quantum model by

$$M_q^{(c)} = (\mathcal{D}^{(c)}(m), L_s^{(c)}(m)).$$

If $m = 1$, then quantum observables are given by real numbers (operators of multiplication by real numbers on the complex plane) and there is only one statistical state $\rho = 1$. Here $\langle A \rangle_\rho = \rho A = A$.

5.3 Prequantum ↦ Quantum Correspondence: Finite-Dimensional Case

5.3.1 *One-Dimensional Case*

We specify the classical statistical model under consideration, see Section 4.3 for the general scheme. States of systems are represented by real numbers, $q \in Q = \mathbf{R}$. Ensembles of such systems are described by probability measures on the real line, *statistical states.* We consider a special class of preparation procedures. They produce ensembles of systems described by Gaussian probability distributions on Q having zero mean value and the dispersion

$$\sigma^2(\mu) = \alpha + O(\alpha^2), \tag{5.12}$$

where, as always, $|O(\alpha^2)| \le C\alpha^2$ for some constant C and sufficiently small α. The crucial point is that α is *a small parameter* of our model. Denote this class of probability distributions by the symbol $S_G^{\approx\alpha}(Q)$. The class of Gaussian probability measures S_G^α considered in Section 4.8 is contained in $S_G^{\approx\alpha}$, whose elements are Gaussian measures with $\sigma^2(\mu) = \alpha$.

For a probability $\mu \in S_G^{\approx\alpha}(Q)$, we have:

$$d\mu(q) = \frac{e^{\frac{-q^2}{2(\alpha+O(\alpha^2))}} dq}{\sqrt{2\pi(\alpha + O(\alpha^2))}}.$$

We recall that, for a probability with the zero mean value, its dispersion is given by

$$\sigma^2(\mu) = \frac{1}{\sqrt{2\pi(\alpha + O(\alpha^2))}} \int_{-\infty}^{\infty} q^2 e^{\frac{-q^2}{2(\alpha+O(\alpha^2))}} dq. \qquad (5.13)$$

As already pointed out, we consider α as a small parameter. Therefore Gaussian probability distributions are very *sharply concentrated around the point* $q_0 = 0$. By using the terminology of functional analysis we say that $\{\mu \equiv \mu(\alpha)\}$ is a δ-family:

$$\lim_{\alpha \to 0} \mu(\alpha) = \delta$$

in the sense of theory of distributions.

In the approximation $\alpha = 0$ all systems are located at a single point, namely, $q_0 = 0$. However, a finer description (in which α cannot be neglected) provides the picture of Gaussian bells concentrated near q_0. We remark that in average a system cannot go far away from q_0. By using the Chebyshov inequality one can obtain for any $C > 0$

$$\mu\{q : |q| > C\} \leq \frac{\alpha + O(\alpha^2)}{C^2} \to 0, \alpha \to 0. \qquad (5.14)$$

However, the probabilistic inequality (5.14) does not exclude a possibility that some systems can move far away from the point q_0 (of course, with a small probability).

We also introduce a class of *physical variables* in the classical statistical model under the following consideration:

(a) $f \in C^\infty(\mathbf{R})$, a smooth function;
(b) $f(0) = 0$;
(c) $|f^{(4)}(q)| \leq c_f e^{r_f|q|}, c_f, r_f \geq 0$.

Denote this functional space by the symbol $\mathcal{V}(Q)$, $Q = \mathbf{R}$.[a]

Lemma 3.1. *Let $f \in C^n$ (so it is n times continuously differentiable) and let its nth derivative have the exponential growth. Then all*

[a]The restriction on the growth of the fourth derivative will be used when we consider the Taylor expansion of f up to the fourth term. The exponential growth implies integrability with respect to any Gaussian measure. Thus conditions a) and c) are of purely mathematical value. Condition b) has a physical interpretation. A physical variable $f(q)$ maps the "vacuum state" $q_0 = 0$ into zero. Thus nothing can be produced from vacuum.

derivatives of orders $k = 0, \ldots, n-1$, also have the exponential growth (in particular, $f(q)$ grows exponentially).

Proof. We use the Taylor expansion with the integral remainder:

$$f(q) = f(0) + f'(0)q + \frac{f''(0)q^2}{2} + \frac{f^{(3)}(0)q^3}{3!} + \ldots$$
$$+ \frac{q^n}{n!} \int_0^1 (1-\theta)^{n-1} f^{(n)}(\theta q) d\theta. \qquad (5.15)$$

Since the growth of any polynomial can be compensated by the decrease of $e^{-r|q|}$, by using the exponential estimate for the nth derivative we obtain:

$$|f(q)| = C_1 e^{r|q|} + C_2 \frac{q^n}{n!} \int_0^1 (1-\theta)^{n-1} e^{r|q\theta|} d\theta \leq C e^{r|q|}. \qquad (5.16)$$

Here all constants depend on f.

This simple exercise from the course of analysis will be useful in our further consideration.

We defined the following classical statistical model on the real line:

(1) *states* of systems are real numbers;
(2) *statistical states* (ensembles of systems) are represented by Gaussian probabilities having zero average and the dispersion $\sigma^2(\mu) = \alpha + O(\alpha^2)$, $\alpha \to 0$;
(3) *physical variables* are smooth functions with the exponentially growing fourth derivative; they map zero into itself, $f(0) = 0$.

We denote this model by

$$M_{\mathrm{cl}}^{\alpha} = (S_{\mathrm{G}}^{\approx\alpha}(Q), \mathcal{V}(Q)).$$

As always in classical statistical physics, the average of a physical variable $f \in \mathcal{V}(Q)$ with respect to an ensemble of systems described by a probability $\mu \in S_{\mathrm{G}}^{\approx\alpha}(Q)$ is given by the integral

$$\langle f \rangle_\mu = \frac{1}{\sqrt{2\pi(\alpha + O(\alpha^2))}} \int_{-\infty}^{\infty} f(q) e^{\frac{-q^2}{2(\alpha+O(\alpha^2))}} dq. \qquad (5.17)$$

Since α is a parameter of the model, we can consider averages as functions of α :

$$\langle f \rangle_\mu \equiv \langle f \rangle_\mu(\alpha).$$

We are interested in the asymptotic expansion of averages when $\alpha \to 0$. In particular, such an asymptotic expansion will give us a possibility to calculate classical averages approximately.

Lemma 3.2. *Let $f \in \mathcal{V}(Q)$ and let $\mu \in S_{\mathrm{G}}^{\approx\alpha}(Q)$. Then*

$$\langle f \rangle_\mu(\alpha) = \frac{\alpha}{2} f''(0) + O(\alpha^2). \tag{5.18}$$

Proof. We start with the scaling of the state variable:

$$q = \sigma(\mu)x \tag{5.19}$$

We have

$$\langle f \rangle_\mu(\alpha) = \frac{1}{\sqrt{2\pi}} \int_{-\infty}^{\infty} f(\sigma(\mu)x) e^{\frac{-x^2}{2}} dx. \tag{5.20}$$

We now expand $f(\sigma(\mu)x)$ by using the fourth-order Taylor formula with the integral remainder, see Lemma 3.1

$$\langle f \rangle_\mu(\alpha) = \frac{\sigma^2(\mu)}{2} f''(0) + \frac{\sigma^4(\mu)}{4!\sqrt{2\pi}} \times \int_{-\infty}^{\infty} x^4 \left(\int_0^1 (1-\theta)^3 f^{(4)}(\sigma(\mu)x\theta) d\theta \right) e^{\frac{-x^2}{2}} dx. \tag{5.21}$$

We recall that for a Gaussian measure with zero mean value all odd momenta are equal to zero. This is an important point of our consideration. This implies that the first nonzero contribution to the classical average is given by the second derivative (quadratic term). Disappearance of the third-order term implies the asymptotics $O(\alpha^2)$. We now estimate the remainder to obtain this asymptotics:

$$|R(f,\mu)| \le \frac{C\sigma^4(\mu)}{4!\sqrt{2\pi}} \int_{-\infty}^{\infty} x^4 \left(\int_0^1 (1-\theta)^3 e^{r\sigma(\mu)|x|\theta} d\theta \right) e^{\frac{-x^2}{2}} dx.$$

Since we consider α as a small parameter, we can assume that $|\sigma(\mu)| \le 1$ in the exponential function. Thus, $|R(f,\mu)| \le C'\sigma^4(\mu) \int_{-\infty}^{\infty} x^4 e^{r|x| - \frac{x^2}{2}} dx$. Since $\sigma^2(\mu) = \alpha + O(\alpha^2)$, we have $R(f,\mu) = O(\alpha^2)$, $\alpha \to 0$.

We consider the dispersion $\sigma^2(\mu)$ as the *intensity of fluctuations* in an ensemble of systems. We define the *relative average* with respect to the intensity of fluctuations by normalizing the average by the main term (namely, α) in the intensity of fluctuations: $\frac{\langle f\rangle_\mu}{\alpha}$. We have

$$\frac{\langle f\rangle_\mu}{\alpha} = \frac{f''(0)}{2} + O(\alpha). \tag{5.22}$$

In particular, $\lim_{\alpha\to 0} \frac{\langle f\rangle_\mu}{\alpha} = \frac{f''(0)}{2}$.

We have shown that $\frac{f''(0)}{2}$ gives the approximation of the (classical) relative average. The precision of such an approximation is α.

If the level of the development of measurement technology were such that in measurements all contributions of the magnitude α were neglected, then averages could be calculated by using the following simple rule:

$$\left[\frac{\langle f\rangle_\mu}{\sigma^2(\mu)}\right]^{\text{approx}} = \frac{f''(0)}{2}. \tag{5.23}$$

At the first sight such averages have nothing to do with classical averages given by integrals. By continuing speculation about such an experimental situation, we may guess that the following interpretation of physics might be elaborated: the rules of classical probability theory are violated and the exotic rule (5.23) for calculating averages is related to special features of systems under consideration (and not to a special approximation procedure for averages). It seems that this is the present situation in QM.

Finally, we remark that calculation of averages with the aid of the formula (5.23) is essentially simpler than that for classical probabilistic averages given by Lebesgue integrals.

5.3.2 *Multidimensional Case*

States are vectors $q \in Q = \mathbf{R}^m$; statistical states are Gaussian distributions with zero mean value and dispersion $\sigma^2(\mu) = \alpha + O(\alpha^2)$. Denote this class of probabilities by the symbol $S_{\mathrm{G}}^{\approx\alpha}(Q)$. We introduce the scalar product and norm on Q : $(\xi, q) = \sum_{j=1}^m \xi_j q_j$ and $\|q\|^2 = \sum_{j=1}^m q_j^2$. If a Gaussian measure μ is nondegenerate (so that the measure of any open set is positive), then

$$d\mu(q) = \frac{e^{-\frac{1}{2}(B^{-1}q,q)}\,dq}{\sqrt{(2\pi)^m \det B}},$$

where $B = B_\mu$ is its covariance operator (we consider everywhere only Gaussian measures with zero mean values). We now define a class of physical variables, $\mathcal{V}(Q)$: (a) $f \in C^\infty(\mathbf{R}^m)$; (b) $f(0) = 0$; (c) $\| f^{(4)}(q)\| \leq c_f e^{r_f \|q\|}$, $c_f, r_f \geq 0$. Thus we have defined the following classical statistical model: $M_{\rm cl} = (S_{\rm G}^{\approx\alpha}(Q), \mathcal{V}(Q))$.

For a function $f : \mathbf{R}^m \to \mathbf{R}$, its nth derivative is a (symmetric) n-linear functional, $f^{(n)}(q) : \mathbf{R}^m \times \ldots \times \mathbf{R}^m \to \mathbf{R}$. The norm of this functional is given by $\| f^{(n)}(q)\| = \sup_{\|h_j\|=1} |f^{(n)}(q)(h_1, \ldots, h_n)|$. It is easy to generalize Lemma 3.1 to the multidimensional case.

Lemma 3.3. *Let $f \in \mathcal{V}(Q)$ and let $\mu \in S_{\rm G}^{\approx\alpha}(Q)$. Then*

$$\langle f\rangle_\mu(\alpha) \equiv \int_{\mathbf{R}^m} f(q)\, d\mu(q) = \frac{\alpha}{2} Tr\, \rho f''(0) + O(\alpha^2), \tag{5.24}$$

where $\rho = B_\mu/\alpha$ is a "density operator," scaling of the covariance operator B_μ of μ.

Proof. By using scaling of the state variable (5.19) and by expanding $f(\sigma(\mu)x)$ using the fourth-order Taylor formula with the integral remainder we obtain

$$\langle f\rangle_\mu(\alpha) = \frac{\sigma^2(\mu)}{2} \mathrm{Tr}\, \rho f''(0) \tag{5.25}$$

$$+\frac{\sigma^4(\mu)}{4!} \int_{\mathbf{R}^m} \left(\int_0^1 (1-\theta)^3\, f^{(4)}(\sigma(\mu)x\theta)(q, q, q, q)\, d\theta \right) d\mu_{\rm scal}(x),$$

where $\mu_{\rm scal}$ is a normalized Gaussian measure, i.e., the image of μ under the scaling (5.19). We now estimate the remainder:

$$|R(f, \mu)| \leq \frac{C\sigma^4(\mu)}{4!} \int_{\mathbf{R}^m} \|x\|^4 \left(\int_0^1 (1-\theta)^3\, e^{r\sigma(\mu)\|x\|\theta}\, d\theta \right) d\mu_{\rm scal}(x).$$

Thus $|R(f, \mu)| \leq C'\sigma^4(\mu) \int_{\mathbf{R}^m} \|x\|^4 \;\; e^{r\|x\|} \;\; d\mu_{\rm scal}(x)$. We have $R(f, \mu) = O(\alpha^2)$, $\alpha \to 0$.

Corollary 3.1. *Let $f \in \mathcal{V}(Q)$ and let $\mu \in S_{\rm G}^{\alpha}(Q)$ be nondegenerate. Then*

$$\frac{1}{\sqrt{(2\pi)^m \det B}} \int_{\mathbf{R}^m} f(q)\, e^{\frac{1}{2}(B^{-1}q, q)}\, dq = \frac{\alpha}{2} Tr\, \rho f''(0) + O(\alpha^2), \tag{5.26}$$

where $\rho = B/\alpha$.

As in the one-dimensional case, we introduce the relative average

$$\frac{\langle f \rangle_\mu}{\alpha} = \frac{\int_{\mathbf{R}^m} f(q)\, d\mu(q)}{\int_{\mathbf{R}^m} \|q\|^2\, d\mu(q)} + O(\alpha).$$

In the case of a nondegenerate Gaussian measure we have

$$\frac{\langle f \rangle_\mu}{\alpha} = \frac{\int_{\mathbf{R}^m} f(q)\, e^{\frac{1}{2}(B^{-1}q,q)}\, dq}{\int_{\mathbf{R}^m} \|q\|^2\, e^{\frac{1}{2}(B^{-1}q,q)}\, dq} + O(\alpha).$$

We have

$$\frac{\langle f \rangle_\mu}{\alpha} = \frac{1}{2}\mathrm{Tr}\, \rho\, f''(0) + O(\alpha). \tag{5.27}$$

We consider maps

$$T : S_{\mathrm{G}}^{\approx\alpha}(Q) \to \mathcal{D}^{(r)}(m),\ \rho = T(\mu) = B_{\mu_{\mathrm{scal}}}; \tag{5.28}$$

$$T : \mathcal{V}(Q) \to L_{\mathrm{s}}^{(r)}(m)),\ A = T(f) = f''(0)/2 \tag{5.29}$$

(we recall that Hessian is always a symmetric matrix).

Theorem 3.1. *The maps (5.28) and (5.29) project the classical statistical model $M_{cl} = (S_{\mathrm{G}}^{\approx\alpha}(Q), \mathcal{V}(Q))$ onto the quantum model $M_q^{(r)} = (\mathcal{D}^{(r)}(m), L_{\mathrm{s}}^{(r)}(m))$ in such a way that classical and quantum averages are coupled by the asymptotic equality*

$$\frac{\langle f \rangle_\mu}{\alpha} = \langle T(f) \rangle_{T(\mu)} + O(\alpha). \tag{5.30}$$

5.3.3 *Prequantum Phase Space in the Two-Dimensional Case*

In the previous sections we considered the prequantum toy model in which the phase space structure was not taken into account. The corresponding quantum model was over the reals. On the other hand, physical reality is described by the classical phase space mechanics and complex quantum mechanics. It is possible to create a prequantum phase space model reproducing the complex quantum mechanics. The crucial point is that classical variables and statistical states, functions and measures on the phase space, have to be *invariant with respect to a special group of transformations of the phase space.* This *fundamental prequantum group* is very simple, it is the *special orthogonal group* $SO(2)$, the group of rotations of the phase space. A trace of the presence of this group was visible already in Chapter 4. The symplectic operator J is the generator of $SO(2)$. For quadratic forms, the J-invariance is equivalent to the $SO(2)$-invariance. However, by considering nonquadratic forms we have to extend the J-invariance to $SO(2)$-invariance. Switching from the real phase space representation to the complex representation induces the use of the unitary group $U(1)$ (see Section 5.13). Invariance with respect to this group is the well-known gauge invariance (in particular, for the classical electromagnetic field). Thus the $U(1)$ gauge invariance plays a fundamental role in determining prequantum physical variables; by some reason (which is not completely clear, but cf. Section 5.13) all functionals of the prequantum field should be $U(1)$ gauge invariant.

States of systems are now represented by points $\phi = (q, p) \in \Omega = Q \times P$, where $Q = P = \mathbf{R}$. Here q is position and p is momentum, so Ω denotes the phase space. Statistical states are represented by Gaussian $SO(2)$-invariant measures having zero mean value and the dispersion

$$\sigma^2(\mu) = \alpha + O(\alpha^2); \tag{5.31}$$

physical variables are given by $SO(2)$-invariant maps, $f : \Omega \to \mathbf{R}$, which satisfy conditions (a), (b), and (c) specifying variables in the real case. Denote these classes of measures and functions, respectively, by $S_G^{\approx\alpha}(\Omega|SO(2))$ and $\mathcal{V}(\Omega|SO(2))$.

We consider the classical model $M_{\rm cl} = (S_G^{\approx\alpha}(\Omega|SO(2)), \mathcal{V}(\Omega|SO(2))$. As in the real case, we can obtain the asymptotic expansion of the classical averages, see (5.27). However, in quantum mechanics we consider the complex structure. We would like to recover it in our classical model. To do this, we shall study in more detail $SO(2)$-invariance properties of classical probabilities and variables.

A measure μ is $SO(2)$-invariant if for any $u \in SO(2)$,

$$\int_{\mathbf{R}^2} f(uq)d\mu(q) = \int_{\mathbf{R}^2} f(q)d\mu(q). \tag{5.32}$$

For a Gaussian measure μ with the covariance matrix $B = B_\mu$, this is equivalent to the condition

$$[u, B] = 0, \ u \in SO(2). \tag{5.33}$$

Let f be a two times differentiable invariant map, so that $f(u\phi) = f(\phi)$, for any $u \in SO(2)$. By representing

$$u = u_\theta = \begin{pmatrix} \cos\theta & -\sin\theta \\ \sin\theta & \cos\theta \end{pmatrix},$$

we have

$$f(\cos\theta q - \sin\theta p, \sin\theta q + \cos\theta p) = f(q, p). \tag{5.34}$$

This is a rather strong constraint determining a very special class of maps. In particular, we obtain: $u^\star\nabla f(u\phi) = \nabla f(\phi)$ and $u^\star f''(u\phi)u = f''(\phi)$. Hence $u^\star\nabla f(0) = \nabla f(0)$ for any rotation, and thus,

$$\nabla f(0) = 0 \tag{5.35}$$

and

$$[f''(0), u] = 0, \;\; u \in SO(2). \tag{5.36}$$

It is convenient to introduce the *commutator* of the set $SO(2)$ in the algebra of all two by two matrices $M^{(r)}(2)$:

$$SO'(2) = \{A \in M^{(r)}(2) : [A, u] = 0, u \in SO(2)\}$$

We remark that a generator of $SO(2)$ can be chosen as the symplectic operator $J = \begin{pmatrix} 0 & 1 \\ -1 & 0 \end{pmatrix}$. Therefore the commutator of $SO'(2)$ coincides with the commutator of J : $\{J\}' = \{A \in M^{(r)}(2) : [A, J] = 0\}$.

Proposition 3.1. *Let $\mu \in S_{\mathrm{G}}^{\approx\alpha}(\Omega|SO(2))$ and let $f \in \mathcal{V}(\Omega|SO(2))$. Then the covariance operator B_μ and Hessian $A = f''(0)$ belong to the commutator $SO'(2)$.*

Hence, $\mu \in S_{\mathrm{G}}^{\approx\alpha}(\Omega|SO(2))$ iff its covariance operator commutes with the symplectic operator J. (If $f \in \mathcal{V}(\Omega|SO(2))$, then its second derivative commutes with J.) Consider now the subspace $S_{\mathrm{G}}^{\alpha}(\Omega|SO(2))$ of $S_{\mathrm{G}}^{\approx\alpha}(\Omega|SO(2))$ consisting of measures having the dispersion which precisely equals α, then $S_{\mathrm{G}}^{\alpha}(\Omega|SO(2))$ coincides with the space $S_{G,\mathrm{sp}}^{\alpha}(\Omega)$, which was used in Section 4.8. In that section we were fine by considering only J-invariance; there was no reason to consider the group $SO(2)$, since for measures and quadratic forms, as we have seen, the J-invariance implies the $SO(2)$-invariance and in Section 4.8 we did not consider nonquadratic functions. In particular, we know that a matrix A belongs to the commutator $SO'(2)$ iff

$$A = \begin{pmatrix} R & -S \\ S & R \end{pmatrix}. \tag{5.37}$$

(In Section 4.1.2 we used slightly different notation to make the present matrix form consistent with the one from that section, $-S$ should be placed instead of S). If A is symmetric, then it is diagonal: $A = \begin{pmatrix} R & 0 \\ 0 & R \end{pmatrix}$. (We remark that we work in the two-dimensional case.) In particular, its trace is given by

$$\mathrm{Tr} A = 2R \tag{5.38}$$

Thus, if $\mu \in S_{\mathrm{G}}^{\approx\alpha}(\Omega|SO(2))$, then its covariance matrix is diagonal, $B = \begin{pmatrix} b & 0 \\ 0 & b \end{pmatrix}$, where $2b = \alpha + O(\alpha^2)$.

Fluctuations of the coordinate q and the momentum p are independent and equally distributed

$$d\mu(q) = \frac{1}{2\pi b} \exp\{-\frac{q^2 + p^2}{2b}\} dq. \tag{5.39}$$

Thus in the two-dimensional case the set of $SO(2)$-invariant Gaussian measures has a simple structure: these are measures of the form (5.39); this is a one-parameter family of Gaussian measures parameterized by the dispersion parameter. By definition, these are *rotation-invariant Gaussian measures.* Thus "prequantum random variables" are rotation invariant.

Denote the marginal distributions of μ by the symbols μ_q and μ_p, respectively. Then

$$\sigma^2(\mu_q) = \frac{1}{\sqrt{2\pi b}} \int_{-\infty}^{+\infty} q^2 \, e^{-\frac{q^2}{2b}} dq = \sigma^2(\mu_p)$$
$$= \frac{1}{\sqrt{2\pi b}} \int_{-\infty}^{+\infty} p^2 \, e^{-\frac{p^2}{2b}} dp.$$

Hence $b = \sigma^2(\mu_q) = \sigma^2(\mu_p) = \frac{1}{2}\sigma^2(\mu) = \frac{\alpha}{2} + O(\alpha^2)$.

Proposition 3.2. *Let $f \in \mathcal{V}(\Omega|SO(2))$. Then all its odd derivatives at the point $q_0 = 0$ are equal to zero.*

For a function $f \in \mathcal{V}(\Omega|SO(2))$, its Hessian has the form $f''(0) = \begin{pmatrix} R & 0 \\ 0 & R \end{pmatrix}$, where $R \in \mathbf{R}$, and hence,

$$f(q, p) = \frac{R(q^2 + p^2)}{2} + O(\alpha^2).$$

Thus "prequantum physical variables" have the quadratic part of a very simple form; by definition, it is rotation invariant. We remark that, in spite of the coincidence of commutators, the $SO(2)$-invariance is not equivalent to the J-invariance.

Example 3.1. Let $f(q, p) = q^3 p - qp^3 = qp(q^2 - p^2)$. Then $f(J\phi) = f(\phi)$, $\phi = (q, p)$. But take $\theta = \pi/4$. Here $u = \frac{1}{2}\begin{pmatrix} 1 & -1 \\ 1 & 1 \end{pmatrix}$. Hence $u\begin{pmatrix} q \\ p \end{pmatrix} = \begin{pmatrix} (q-p)/2 \\ (q+p)/2 \end{pmatrix}$. Thus $f(u\Psi) = (q-p)(q+p)(q-p-q-p)(q-p+q+p)/16 = -qp((q^2 - p^2)/4$.

We are now ready to recover in this simple case the complex structure of quantum mechanics (see Section 4.1 for the general scheme). Any matrix belonging to $SO'(2)$ can be represented in the form $A = R\,I + S\,(-J)$, where I and J are the unit and symplectic matrices, respectively. By mapping I into 1 and $(-J)$ into i we obtain a map of the commutator $SO'(2)$ onto the set of complex numbers $\mathbf{C}$:

$$j : SO'(2) \to \mathbf{C},\, z = j(A) = R + iS. \tag{5.40}$$

This is an isomorphism of two fields. In particular, a symmetric matrix $A = \begin{pmatrix} R & 0 \\ 0 & R \end{pmatrix}$ is represented by the real number $j(A) = R$. This is the operator of multiplication by R. The trace of this operator in the one-dimensional complex space $\mathbf{C}$ (with the scalar product, $(z, w) = z\bar{w}$) equals R. By (5.38) we obtain

$$\mathrm{Tr}\, A = 2\mathrm{Tr}\, j(A), \tag{5.41}$$

where on the left-hand side we have the real trace and on the right-hand side the complex trace (see Section 4.8 for general discussion).

Now we can write the basic asymptotic equality for averages in the complex form.

Lemma 3.4. *Let $f \in \mathcal{V}(\Omega|SO(2))$ and let $\mu \in S_{\mathrm{G}}^{\approx\alpha}(\Omega|SO(2))$. Then*

$$\langle f\rangle_\mu(\alpha) \equiv \int_{\mathbf{R}^2} f(q, p)\, d\mu(q, p) = \alpha j(f''(0)/2) + O(\alpha^2). \tag{5.42}$$

Proof. We make the scaling of the state variable

$$\phi = \sigma(\mu)\Phi = (\sqrt{\alpha} + O(\alpha))\Phi, \tag{5.43}$$

where $\phi = (q, p)$ and $\Phi = (x, y)$.

Then the image of μ is again a Gaussian measure, say μ_{scal}, having the dispersion $\sigma^2(\mu_{\mathrm{scal}}) = 1$. Set $B = B_{\mu_{\mathrm{scal}}}$. In the two-dimensional case $B = \begin{pmatrix} 1/2 & 0 \\ 0 & 1/2 \end{pmatrix}$ and Tr $B = 1$. We now have

$$\langle f \rangle_\mu(\alpha) = \frac{\sigma^2(\mu)}{2} \mathrm{Tr}\ B\ f''(0) + O(\alpha^2), \tag{5.44}$$

where Tr is the real trace. By using (5.41) we obtain

$$\langle f \rangle_\mu(\alpha) = \sigma^2(\mu)\mathrm{Tr}\ j(B\ f''(0)) + O(\alpha^2)$$

$$= \alpha \mathrm{Tr}\ j(B)\ j(f''(0))) + O(\alpha^2) = \alpha j(B)\ j(f''(0))) + O(\alpha^2),$$

since in the two-dimensional case $j(A) \in \mathbf{C}$ for any matrix A (from the domain of definition of j). We remark that $j(B) = 1/2$. Thus $\rho = 2j(B) = 1$. Hence, we can write

$$\langle f \rangle_\mu(\alpha) = \alpha\rho\ j(f''(0))/2) + O(\alpha^2) \tag{5.45}$$

and hence (5.42).

We recall that in the one-dimensional quantum mechanics there is only one "density matrix," namely, $\rho = 1 \in \mathbf{R}$. This fact trivializes the previous considerations, but this trivial presentation is useful as a ground of coming multidimensional presentation.

As always, it is convenient to consider the renormalization of averages by the main term in the intensities of fluctuations of the coordinates and momenta: $\frac{\langle f \rangle_\mu}{\alpha}$. Then we get

$$\frac{\langle f \rangle_\mu}{\alpha} = j(f''(0)/2) + O(\alpha). \tag{5.46}$$

5.3.4 *Prequantum Phase Space in Multidimensional Case*

States of systems are now represented by points $\phi = (q, p) \in \Omega = Q \times P$, where $Q = P = \mathbf{R}^m$. Here $q = (q_1, \ldots, q_m)$ is the position and $p = (p_1, \ldots, p_m)$ is momentum, so Ω denotes the phase space. Let us consider the canonical representation of the group $SO(2)$ in the phase space $\Omega = Q \times P$

$$u = u_\theta = \begin{pmatrix} \cos\theta\ I & -\sin\theta\ I \\ \sin\theta\ I & \cos\theta\ I \end{pmatrix}, \tag{5.47}$$

where I is the unit matrix from $M^{(r)}(m)$. The corresponding group of **R**-linear operators (or $2m \times 2m$ matrices) we denote by the symbol $SO_m(2)$. The classical model $M_{cl} = (S_G^{\approx\alpha}(\Omega|SO_m(2)), \mathcal{V}(\Omega|SO_m(2))$ is defined in the same way as in the two-dimensional case.

A Gaussian measure is $SO_m(2)$-invariant iff its covariance operator belongs to the commutator $SO'_m(2) = \{A \in M^{(r)}(2m) : [A, u] = 0, u \in SO_m(2)\}$. We remark that it is equivalent to commuting with the symplectic operator J. Thus, the space $S^{\alpha}_{G,\text{sp}}$ (see Section 4.8) coincides with the subspace of $S_G^{\approx\alpha}(\Omega|SO_m(2))$ consisting of measures with the dispersion α.

If a smooth function f is $SO(2)$-invariant, then all its odd derivatives are equal to zero and the second derivative belongs to $SO'_m(2)$. A matrix $A \in SO'_m(2)$ if it has the form (5.37), where $R, S \in M^{(r)}(m)$. In contrast to the two-dimensional case, a symmetric matrix from $SO'_m(2)$ can be nondiagonal. It has the form (5.37), where $R^* = R$ and $S^* = -S$ (see Section 4.1 for general consideration).

We consider a natural map generalizing the map $j : SO'_m(2) \to \mathbf{C}$. It maps the commutator $SO'_m(2)$ onto the set of complex matrices $M^{(c)}(m)$:

$$j : SO'_m(2) \to M^{(c)}(m), z = j(A) = R + iS. \tag{5.48}$$

This is an isomorphism of two rings. Symmetric matrices are mapped onto Hermitian matrices. Let us denote real and complex conjugations by $*$ and $\star$, respectively. We have $(R+iS)^\star = R^* - iS^* = R + iS$. We also remark that for a symmetric matrix A :

$$\text{Tr } j(A) = \text{Tr } (R + iS) = \text{Tr } R = \frac{1}{2} \text{ Tr } A. \tag{5.49}$$

Lemma 3.5. *Let $f \in \mathcal{V}(\Omega|SO_m(2))$ and let $\mu \in S_G^{\approx\alpha}(\Omega|SO_m(2))$. Then*

$$\langle f \rangle_\mu(\alpha) \equiv \int_{\mathbf{R}^{2m}} f(q, p)\, d\mu(q, p) = \alpha\ Tr\rho\ j(f''(0)/2) + O(\alpha^2), \tag{5.50}$$

where $\rho \in \mathcal{D}^{(c)}(m)$ and ρ is the complex covariance operator of $\sqrt{\alpha}$-scaling of μ.

Proof. We make scaling (5.43) and obtain the measure μ_{scal} with the covariance operator $B = B_{\mu_{\text{scal}}}$, and $\text{Tr } B = 1$. We set $\rho =$

$2j(B)$, here $\mathrm{Tr}\,\rho = 2(\mathrm{Tr}\,B)/2 = 1$ and hence $\rho \in \mathcal{D}^{(c)}(m)$. We note the relation coupling the real representation of the complex covariance operator $D \equiv D_{\mu_{\mathrm{scal}}}$ and the real covariance operator of this measure, namely, B (see Section 4.8) equality (4.63): $D^{\mathrm{real}} = 2B$. Since $j(D^{\mathrm{real}}) = D$, we have $\rho = D$. Finally, we can obtain the multidimensional version of (5.45):

$$\langle f \rangle_\mu(\alpha) = \alpha \mathrm{Tr} \rho\; j(f''(0))/2) + O(\alpha^2). \qquad (5.51)$$

We now modify the classical → quantum projections, (5.28) and (5.29), to make them consistent with the complex structure

$$T : S_{\mathrm{G}}^{\approx\alpha}(\Omega|SO_m(2))) \to \mathcal{D}^{(c)}(m),\; \rho = T(\mu) = D_{\mu_{\mathrm{scal}}}; \qquad (5.52)$$

$$T : \mathcal{V}(Q) \to L_{\mathrm{s}}^{(c)}(m),\; \widehat{A} = T(f) = j(f''(0)/2) \qquad (5.53)$$

(We recall that $\rho = 2j(B_{\mu_{\mathrm{scal}}})$.)

The maps (5.52) and (5.53) project the classical statistical model

$$M_{\mathrm{cl}} = (S_{\mathrm{G}}^{\approx\alpha}(\Omega|SO_m(2)), \mathcal{V}(\Omega|SO_m(2)))$$

onto the quantum model $M_{\mathrm{q}}^{(c)} = (\mathcal{D}^{(c)}(m), L_{\mathrm{s}}^{(c)}(m))$ in such a way that classical and quantum averages are coupled by the asymptotic equality

$$\frac{\langle f \rangle_\mu}{\alpha} = \langle T(f) \rangle_{T(\mu)} + O(\alpha). \qquad (5.54)$$

5.4 Prequantum Phase Space: Infinite-Dimensional Case

States of systems are now represented by points $\phi = (q, p) \in H = Q \times P$, where Q and P are two copies of a real separable infinite-dimensional Hilbert space with the scalar product $(\cdot, \cdot)$ and the corresponding norm $\|\cdot\|$. Here $q \in Q$ is a field analog of the position variable and $p \in P$ is a field analog of the momentum variable. H denotes the field phase space. The real Hilbert space structure on H is given by the scalar product

$$(\phi_1, \phi_2) = (q_1, q_2) + (p_1, p_2). \qquad (5.55)$$

In physics $Q = P = L_2(\mathbf{R}^3; \mathbf{R})$ is the space of real-valued square-integrable functions. Thus both "position" and "momentum" are functions of $x \in \mathbf{R}^3$. These are simply classical fields. A point of such phase space is a classical vector field $\phi(x) = (q(x), p(x))$.

Let us consider the canonical representation of the group $SO(2)$ in the phase space $H = Q \times P$, see (5.47). The corresponding group of continuous **R**-linear operators in H we denote by the symbol $SO_\infty(2)$.

The classical model

$$M_{\rm cl} = (S_{\rm G}^{\approx\alpha}(H|SO_\infty(2)), \mathcal{V}(H|SO_\infty(2)))$$

is defined in the same way as in the finite-dimensional case. We just recall a few basic notions from theory of differentiable functions on infinite-dimensional spaces.

The differential calculus for maps $f : H \to \mathbf{R}$ does not differ so much from the differential calculus in the finite-dimensional case, $f : \mathbf{R}^n \to \mathbf{R}$ [130, 138]. Instead of the norm on $\mathbf{R}^n$, one should use the norm on H. We consider the so-called Frechet differentiability. Here a function f is differentiable if it can be represented as

$$f(\phi_0 + \Delta\phi) = f(\phi_0) + f'(\phi_0)(\Delta\phi) + o(\Delta\phi),$$

where $\lim_{\|\Delta\phi\|\to 0} \frac{\|o(\Delta\phi)\|}{\|\Delta\phi\|} = 0$. Here at each point ϕ the derivative $f'(\phi)$ is a continuous linear functional on H; so it can be identified with the element $f'(\phi) \in H$. Then we can define the second derivative as the derivative of the map $\phi \to f'(\phi)$ and so on. A map f is differentiable n-times iff:

$$f(\phi_0 + \Delta\phi) = f(\phi_0) + f'(\phi_0)(\Delta\phi) + \frac{1}{2} f''(\phi_0)(\Delta\phi, \Delta\phi) + \dots$$
$$+ \frac{1}{n!} f^{(n)}(\phi_0)(\Delta\phi, \dots, \Delta\phi) + o_n(\Delta\phi),$$

where $f^{(n)}(\phi_0)$ is a symmetric continuous n-linear form on H and

$$\lim_{\|\Delta\phi\|\to 0} \frac{\|o_n(\Delta\phi)\|}{\|\Delta\phi\|^n} = 0.$$

It is important for us that the second derivative $f''(\phi_0)$ can be represented by a self-adjoint operator $f''(\phi_0)(u, v) = (f''(\phi_0)u, v)$, $u, v \in H$. We remark that for $\phi_0 = 0$ we have

$$f(\phi) = f(0) + f'(0)(\phi) + \frac{1}{2} f''(0)(\phi, \phi) + \cdots$$
$$+ \frac{1}{n!} f^{(n)}(0)(\phi, \dots, \phi) + o_n(\phi).$$

As in the finite-dimensional case, the reminder can be represented in the integral form.

A Gaussian measure is $SO_\infty(2)$-invariant iff its covariance operator belongs to the commutator

$$SO'_\infty(2) = \{A \in \mathcal{L}(H) : [A, u] = 0, u \in SO_\infty(2)\}.$$

It is equivalent to commuting with the symplectic operator. We remark that in the notation of Section 4.1 $SO'_\infty(2) = \mathcal{L}_{\rm sp}(H)$.

As in the finite-dimensional case, if a smooth function $f : H \to \mathbf{R}$ is $SO_\infty(2)$-invariant, then all its odd derivatives are equal to zero and the second derivative belongs to $SO'_\infty(2)$.

Example 4.1. Let $\widehat{H} \in \mathcal{L}_{\rm sp,s}(H)$, i.e., $[\widehat{H}, J] = 0$. Then any polynomial function

$$f(\phi) = \sum_{k=1}^{N} a_k(\widehat{H}\phi, \phi)^k, \; a_k \in \mathbf{R},$$

is $SO'_\infty(2)$-invariant. In this case this is equivalent to the J-invariance.

Let us now consider the complexification of the Hilbert space H : $H_{\rm c} = Q \oplus iP$.

There is a natural map (generalizing the map $j : SO'_m(2) \to \mathbf{C}^m$) of the commutator $SO'_\infty(2)$ onto $\mathcal{L}(H_{\rm c})$:

$$j : SO'_\infty(2) \to \mathcal{L}(H_{\rm c}), \; z = j(A) = R + iS. \tag{5.56}$$

This is an isomorphism of two rings. Self-adjoint operators (with respect to the real scalar product) are mapped onto self-adjoint operators (with respect to the complex scalar product).

In the same way as in the finite-dimensional case, we prove that for $f \in \mathcal{V}(H|SO_\infty(2))$ and $\mu \in S_{\rm G}^{\approx\alpha}(H|SO_\infty(2))$, the asymptotic equality (5.50), where $\rho \in \mathcal{D}(H_{\rm c})$, holds. We also consider the infinite-dimensional generalization of the classical $\to$ quantum projections, (5.52) and (5.53) $T : S_{\rm G}^{\approx\alpha}(H|SO_\infty(2)) \to \mathcal{D}(H_{\rm c})$, $T : \mathcal{V}(H|SO_\infty(2)) \to \mathcal{L}_s(H_{\rm c})$. The map T projects the classical statistical model

$$M_{\rm cl} = (S_{\rm G}^{\approx\alpha}(H|SO_\infty(2)), \mathcal{V}(H|SO_\infty(2)))$$

onto the quantum model

$$M_{\rm q}(H_{\rm c}) = (\mathcal{D}(H_{\rm c}), \mathcal{L}_s(H_{\rm c}))$$

in such a way that classical and quantum averages are coupled by the asymptotic equality (5.54).

We remark that if a prequantum variable is a quadratic form, i.e., $f_A(\phi) = (\widehat{A}\phi, \phi)$, then it has the quantum image $T(f_A) = \widehat{A}$, see also Eq. (3.4), and the asymptotic equality (5.54) is reduced to the equality $\langle f_A \rangle_\mu / \sigma^2(\mu) = \mathrm{Tr} \rho \widehat{A}$; in particular, if μ has the unit dispersion, we have $\langle f_A \rangle_\mu = \mathrm{Tr} \rho \widehat{A} = \mathrm{Tr} D_\mu \widehat{A}$, where D_μ is the complex covariance operator of μ, cf. Section 3.1.3. It is also clear that the restriction of the map T to the subspace of quadratic forms $V_{\rm quad,sp}(H)$ (see Section 4.7) of the functional space $\mathcal{V}(H|SO_\infty(2))$ is one-to-one.[a] Consider now the classical statistical model $M_{\rm cl} = (S_{\rm G}^{\approx\alpha}(H|SO_\infty(2)), V_{\rm quad,sp}(H))$. Then the map T establishes the one-to-one correspondence of this model with QM.

We remark that our projection map $T : \mathcal{V}_{\rm sp}(H) \to \mathcal{L}_{\rm s}(H_{\rm c})$ fulfills the important postulate for classical $\to$ quantum correspondence which was used by J. von Neumann [296] in his no-go theorem

$$T(\sum \lambda_j f_j) = \sum \lambda_j T(f_j), \;\; \lambda_j \in \mathbf{R}, \; f_j \in \mathcal{V}_{\rm sp}(H). \qquad (5.57)$$

Here quantum observables $\widehat{A}_j = T(f_j)$ can be incompatible, so these operators can be noncommuting. (This postulate was strongly criticized by Bell [31] and Ballentine [25] as nonphysical: what is a physical meaning of a linear combination of incompatible observables?)

Conclusion. *Averages of quantum mechanics can be considered as Taylor approximations of averages of CSM with the infinite-dimensional phase space. In the quantum formalism a classical physical variable (a smooth function on the classical phase space) is represented by its second derivative. A classical statistical state, Gaussian probability measure on the phase space, is represented by its covariance operator (with corresponding normalization).*

---

[a]We recall the evident fact that, for quadratic forms, the $SO(2)$-invariance is equivalent to the J-invariance.

5.5 Gaussian Measures Corresponding to Pure Quantum States

Let Ψ be a pure quantum state: $\Psi \in H_c$, $\|\Psi\| = 1$ and let $\rho_\Psi = \Psi \otimes \Psi$ be the corresponding density operator; in Dirac's notation $\rho_\Psi = |\Psi\rangle\langle\Psi|$.

We consider the Gaussian measure $\mu_{\Psi,\alpha} \in S_G^\alpha(H|SO_\infty(2))$ with the complex covariance operator $D_{\mu_{\Psi,\alpha}} = \alpha\rho_\Psi$. It is easy to see that it is concentrated on the one-dimensional complex space (the real plane) $\Pi_\Psi = \{\phi \in H_c : \phi = c\Psi, c \in \mathbf{C}\}$ and $T(\mu_{\Psi,\alpha}) = \rho_\Psi$. For any prequantum physical variable $f(\phi)$, the prequantum and quantum averages are coupled by the following asymptotic equality:

$$\langle f \rangle_\mu/\alpha = \langle j(f''(0)/2)\Psi, \Psi\rangle + O(\alpha).$$

By selecting $\alpha = 1$, we obtain the Gaussian measure on Π_Ψ with the complex covariance operator ρ_Ψ. Here

$$\langle f \rangle_\mu = \langle \widehat{A}\Psi, \Psi\rangle + O(1), \ \ f \in \mathcal{V}(H|SO_\infty(2)),$$

where $\widehat{A} = T(f) \equiv j(f''(0)/2$.

Physical interpretation. (cf. Margenau [239, 240] and Ballentine [25, 28]) *In PCSFT there are no "pure quantum states." The states which are interpreted (in the conventional quantum formalism) as pure, in fact, represent J-invariant Gaussian measures having two-dimensional supports. Such states can be imagined as fluctuations of fields concentrated on two-dimensional real planes of the infinite-dimensional state phase space.*[a]

5.6 Illustration of the Prequantum $\mapsto$ Quantum Coupling in the Case of Qubit Mechanics

5.6.1 *Classical Representation for Spin Operators*

The Pauli matrices are a set of 2×2 complex Hermitian and unitary matrices. They are $\sigma_1 = \begin{pmatrix} 0 & 1 \\ 1 & 0 \end{pmatrix}$, $\sigma_2 = \begin{pmatrix} 0 & -i \\ i & 0 \end{pmatrix}$, $\sigma_3 = \begin{pmatrix} 1 & 0 \\ 0 & -1 \end{pmatrix}$.

[a] *We recall that we still work in the idealized situation where vacuum fluctuations are not taken into account.*

Let $H_c = C^2$ with the complex coordinates $z = (z_1, z_2)$, $z_j = q_j + ip_j$, $j = 1, 2$, and $\Omega = \mathbf{R}^2 \times \mathbf{R}^2$ with the real coordinates $\phi = (q_1, q_2, p_1, p_2)$. We consider spin operators $\sigma(a) = \sum_j^3 a_j \sigma_j : \mathbf{C}^2 \to \mathbf{C}^2$, $a = (a_1, a_2, a_3)$. Let us consider real matrices $\sigma_j^{(r)} = j^{-1}(\sigma_j)$:

$$\sigma_1^{(r)} = \begin{pmatrix} \sigma_1 & 0 \\ 0 & \sigma_1 \end{pmatrix} = \begin{pmatrix} 0 & 1 & 0 & 0 \\ 1 & 0 & 0 & 0 \\ 0 & 0 & 0 & 1 \\ 0 & 0 & 1 & 0 \end{pmatrix}, \sigma_2^{(r)} = \begin{pmatrix} 0 & -i\sigma_2 \\ i\sigma_2 & 0 \end{pmatrix} = \begin{pmatrix} 0 & 0 & 0 & -1 \\ 0 & 0 & 1 & 0 \\ 0 & 1 & 0 & 0 \\ -1 & 0 & 0 & 0 \end{pmatrix},$$

$$\sigma_3^{(r)} = \begin{pmatrix} \sigma_3 & 0 \\ 0 & \sigma_3 \end{pmatrix} = \begin{pmatrix} 1 & 0 & 0 & 0 \\ 0 & -1 & 0 & 0 \\ 0 & 0 & 1 & 0 \\ 0 & 0 & 0 & -1 \end{pmatrix}.$$

We set

$$\sigma^{(r)}(a) \equiv j^{-1}(\sigma(a)) = \sum_j^3 a_j \sigma_j^{(r)} : \mathbf{R}^4 \to \mathbf{R}^4, \quad a = (a_1, a_2, a_3).$$

and consider *classical random spin variables*

$$f_a(\phi) = (\sigma^{(r)}(a)\phi, \phi) = \langle \sigma(a)\phi, \phi \rangle.$$

Then, for the observable part of prequantum$\mapsto$quantum correspondence map T, we have

$$T(f_a) = \sigma(a)$$

and for any $SO_2(2)$-invariant Gaussian measure μ, on $\Omega = \mathbf{R}^2 \times \mathbf{R}^2$ with dispersion 1, we have[a]

$$\int_{\mathbf{R}^4} f_a(\phi) d\mu(\phi) = \mathrm{Tr}\, \rho\sigma(a),$$

where ρ is the complex covariance operator of μ. For example,

$$\int_{\mathbf{R}^4} (q_1 q_2 + p_1 p_2) d\mu(q_1, q_2, p_1, p_2) = \mathrm{Tr}\, \rho\sigma_1 = \langle \sigma_1 \rangle_\rho,$$

$$\int_{\mathbf{R}^4} (p_1 q_2 - p_2 q_1) d\mu(q_1, q_2, p_1, p_2) = \mathrm{Tr}\, \rho\sigma_2 = \langle \sigma_2 \rangle_\rho,$$

[a]For quadratic forms, we have the precise equality of quantum and prequantum averages; therefore, one can always eliminate the small parameter α and work with normalized measures.

$$\int_{\mathbf{R}^4} (q_1^2 - q_2^2 + p_1^2 - p_2^2) d\mu(q_1, q_2, p_1, p_2) = \mathrm{Tr}\ \rho\sigma_3 = \langle\sigma_3\rangle_\rho.$$

We also remark that

$$\int_{\mathbf{R}^4} (q_1^2 + q_2^2 + p_1^2 + p_2^2) d\mu(q_1, q_2, p_1, p_2) = \mathrm{Tr}\ \rho I = 1.$$

By introducing vectors $\phi_1 = (q_1, p_1)$ and $\phi_2 = (q_2, p_2)$ we rewrite these equalities in a shorter way:

$$\int_{\mathbf{R}^4} (\phi_1, \phi_2) d\mu(\phi_1, \phi_2) = \mathrm{Tr}\ \rho\sigma_1; \int_{\mathbf{R}^4} (J\phi_1, \phi_2) d\mu(\phi_1, \phi_2) = \mathrm{Tr}\ \rho\sigma_2,$$

and, finally,

$$\int_{\mathbf{R}^4} (\|\phi_1\|^2 - \|\phi_2\|^2) d\mu(\phi_1, \phi_2) = \mathrm{Tr}\ \rho\sigma_3.$$

We remark that it is easy to pass to the complex representation, i.e., to integration over $\mathbf{C}^2$.

5.6.2 *Prequantum Gaussian Measures Corresponding to Pure States*

Let us now consider Gaussian measures on $\Omega = \mathbf{R}^4$ corresponding to pure quantum states. These are singular Gaussian measures which are concentrated on $SO_2(2)$-invariant planes in $\mathbf{R}^4$. To determine such a measure, we find its covariance operator.

Let $\Psi = u + iv$, $u = (u_1, u_2) \in \mathbf{R}^2$, $v = (v_1, v_2) \in \mathbf{R}^2$ be a pure quantum state and let $\rho_\Psi = \Psi \otimes \Psi$. Then $T^{-1}(\rho_\Psi) = \mu_\Psi$ (we choose $\alpha = 1$, i.e., the unit dispersion of the prequantum measure), where the Gaussian measure μ_Ψ has the real covariance operator

$$B_\Psi = \frac{1}{2} \begin{pmatrix} \|g_1\|^2 & (g_1, g_2) & 0 & (J g_1, g_2) \\ (g_1, g_2) & \|g_2\|^2 & (g_1, J g_2) & 0 \\ 0 & (g_1, J g_2) & \|g_1\|^2 & (g_1, g_2) \\ (J g_1, g_2) & 0 & (g_1, g_2) & \|g_2\|^2 \end{pmatrix}. \quad (5.58)$$

Here $g_1 = (u_1, v_1)$ and $g_2 = (u_2, v_2)$ are variables which are conjugate to $\phi_1 = (q_1, p_1)$ and $\phi_2 = (q_2, p_2)$.

We obtain representation (5.58). The real space realization of ρ_Ψ is given by the operator

$$j^{-1}(\rho_\Psi) = \begin{pmatrix} u\otimes u + v\otimes v & v\otimes u - u\otimes v \\ u\otimes v - v\otimes u & u\otimes u + v\otimes v \end{pmatrix}.$$

In the chosen system of coordinates on the phase space we have

$$u\otimes u = \begin{pmatrix} u_1^2 & u_1u_2 \\ u_1u_2 & u_2^2 \end{pmatrix},\ v\otimes v = \begin{pmatrix} v_1^2 & v_1v_2 \\ v_1v_2 & v_2^2 \end{pmatrix},$$

$$v\otimes u = \begin{pmatrix} u_1v_1 & u_2v_1 \\ u_1v_2 & u_2v_2 \end{pmatrix},\ u\otimes v = \begin{pmatrix} v_1u_1 & v_2u_1 \\ v_1u_2 & v_2u_2 \end{pmatrix}.$$

Hence,

$$u\otimes u + v\otimes v = \begin{pmatrix} u_1^2+v_1^2 & u_1u_2+v_1v_2 \\ u_1u_2+v_1v_2 & u_2^2+v_2^2 \end{pmatrix},$$

$$v\otimes u - u\otimes v = \begin{pmatrix} 0 & u_2v_1 - v_2u_1 \\ u_1v_2 - v_1u_2 & 0 \end{pmatrix}.$$

To illustrate the correspondence between the real and complex state spaces, we now show directly that, for B_Ψ given by (5.58), $2j(B_\Psi) = \Psi\otimes\Psi$. We have

$$2j(B_\Psi) = \begin{pmatrix} \|g_1\|^2 & (g_1, g_2) \\ (g_1, g_2) & \|g_2\|^2 \end{pmatrix} + i\begin{pmatrix} 0 & (Jg_1, g_2) \\ (g_1, Jg_2) & 0 \end{pmatrix}.$$

This operator acts on a complex vector $z = (z_1, z_2)$ in the following way:

$$z_1' \equiv (2j(B_\Psi)z)_1 = \|g_1\|^2 z_1 + [(g_1, g_2) + i(Jg_1, g_2)]z_2,$$
$$z_2' \equiv (2j(B_\Psi)z)_2 = [(g_1, g_2) + i(g_1, Jg_2)]z_1 + \|g_2\|^2 z_2.$$

On the other hand, $\Psi\otimes\Psi(z) = \langle z, \Psi\rangle\Psi = (z_1\overline{\Psi}_1 + z_2\overline{\Psi}_2)\Psi$. In Dirac's notation $\Psi\otimes\Psi(z) = \langle z|\Psi\rangle|\Psi\rangle$.

Here

$$z_1' = (u_1 - iv_1)(u_1 + iv_1)z_1 + (u_2 - iv_2)(u_1 + iv_1)z_2,$$
$$z_2' = (u_1 - iv_1)(u_2 + iv_2)z_1 + (u_2 - iv_2)(u_2 + iv_2)z_2.$$

Thus,

$$z_1' = (u_1^2 + v_1^2)z_1 + [(u_1u_2 + v_1v_2) + i(u_2v_1 - u_1v_2)]z_2,$$
$$z_2' = (u_2^2 + v_2^2)z_2 + [(u_1u_2 + v_1v_2) + i(u_1v_2 - u_2v_1)]z_2.$$

Let us consider the Gaussian measures corresponding to pure states for spin up and spin down, $|1\rangle = \begin{pmatrix} 1 \\ 0 \end{pmatrix}$ and $|0\rangle = \begin{pmatrix} 0 \\ 1 \end{pmatrix}$. For the vector $|1\rangle$, we have: $u_1 = 1$, $u_2 = v_1 = v_2 = 0$. Thus,

$$2B_{|1\rangle} = \begin{pmatrix} 1 & 0 & 0 & 0 \\ 0 & 0 & 0 & 0 \\ 0 & 0 & 1 & 0 \\ 0 & 0 & 0 & 0 \end{pmatrix}$$

and the Fourier transform of the measure $\mu_{|1\rangle}$ is given by

$$\tilde{\mu}_{|1\rangle}(\xi_1, \xi_2, \eta_1, \eta_2) = e^{-\frac{1}{4}(\xi_1^2+\eta_1^2)}.$$

This is the standard Gaussian measure on the plane $q_2 = 0,\ p_2 = 0$ having the density $\frac{1}{\pi}e^{-(q_1^2+p_1^2)}$.

In the same way $\mu_{|0\rangle}$ is the standard Gaussian measure on the plane $q_1 = 0,\ p_1 = 0$ having the density $\frac{1}{\pi}e^{-(q_2^2+p_2^2)}$. Let us now consider the Gaussian measure corresponding to the superposition of spin up and spin down states:

$$\Psi_\theta = \frac{1}{\sqrt{2}}(|0\rangle + e^{i\theta}|1\rangle).$$

Here $u_1 = \cos\theta,\ v_1 = \sin\theta$. Hence,

$$B_{\Psi_\theta} = \frac{1}{2}\begin{pmatrix} \cos^2\theta & \cos\theta\sin\theta & 0 & 0 \\ \cos\theta\sin\theta & \sin^2\theta & 0 & 0 \\ 0 & 0 & \cos^2\theta & \cos\theta\sin\theta \\ 0 & 0 & \cos\theta\sin\theta & \sin^2\theta \end{pmatrix}$$

and the Fourier transform of μ_{Ψ_θ} is given by

$$\tilde{\mu}_{\Psi_\theta}(\xi_1, \xi_2, \eta_1, \eta_2) = e^{-\frac{1}{4}[(\cos\theta\xi_1+\sin\theta\xi_2)^2+(\cos\theta\eta_1+\sin\theta\eta_2)^2]}.$$

Thus pure states Ψ_θ correspond to the standard Gaussian measures concentrated on planes obtained by rotations.

5.7 Prequantum Classical Statistical Field Theory

Let $Q = P = L_2(\mathbf{R}^3; \mathbf{R})$ be the Hilbert space of real-valued square-integrable functions $\phi : \mathbf{R}^3 \to \mathbf{R}$ with the scalar product

$$(\phi_1, \phi_2) = \int_{\mathbf{R}^3} \phi_1(x)\phi_2(x)dx.$$

Our classical phase space $H = L_2(\mathbf{R}^3; \mathbf{R}) \times L_2(\mathbf{R}^3; \mathbf{R})$ consists of vector functions $\phi(x) = \begin{pmatrix} q(x) \\ p(x) \end{pmatrix}$. The symplectic operator J on this phase space has the form

$$q_1(x) = p(x),\ p_1(x) = -q(x) \qquad (5.59)$$

and the symplectic form on H is defined by

$$w(\phi_1, \phi_2) = \int_{\mathbf{R}^3} (p_2(x)q_1(x) - p_1(x)q_2(x))dx.$$

The fundamental law of PCSFT is the $SO(2)$-invariance of physical variables.[a] By introducing the canonical complex structure on H, we obtain the space $H_c = L_2(\mathbf{R}^3)$, the complex Hilbert space of square-integrable functions $\phi : \mathbf{R}^3 \to \mathbf{C}$, $\phi = q(x) + ip(x))$ with the scalar product $\langle \phi_1, \phi_2 \rangle = \int_{\mathbf{R}}^3 \phi_1(x)\overline{\phi_2(x)}dx$. Let us consider an integral operator

$$\widehat{A} : H \to H, \ \widehat{A}\phi(x) = \int_{\mathbf{R}^3} A(x, y)\phi(y)dy.$$

The kernel $A(x, y)$ of such an operator has the block structure. This operator is J-invariant iff $A_{11}(x, y) = A_{22}(x, y)$, $A_{12}(x, y) = -A_{21}(x, y)$, and it is symmetric iff $A_{11}(x, y) = A_{11}(y, x)$, $A_{12}(y, x) = A_{21}(x, y) = -A_{12}(x, y)$. The corresponding quadratic form is given by

$$f(\phi) = \int A_{11}(x, y)\phi_1(x)\phi_2(y)dxdy$$
$$+2\int A_{12}(x, y)\phi_2(x)\phi_1(y) + \int A_{11}(x, y)\phi_2(x)\phi_2(y)dxdy.$$

Let μ be a J-invariant measure on $H = L_2(\mathbf{R}^3; \mathbf{R}) \times L_2(\mathbf{R}^3; \mathbf{R})$. (For a Gaussian measure, this is equivalent to the $SO(2)$-invariance.) Its complex covariance is defined as

$$\langle D_\mu \phi_1, \phi_2 \rangle = \int_{L_2(\mathbf{R}^3;\mathbf{R}) \times L_2(\mathbf{R}^3;\mathbf{R})} \left(\int_{\mathbf{R}^3} \phi_1(x)\overline{\phi(x)}dx \int_{\mathbf{R}^3} \phi(x)\overline{\phi_2(x)}dx \right) d\mu(\phi).$$

[a] Primarily, PCSFT was created to reproduce probabilistic predictions of QM. (Therefore it was called "prequantum.") However, by considering nonquadratic functionals of the prequantum field we moved beyond QM. It is not clear whether $SO(2)$-invariance is important for the "beyond quantum model" as well. We may try to proceed by eliminating this constraint (see Section 5.13). In this case we have to consider even quadratic forms which do not have the QM counterpart. On the other hand, as it will be shown in Section 5.13.1, any quadratic form on the phase space induces (in a natural way, at least from the mathematical viewpoint) the $SO(2)$-invariant quadratic form. (Any **R**-linear operator on the phase space induces the corresponding **C**-linear operator, see (5.93).) It is not clear whether the map (5.93) is just a mathematical trick or it has a physical realization, i.e., can be physically realized as an indirect measurement of the quadratic variable which is not $SO(2)$-invariant.

Let μ have the unit dispersion

$$\sigma^2(\mu) = \int_{L_2(\mathbf{R}^3;\mathbf{R})\times L_2(\mathbf{R}^3;\mathbf{R})} \left(\int_{\mathbf{R}^3} |\phi(x)|^2 dx\right) d\mu(\phi) = 1.$$

We find the average of the quadratic physical variable f in the statistical state μ:

$$\begin{aligned}\langle f\rangle_\mu &= \int_{L_2(\mathbf{R}^3;\mathbf{R})\times L_2(\mathbf{R}^3;\mathbf{R})} \Big(\int A_{11}(x,y)q(x)q(y)dxdy \\ &\quad + 2\int A_{12}(x,y)p(x)q(y)dxdy \\ &\quad + \int A_{11}(x,y)p(x)p(y)dxdy\Big) d\mu_B(q,p) \\ &= \mathrm{Tr}\rho\widehat{A},\end{aligned}$$

where $\rho = D_\mu$.

Let us forget for a moment about mathematical difficulties and consider a singular integral operator, the differential operator

$$\widehat{H} = -\frac{\Delta}{2m} + V(x)$$

In the phase space H we consider the diagonal operator $\widehat{H}_{11} = \widehat{H}_{22} = \widehat{H}$, $\widehat{H}_{12} = \widehat{H}_{21} = 0$ or we can directly consider $\widehat{H}$ as acting in the complex Hilbert space $H_c = L_2(\mathbf{R}^3)$. Consider the corresponding quadratic functional

$$\phi \mapsto \langle \widehat{H}\phi, \phi\rangle = \int_{\mathbf{R}^3} \left(\frac{|\nabla\phi(x)|^2}{2m} + V(x)|\phi(x)|^2\right) dx.$$

This is the classical energy of a system with the infinite number of degrees of freedom, the system of coupled fields $q(x)$ and $p(x)$. This system is a classical vector field; the parameter m (mass) is one of the characteristics of this field.

Remark 6.1. In nonrelativistic QM all energy quadratic forms have the diagonal form: $f_H(q,p) = \langle \widehat{H}\phi, \phi\rangle = (\widehat{H}q, q) + (\widehat{H}p, p)$, where $\widehat{H} = \widehat{H}^*$ and $\phi = q + ip$. This is a rather special class of quadratic $SO(2)$-invariant functionals. In general, we can expect functionals $f(q,p) = \langle(\widehat{H} + i\widehat{R})\phi, \phi\rangle = (\widehat{H}q, q) + 2(\widehat{R}q, p) + (\widehat{H}p, p)$, where $\widehat{H} = \widehat{H}^*$, $\widehat{R} = -\widehat{R}^*$, cf. Chapter 3.

We can find the average of the prequantum field energy:

$$\int_{L_2(\mathbf{R}^3;\mathbf{R})\times L_2(\mathbf{R}^3;\mathbf{R})} \langle \widehat{H}\phi, \phi\rangle d\mu(\phi)$$

$$= \int_{L_2(\mathbf{R}^3;\mathbf{R})\times L_2(\mathbf{R}^3;\mathbf{R})} \left(\int_{\mathbf{R}^3} (\frac{|\nabla\phi(x)|^2}{2m} + V(x)|\phi(x)|^2)dx \right) d\mu(\phi)$$

$$= \text{Tr } D_\mu \left(\frac{-\Delta}{2m} + V(x)\right) = \text{Tr } \rho\widehat{H}.$$

Since the operator $\widehat{H}$ is unbounded, the Tr $\rho\widehat{H}$ is not well defined for an arbitrary Gaussian measure.

5.7.1 *Interpretation of Mass*

We interpret m as a parameter, *mass,* determining a scalar-complex field (or a pair of self-inducing real fields); the term $-\frac{\Delta}{2m}$ describes cross-inducing of the field components encoded in the system of Hamilton equations. In the complex representation (with the Schrödinger equation) this cross-inducing is not visible very much. In the Hamiltonian formalism, we see that components of a prequantum field induce each other in the same way as components of the electromagnetic field described by the Maxwell equation. This is a special system of Hamiltonian equations on an infinite-dimensional phase space. Thus, depending on the mass parameter, we obtain a family of different prequantum fields corresponding to elementary particles: the electron field, proton field, and neutron field. Each of these fields can be represented as a classical Gaussian random signal with the covariance operator determined by the quantum state of the corresponding particle. This interpretation is close to that given by QFT: each quantum particle is an excitation of a special quantum field. In QFT there are as many different types of quantum fields as there are types of quantum particles: a quantum field for electron, for proton, for neutron, ... The "only" distinguishing feature of the PCSFT approach is that all fields are classical. Different fields interact with each other and external potentials (both in QFT and PCSFT).

PCSFT does not guarantee that the quantum observable of energy $\widehat{H}$ really corresponds to a quadratic classical variable of energy $\phi \mapsto \langle \widehat{H}\phi, \phi\rangle$. For nonquadratic functions, the prequantum$\mapsto$quantum map is not one-to-one. We illustrate this feature of the model by an example. Departure from quadratic functionals implies the asymptotic coupling between classical (prequantum) and quantum averages. Consider a measure with the dispersion $\sigma^2(\mu) = \alpha$, where α is a small parameter of the model, the magnitude of fluctuations of the prequantum random field.

Take, e.g., a classical energy variable of the form

$$\mathcal{F}(\phi) = \int_{\mathbf{R}^3} \left(\frac{|\nabla\phi(x)|^2}{2m} + V(x)|\phi(x)|^2 \right) dx + g \int_{\mathbf{R}^3} |\phi(x)|^4 dx, \ g > 0$$

(which is J-invariant). Then it produces approximately the same quantum average as the quadratic energy variable

$$\frac{\langle \mathcal{F} \rangle_\mu}{\alpha} = \mathrm{Tr}\, \rho \widehat{H} + \alpha g \int_{L_2(\mathbf{R}^3;\mathbf{R}) \times L_2(\mathbf{R}^3;\mathbf{R})} \left(\int_{\mathbf{R}^3} |\phi(x)|^4 dx \right) d\mu(\phi)$$

Of course, such nonquadratic Hamilton function induces the nonlinear Hamiltonian dynamics in the infinite-dimensional phase space H, and, in principle, it could be distinguished from the standard linear dynamics corresponding to quadratic Hamilton functions.

We now consider the quadratic classical variables inducing quantum observables of the position $\mathbf{x}_j$ and the momentum $\mathbf{p}_j (j = 1, 2, 3)$:

$$f_{\mathbf{x}_j}(\phi) = \langle \mathbf{x}_j \phi, \phi \rangle = \int_{\mathbf{R}^3} x_j |\phi(x)|^2 dx.$$

$$f_{\mathbf{p}_j}(\phi) = \langle \mathbf{p}_j \phi, \phi \rangle = \int_{\mathbf{R}^3} y_j |\tilde{\phi}(y)|^2 dy,$$

where $\tilde{\phi}(y)$ is the Fourier transform of the L_2-function $\phi(x)$. We can also consider the quadratic classical variables inducing the angular momentum operators, e.g.,

$$f_{\mathbf{J}_z}(\phi) = \langle \mathbf{J}_z \phi, \phi \rangle = -i \int_{\mathbf{R}^3} \left(x \frac{\partial \phi}{\partial y} - y \frac{\partial \phi}{\partial x} \right) \bar{\phi}\, dxdydz$$

We remark that, however, the same quantum operator can also be induced, e.g., by the classical variable

$$f_{\mathbf{J}_z}(\phi) = \langle \mathbf{J}_z \phi, \phi \rangle + \langle \mathbf{J}_z \phi, \phi \rangle^2.$$

5.8 PCSFT Formalism for Classical Electromagnetic Field-2

The PCSFT formalism provides not only a possibility to reproduce the conventional QM starting with "prequantum classical fields," but also to create an approximative probabilistic description of any classical field theory. We have already started to create such a description for the classical (macroscopic) electromagnetic field (Section 3.5). We now develop this viewpoint in a more general framework: in Chapter 3 and, in particular, in Section 3.5, we considered quadratic functionals of classical fields. Later we generalized PCSFT to include nonquadratic field functionals. In this section we consider such functionals for the model of Section 3.5. We also represent the system of Maxwell equations (in vacuum) as the system of the Hamiltonian equations (with the L_2-phase space) and then by using the general scheme connecting the Hamiltonian equations with the Schrödinger equation (see Chapter 4) we represent Maxwell equations as an equation of Schrödinger type.

Let us consider statistical ensembles of electromagnetic fields[a] $\phi(x) = (E(x), B(x))$. Here $x \in \mathbf{R}^3$ and $E(x) = (E_1(x), E_2(x), E_3(x))$ and $B(x) = (B_1(x), B_2(x), B_3(x))$ are electric and magnetic components of the field. "Physical variables" are functionals $f(\phi) \equiv f(E, B)$ of the electromagnetic field; for example, the energy variable is given by the quadratic form (Section 3.5)

$$\mathcal{E}(E, B) = \int_{\mathbf{R}^3} (E^2(x) + B^2(x))dx.$$

Statistical ensembles of fields (random fields) are mathematically described by probability measures on the space of fields which is denoted by Ω. In the previous consideration we were fine with choosing Ω as the L_2-space of scalar or vector valued fields on $\mathbf{R}^3$. However, the coordinates of the electric field are constrained, the same is valid for the coordinates of the magnetic field. They have to satisfy the system of differential equations

$$\nabla E(x) = 0, \ \nabla B(x) = 0. \tag{5.60}$$

---

[a]Such ensembles represent random fields in the set-theoretic model, the Kolmogorov's model. As was stated many times, on the basis of the ergodicity hypothesis we can switch from the set-theoretical model dealing with ensembles to the time series model dealing with trajectories (in the L_2-space).

According to the general ideology of construction of functional spaces which is used in theory of partial differential equations, these constraints have to be incorporated into the space structure. Consider the space of smooth quickly decreasing vector functions satisfying this system of constraints. Then take the norm induced from the L_2-space and complete our space with respect to the L_2-norm. The resulting functional space is denoted by Ω. This is the state space of the classical electromagnetic field. Our aim is to present a procedure of approximative calculation of averages with respect to such random fields which would mathematically coincide with the standard quantum formalism with the state space Ω. We state again that, since Ω has a natural phase space structure generated by duality between electric and magnetic fields, we can introduce the canonical complex structure on it and proceed with a complex Hilbert state space. However, for the classical electromagnetic field this is just a special mathematical representation. In any event, there is nothing physically special in using the Riemann–Silberstein representation. Therefore we prefer to proceed in the real representation.

We continue the presentation of the theory of classical random electromagnetic fields, i.e., signals. The average with respect to an ensemble of fields represented by a measure μ is given by

$$\langle f\rangle_\mu = \int_\Omega f(E, B)d\mu(E, B).$$

For example, the average of the field energy variable $\mathcal{E}(E, B)$ equals

$$\langle \mathcal{E}\rangle_\mu = \int_\Omega \Big(\int_{\mathbf{R}^3} (E^2(x) + B^2(x))dx\Big)d\mu(E, B).$$

As was pointed out in Section 3.1.5, under the ergodicity assumption ensemble averages can be represented as time averages; for example, the energy average can be found as

$$\langle f\rangle_\mu \equiv \lim_{T\to\infty} \frac{1}{T}\int_0^T \Big(\int_{\mathbf{R}^3} (E^2(s, x) + B^2(s, x))dx\Big)ds.$$

The regime $T \to \infty$ corresponds to the time scale of measurement of classical electromagnetic signals under consideration. This is a classical analog of the time scale of quantum measurements. We also use the symbol t to denote this scale. It is assumed that the t-scale is essentially coarser than the time scale of classical field dynamics. The latter is based on the time parameter denoted by s.

For general functionals $f(E, B)$, calculation of infinite-dimensional integrals and asymptotic time averages are very complicated problems. Therefore some approximative methods for calculation may be useful. (Here we forget about QM at all! We are interested in a purely computational problem for averages with respect to classical random electromagnetic signals.) In such approximative approaches the crucial role is played by the choice of a small parameter. The essence of our story is that we consider ensembles of fields which have *small dispersion* with respect to the mean field value $\bar{\phi} = (\bar{E}, \bar{B})$

$$\sigma^2(\mu) = \int \left(\int_{\mathbf{R}^3} [(E(x) - \bar{E}(x))^2 + (B(x) - \bar{B}(x))^2] dx \right) d\mu(E, B) = \alpha. \tag{5.61}$$

Here α is the small parameter of the model under consideration. We remark that the dispersion of a random field can be represented as the average of its centered energy $\sigma^2(\mu) = \langle \mathcal{E}(E - \bar{E}, B - \bar{B}) \rangle_\mu$. Thus this energy average is a small parameter. In our model (for approximate computing of averages with respect to fluctuations of macroscopic, i.e., having macroscopic energy, electromagnetic field) quantities of the magnitude $O(\alpha^2)$ are considered as negligibly small. We have a technique of expansion of integrals over the L_2-space with respect to the parameter α (see Section 5.7). By neglecting terms of the magnitude $O(\alpha^2)$, $\alpha \to 0$, we obtain the approximation of averages with respect to fluctuations of the classical electromagnetic random field. We know that such an approximation coincides with the trace formula for quantum averages. In the latent form the quantum-like statistical structure is present in the theory of classical random electromagnetic waves (and any theory of the wave type). This structure can be used in a purely pragmatic way: to simplify the calculation of complicated averages. It may happen that this approach to the approximate calculation of averages of classical random signals can be used in radiophysics. (We cannot exclude the possibility that it has already been used, cf. Section 5.1.)

Let a random field have the average

$$\int E d\mu(E, B) = \bar{E} \text{ and } \int B d\mu(E, B) = \bar{B} \tag{5.62}$$

In reality these averages are typically calculated as the time averages

$$\bar{E} = \lim_{T\to\infty} \frac{1}{T}\int_0^T E(s,x)ds, \ \bar{B} = \lim_{T\to\infty} \frac{1}{T}\int_0^T B(s,x)ds. \quad (5.63)$$

Here the s-time is the ordinary physical time of the propagation of a signal and T is the time window of signal averaging. One of the reasons for using a time window is practical impossibility to monitor a signal on the s-time scale. Another reason is just an attempt to save monitoring resources and to work with signal averages, instead of the complete signal measured at each instance of time s.

We assume that fluctuations of the macroscopic electromagnetic field are *symmetric with respect to the mean value field* $\bar{\phi} = (\bar{E}, \bar{B})$ and that $\sigma^2(\mu) = \alpha$. We also need a mathematical integration restriction which guarantees the existence of all integrals in coming calculations

$$\int_\Omega \exp\{r\sqrt{\mathcal{E}(E,B)}\}d\mu(E,B) < \infty, \quad (5.64)$$

for any $r > 0$. Examples are provided by Gaussian measures on Ω. However, we do not restrict our consideration to Gaussian random signals.

In the infinite-dimensional case the easiest way is to define a measure by its Fourier transform:

$$\tilde{\mu}(\xi,\eta) = \int_\Omega \exp\{i\int_{\mathbf{R}^3}[(\xi(x), E(x)) + (\eta(x), B(x))]dx\}d\mu(E,B).$$

For a Gaussian measure its Fourier transform is the quadratic exponent

$$\tilde{\mu}(u) = \exp\{-\frac{1}{2}(Cu, u) + i(u, \bar{\phi})\}.$$

Here the quadratic form in the exponent is *the correlation functional* of μ and $u = (\xi, \eta) \in \Omega$, $\bar{\phi} = (\bar{E}, \bar{B})$ is the mean value of the Gaussian random field. The operator $C : \Omega \to \Omega$ is the covariance operator of the Gaussian measure μ. Formally the Gaussian measure μ can be written as

$$d\mu(\phi) = \exp\{-\frac{1}{2}(C^{-1}(\phi - \bar{\phi}), \phi - \bar{\phi})\}d\phi,$$

where $\phi = (E, B)$, $\bar{\phi} = (\bar{E}, \bar{B})$, and $d\phi$ is the "normalized volume of the field space." However, $d\phi$ has no mathematical meaning by itself. For an arbitrary measure μ such that $\langle \mathcal{E} \rangle_\mu < \infty$, its correlation functional is defined as

$$(Cu, u) = \int_\Omega (u, \phi - \bar{\phi})^2 d\mu(\phi),$$

or

$$(Cu, u) = \int_\Omega \left(\int_{\mathbf{R}^3} [(\xi(x), E(x) - \bar{E}(x)) + (\eta(x), B(x) - \bar{B}(x))] dx \right)^2 d\mu(E, B).$$

By using distribution kernels we can write this quadratic form as

$$(Cu, u) = \int_{\mathbf{R}^3} [C_{11}^{ij}(x, y)\xi_i(x)\xi_j(x) + 2C_{12}^{ij}(x, y)\xi_i(x)\eta_j(y) + C_{22}^{ij}(x, y)\eta_i(x)\eta_j(y)] dx dy,$$

where C_{nm} are blocks of the covariance matrix C.

Let us now consider some quadratic form of the electromagnetic field: $f_A(E, B) \equiv (\widehat{A}\phi, \phi) = \int_{\mathbf{R}^6} [A_{11}^{ij}(x, y)E_i(x)E_j(y) + 2A_{12}^{ij}E_i(x)B_j(y) + A_{22}^{ij}(x, y)B_i(x)B_j(y)] dx dy$. We know that, for quadratic forms, the following equality holds:

$$\int f_A(E - \bar{E}, B - \bar{B}) d\mu(E, B) = \mathrm{Tr}\, C\widehat{A},$$

where C is the covariance operator of μ.

We consider the following class of field functionals. Each $f(E, B)$ is an analytic functional of fields and it has the exponential growth, $|f(E, B)| \le ce^{r\sqrt{\mathcal{E}(E, B)}}$ for some $c, r > 0$ and all E and B.

In principle, to reproduce the complete analog of the QM formalism, we have to operate only with $SO(2)$-invariant field functionals, $f(E, B) = f(E', B')$, where the $SO(2)$-group acts on the electromagnetic field as

$$E' = \cos\theta E + \sin\theta B, \;\; B' = -\sin\theta E + \cos\theta B. \tag{5.65}$$

However, at the moment I do not have any idea whether the transformation (5.65) of the classical electromagnetic field plays any significant role in the theory of classical signals. Therefore I prefer to proceed without the constraint

$$f(E, B) = f(E', B'). \tag{5.66}$$

I also do not know whether the $SO(2)$-invariance of probability distributions is important in the theory of classical random signals. Therefore I prefer to proceed without $SO(2)$-invariance constraint for probability distributions.

5.9 Asymptotic Expansion of Averages with Respect to Electromagnetic Random Field

Let f and μ belong to the classes of field functionals and measures, respectively, which were described in Section 5.8. In the integral giving the average of f we make the change of variables

$$E \to \bar{E} + \sqrt{\alpha}E, \quad B \to \bar{B} + \sqrt{\alpha}B. \tag{5.67}$$

We have

$$\langle f \rangle_\mu = \int_\Omega f(\bar{E} + \sqrt{\alpha}E, \bar{B} + \sqrt{\alpha}B) d\mu_{\text{norm}}(E, B), \tag{5.68}$$

where μ_{norm} is the image of μ under the field-coordinate transformation (5.67). We remark that the measure μ_{norm} has zero mean value. Its covariance operator $\rho = C/\alpha$. Thus $\text{Tr}\rho = 1$. Since $C^* = C$ and $C \geq 0$ for any covariance operator, the operator ρ has all the properties of a density operator. We emphasize that we study a purely classical system, the classical electromagnetic field!

As usual in this chapter, we expand a function $f(\phi)$ in the integral with respect to the probability distribution of a classical random signal into the Taylor series. We set $\phi = (E, B)$ and $\bar{\phi} = (\bar{E}, \bar{B})$

$$\begin{aligned}\langle f \rangle_\mu = \int_\Omega (f(\bar{\phi}) &+ \sqrt{\alpha}(f'(\bar{\phi}), \phi) + \frac{\alpha}{2}(f''(\bar{\phi})\phi, \phi) \\ &+ \frac{\alpha^{3/2}}{3!} f^{(3)}(\bar{\phi})(\phi, \phi, \phi) \\ &+ \frac{\alpha^2}{4!} f^{(4)}(\bar{\phi})(\phi, \ldots, \phi) + \ldots) d\mu_{\text{norm}}(\phi).\end{aligned}$$

Since μ was chosen symmetric with respect to its mean value, the μ_{norm} is symmetric (so it is not changed under the transformation, $(E, B) \to (-\bar{E}, -\bar{B})$.) Thus all odd integrals are equal to zero. We obtain

$$\langle f \rangle_\mu = f(\bar{E}, \bar{B}) + \frac{\alpha}{2} \text{Tr } \rho \ f''(\bar{E}, \bar{B}) + O(\alpha^2), \alpha \to 0. \tag{5.69}$$

5.10 Interpretation

The expansion (5.69) induces a hierarchy of models describing approximately averages with respect to a classical random signal.

5.10.1 *Model with the* $[\alpha = 0]$*-Approximation*

First assume that we are able to measure only averages $\bar{E}(x)$, $\bar{B}(x)$ and not fields $E(x, \omega)$, $B(x, \omega)$. Here ω is the random parameter.

To simulate such a situation, consider, e.g., two time scales:

(a) t is the time scale of measurements,
(b) s is a finer time scale.[a]

Fluctuations at the s-scale induce a noisy electromagnetic signal. In the present model such "fine scale fluctuations" are not observable or they are simply eliminated by averaging the signal on the s-scale. Only averages $\bar{E}(t, x)$ and $\bar{B}(t, x)$ are considered as physical quantities.

In such situation we completely neglect contributions to averages having the α-magnitude. Roughly speaking, we set equal to zero the dispersion of a signal. The asymptotic equality (5.112) for averages is reduced to a simple rule, cf. Ref. [290] and Section 5.1,

$$\langle f \rangle_\mu \approx f(\bar{E}, \bar{B}). \tag{5.70}$$

Instead of averages of functionals of the random electromagnetic field, we can consider just functionals of the averages $(\bar{E}, \bar{B})$. These are the only physical variables on this level of approximation.

We now increase the precision of averaging of functionals of classical signals.

5.10.2 *Model with the* α*-Approximation*

Suppose now that we approached a higher precision of measurements and the first order of random fluctuations can also be taken into account. We consider the dispersion α of the field-fluctuations as a small, but not negligibly small parameter. We do not restrict the probabilistic description of random fields to the average variables

[a]In the case of conventional QM (and not a quantum-like description of the classical electromagnetic field presented in this section) this fine time scale is unapproachable with the aid of existing measurement devices. In the present quantum-like model this scale is either unapproachable or the use of the coarser time scale t can simply save resources for monitoring of signals. In the latter case we may, in principle, monitor a signal on the s-scale, but we are fine with a coarser description.

$(\bar{E}, \bar{B})$. At the same time we consider a situation such that the quantities of the magnitude $O(\alpha^2)$ are still not approached (or we need not such a high precision in our statistical analysis of random signals). In this situation the approximation of the form

$$\langle f \rangle_\mu \approx f(\bar{E}, \bar{B}) + \frac{\alpha}{2} \mathrm{Tr} \rho f''(\bar{E}, \bar{B})$$

matches the experimental statistical data. If we restrict our consideration to field functionals such that $f(\bar{E}, \bar{B}) = 0$, then we obtain precisely the quantum formula for averages, $\langle f \rangle_\mu \approx \frac{\alpha}{2} \mathrm{Tr} \rho f''(\bar{E}, \bar{B})$.

As we have seen, it is possible to construct a quantum-like representation of our purely classical field model. We start with scaling of all classical variables by amplification of them: $f_\alpha(E, B) = \frac{f(E,B)}{\alpha}$. This scaling is useful in the situation, where a classical signal is too weak. To be detected, functionals of this signal, e.g., its energy, have to be amplified with the aid of normalization by the dispersion of the signal. In the present quantum-like model for the classical electromagnetic field the dispersion coincides with the average of the signal energy. We rewrite the basic equality for approximate calculation of averages as

$$\langle f_\alpha \rangle_\mu \approx \frac{1}{2} \mathrm{Tr} \rho f''(\bar{E}, \bar{B}). \tag{5.71}$$

The amplified functional $f_\alpha(E, B)$ (classical physical variable, e.g., amplified field energy) is mapped into the second derivative of f at the mean field point $\bar{\phi} = (\bar{E}, \bar{E})$:

$$f_\alpha \to \hat{A} \equiv T(f) = \frac{1}{2} f''(\bar{E}, \bar{B}). \tag{5.72}$$

We note that the second derivative is always a symmetric operator. Here α is the scaling factor of the quantum-like approximative representation of the classical random electromagnetic field model. The operator $\hat{A}$ is considered as the quantum-like variable representing the classical variable f_α.

A measure μ (a random field $\phi(x, \omega) = (E(x, \omega), B(x, \omega))$ is represented by the scaling of its covariance operator $\mu \to \rho \equiv T(\mu) = \frac{C}{\alpha}$. Thus the classical average $\langle f_\alpha \rangle_\mu$ can be approximated by the quantum-like average $\langle T(f) \rangle_{T(\mu)} \equiv \mathrm{Tr} \rho \hat{A}$ up to terms of the magnitude $O(\alpha)$.

Of course, by increasing the precision of the monitoring of a signal we shall violate the quantum-like von Neumann law. The standard quantum approximation will become not useful. To obtain a better (but still approximative) description, the derivatives of orders higher than 2 have to be taken into account (see Section 5.14).

5.11 Simulation of Quantum-Like Behavior for the Classical Electromagnetic Field

Our model provides a good possibility to describe the quantum probabilistic behavior in classical field-systems. Let us consider a classical random field having a small dispersion of energy fluctuations, α. Our aim was to show that the quantum-like probabilistic description arises as a consequence of neglecting small energies of the magnitude $O(\alpha)$.

We assume that $\bar{E} = 0$ and $\bar{B} = 0$. We collect experimental data which are measured with the precision such that energies of the order α^2 are neglected. We find the experimental average $\langle f_\alpha \rangle_{\text{exp}}$ of the amplification $f_\alpha(E, B)$ of some functional of $f(E, B)$. Then we shall check whether the quantum-like approximation works well, $\langle f_\alpha \rangle_{\text{exp}} \approx \text{Tr}\rho\hat{A}$, where $\hat{A} = T(f_\alpha)$ and $\rho = T(\mu)$.

To find the deviations from the quantum-like approximation, we have to consider polynomials of the classical field having at least fourth degree. For example, let us consider the square-energy functional

$$f(E, B) = \mathcal{E}^2(E, B) = \Big(\int_{\mathbf{R}^3} (E^2(x) + B^2(x))dx\Big)^2 \tag{5.73}$$

and its amplification

$$f_\alpha(E, B) = \frac{1}{\alpha}\mathcal{E}^2(E, B) = \frac{1}{\alpha}\Big(\int_{\mathbf{R}^3} (E^2(x) + B^2(x))dx\Big)^2 \tag{5.74}$$

For the form (5.73), the classical average

$$\langle \mathcal{E}^2 \rangle_\mu \neq 0$$

(in fact, it is of the magnitude $O(\alpha^2)$), but the quantum-like average given by the trace formula is equal to zero.

By approaching a higher precision in the measurement of field fluctuations one will see the recovering of the classical statistical behavior. However, if a higher precision could not be approached, one might present a *conjecture of the violation of classical statistical laws.* This would mimic the present situation in QM.

We remark again that before approaching the α-precision for the energy of fluctuations, we can operate with averages $\bar{E}, \bar{B}$ (we note that they satisfy the Maxwell equation, see Section 5.12) and consider the statistical mechanics of classical fields which are represented by their averages as deterministic mechanics. Instead of classical averages $\langle f_\alpha \rangle_\mu$ or quantum-like averages, one can operate with physical variables of classical average-fields, namely, with $f(\bar{E}, \bar{B})$.

By approaching the α-precision for the energy of fluctuations we shall recover the quantum-like averages. At this stage we can use the quantum-like model in which variables are represented by symmetric operators and random fields by density operators. The classical statistical structure would not be completely recovered yet.

Even by approaching the α^2-precision for the energy of fluctuations we would not completely recover the classical statistical structure.

We have seen that it is possible to proceed on different levels of approximate calculation of averages of classical signals. In Ref. [207] this ideology was applied to a model functioning of the brain based on processing of classical electromagnetic signals.

5.12 Maxwell Equations as Hamilton Equations or as Schrödinger Equations

We consider the Maxwell equations

$$\frac{1}{c}\frac{\partial E}{\partial t}(t, x) = \nabla \times B(t, x), \tag{5.75}$$

$$\frac{1}{c}\frac{\delta B}{\delta t}(t, x) = -\nabla \times E(t, x), \tag{5.76}$$

in empty space. This system of dynamical equations has to be constrained by system (5.60). As already mentioned, the standard functional analysis way to do this is to incorporate these constraints into a proper functional space. We have already constructed such a space Ω. Thus we consider the system of Maxwell dynamical equations in the functional space Ω.

We shall work with the vector field $\phi(t, x) = (E(t, x), B(t, x)) \in \Omega$ for any t. The simplest random field evolution is produced by choosing random initial conditions:

$$E(t_0, x, \omega) = E_0(x, \omega),\ B(t_0, x, \omega) = B_0(x, \omega), \tag{5.77}$$

where $\phi_0(x, \omega) = (E_0(x, \omega), B_0(x, \omega))$ has the probability distribution μ.

Since equations (5.75) and (5.76) are linear, the mean field satisfies the same system of equations

$$\frac{1}{c}\frac{\partial \bar{E}}{\partial t}(t, x) = \nabla \times \bar{B}(t, x),\ \frac{1}{c}\frac{\partial \bar{B}}{\partial t}(t, x) = -\nabla \times \bar{E}(t, x), \tag{5.78}$$

$$\bar{E}(t_0, x) = \bar{E}_0(x),\ \bar{B}(t_0, x) = \bar{B}_0(x). \tag{5.79}$$

Thus we can consistently work in the model A which is based on the $[\alpha = 0]$-approximation, i.e., we can forget about fluctuations of random electromagnetic signals and work with their averages.

We recall that the Maxwell equations are infinite-dimensional Hamiltonian equations. They can be represented in the form

$$\frac{\partial E}{\partial t} = \frac{\delta \mathcal{H}(E, B)}{\delta B} \tag{5.80}$$

$$\frac{\partial B}{\partial t} = -\frac{\delta \mathcal{H}(E, B)}{\delta E}, \tag{5.81}$$

where the Hamilton function has the form

$$\mathcal{H}(E, B) = \frac{c}{2}\int_{\mathbf{R}^3} [(\nabla \times E, E) + (\nabla \times B, B)]dx. \tag{5.82}$$

By using the symplectic operator $J = \begin{pmatrix} 0 & I \\ -I & 0 \end{pmatrix}$ in Ω, so that $J(E, B) = (B, -E)$, we can write any Hamiltonian system as

$$\frac{\partial}{\partial t}(E, B) = J\,\nabla \mathcal{H}(E, B). \tag{5.83}$$

(Here (E, B) denotes the vector field with electric and magnetic vector components.) In our case, since $\mathcal{H}(E, B)$ is quadratic, its gradient is a linear operator, namely,

$$L = c\nabla \times . \tag{5.84}$$

Finally, the Maxwell equation can be written as

$$\frac{\partial \phi}{\partial t} = J\, L\phi, \phi = (E, B). \tag{5.85}$$

We now introduce the complex representation of the classical electromagnetic field $\phi(t, x) = E(t, x) + i B(t, x)$. Since $J^2 = -1$, we obtain the Schrödinger-type equation

$$i \frac{\partial \phi}{\partial t} = L\phi \tag{5.86}$$

for the vector-valued "wave function" ϕ. Therefore the classical random electromagnetic field is the solution of the Schrödinger equation (5.86) with the random initial condition $\phi_0(x, \omega) = (E_0(x, \omega), B_0(x, \omega))$. This is the quantum-like representation of the classical stochastic electrodynamics. We repeat shortly our story about hierarchies of approximation of averages.

The coarsest approximation was generated by neglecting the quantities of the magnitude α. In this model we use the classical Maxwell dynamics (5.78) and (5.79) for averages. These averages can be considered as "classical electromagnetic fields."

Then by taking into account the dispersion of fluctuations α we obtain the quantum-like dynamics given by Schrödinger equation with random initial conditions.

5.13 Quadratic Variables without Quantum Counterpart

5.13.1 *Examples*

As we have seen, there can be established a one-to-one correspondence between quantum observables, self-adjoint bounded operators[a], and prequantum variables of the form $f_A(\phi) = \langle \widehat{A}\phi, \phi \rangle$, where $\widehat{A} \in \mathcal{L}_s(H_c)$ and H_c is the complex representation of the phase space $H = Q \times P$, i.e., $H_c = Q \oplus i P$. In the phase space formalism these are quadratic forms $f_A(\phi) = (\widehat{A}\phi, \phi)$, where $\widehat{A} \in \mathcal{L}_{sp,s}(H)$.

Take now $\widehat{A} \in \mathcal{L}_s(H) \setminus \mathcal{L}_{sp,s}(H)$, i.e., $[\widehat{A}, J] \neq 0$. Such operator cannot be represented as a **C**-linear operator in H_c and hence it has no counterpart in QM. However, the corresponding quadratic form of the prequantum field $f_A(\phi)$ is well defined.

[a]We proceed with bounded operators (except Chapter 6).

For example, consider the two-dimensional phase space, i.e., $Q = P = \mathbf{R}$, $H = \mathbf{R}^2$. Here $H_c = \mathbf{C}$ and all "quantum observables" are simply real multipliers in $\mathbf{C}$. Take now

$$A = \frac{1}{2}\begin{pmatrix} 0 & 1 \\ 1 & 0 \end{pmatrix}. \tag{5.87}$$

Then $[A, J] \neq 0$. Nevertheless, we can consider the quadratic form $f_A(\phi) \equiv f_A(q, p) = qp$. This is *a prequantum variable without a quantum counterpart.*

This example can be generalized to the multidimensional case

$$f_{ij}(\phi) \equiv f_{ij}(q, p) = q_i p_j \tag{5.88}$$

or even to the infinite-dimensional case. In the latter case we work with modes of the classical prequantum field $\phi(x) = (q(x), p(x))$ corresponding to an orthonormal basis in $L_2(\mathbf{R}^3; \mathbf{R})$, say $\{e_j(x)\}$. Set

$$q_j = \int_{\mathbf{R}^3} q(x)e_j(x)dx,\ p_j = \int_{\mathbf{R}^3} p_j(x)e_j(e)dx. \tag{5.89}$$

Then $f_{ij}(\phi)$ is a prequantum variable, a functional of the classical field $\phi(x)$, which has no counterpart in QM. Take now two functions $u, v \in L_2(\mathbf{R}; \mathbf{R})$ and set

$$f_{u,v}(\phi) = (q, u)(p, v) = \int_{\mathbf{R}^3} q(x)u(x)dx \int_{\mathbf{R}^3} p(x)v(x)dx.$$

We remark that a quadratic form $G(\phi, \phi)$ corresponds to the operator

$$\widehat{G} \in \mathcal{L}_{\mathrm{sp,s}}(H) \text{ if } G(J\phi, J\phi) = G(\phi, \phi). \tag{5.90}$$

Indeed, if $\widehat{G} \in \mathcal{L}_{sp,s}(H)$, then $G(J\phi, J\phi) = (\widehat{G}J\phi, J\phi) = (J\widehat{G}\phi, J\phi) = (J^\star J\widehat{G}\phi, \phi) = (\widehat{G}\phi, \phi) = G(\phi, \phi)$.

However, we see that in our example $f_{u,v}(J\phi) = f_{u,v}(p, -q) = -f_{u,v}(\phi)$.

Consider now the classical electromagnetic field $\phi(x) = (E(x), B(x))$, where $E = (E_1, E_2, E_3)$, $B = (B_1, B_2, B_3)$. In Sections 3.5 and 5.8–5.11 we presented the quantum-like model for a random electromagnetic fields. Prequantum variables were given by quadratic functionals $f_A(E, B)$, where $[\widehat{A}, J] = 0$ and J was a symplectic operator in the (E, B)-phase space.

$$J\begin{pmatrix} E \\ B \end{pmatrix} = \begin{pmatrix} B \\ -E \end{pmatrix}.$$

For such quadratic forms, we have $f_A(E, B) = f_A(B, -E)$. However, for example, the form $f_{x,y,i,j}(E, B) = E_i(x)B_j(y)$, $x, y \in \mathbf{R}^3$, $i, j = 1, 2, 3$, does not belong to this class and neither does the form

$$f(E, B) = \int_{\mathbf{R}^3} E(x)B(x)dx. \tag{5.91}$$

It is not clear why only J-invariant forms of the prequantum field lead to quantum observables. Consider now an electron field $\phi(x)$, i.e., a prequantum field corresponding to an electron states of QM. In the complex representation a quadratic form $G(\phi, \bar{\phi})$ corresponds to a quantum observable if it is invariant with respect to the gauge transformations

$$\phi \to e^{i\theta}\phi \tag{5.92}$$

i.e., $G(e^{i\theta}\phi, e^{-i\theta}\bar{\phi}) = G(\phi, \bar{\phi})$.

Thus, from the PCSFT viewpoint, QM describes only $U(1)$-gauge invariant[a] quadratic forms of the classical prequantum field. Why does it happen? My conjecture is that in the QM formalism a quantum state is determined up to the phase. Roughly speaking, this elegant feature of the theory, the invariance with respect to the $U(1)$-gauge group, is a consequence of the impossibility to describe (and measure) precisely field components and, in particular, phases.

PCSFT describes the field components. It seems that the condition of $U(1)$-invariance (J-commutativity for operators) might be omitted. Any functional of the form $f_A(\phi) = (\widehat{A}\phi, \phi)$, $\widehat{A} \in \mathcal{L}_s(H)$, has the physical meaning. We hope that experimenters will be able to measure such functionals in the future.

5.13.2 *A Mapping from **R**-Linear to **C**-Linear Operators*

Take an arbitrary operator $\widehat{A} \in \mathcal{L}_s(H)$, $H = Q \times P$, i.e.,

$$\widehat{A} = \begin{pmatrix} \widehat{A}_{11} & \widehat{A}_{12} \\ \widehat{A}_{21} & \widehat{A}_{22} \end{pmatrix},$$

---

[a]This invariance was studied in detail in Sections 5.3.4, 5.4, where the real representation of the Hilbert state space as a phase space was used; instead of the group $U(1)$, we considered the real group $SO(2)$ and its representations in phase spaces of dimensions $m = 1, 2, \ldots, \infty$, the groups $SO_m(2)$.

where $\widehat{A}_{11}^{\star} = \widehat{A}_{ij}$, $\widehat{A}_{12}^{\star} = \widehat{A}_{21}$. Consider the following mapping

$$\widehat{A} \to \widehat{A}_c = \frac{1}{2}\begin{pmatrix} \widehat{A}_{11} + \widehat{A}_{22} & \widehat{A}_{12} - \widehat{A}_{21} \\ \widehat{A}_{21} - \widehat{A}_{12} & \widehat{A}_{11} + \widehat{A}_{22} \end{pmatrix}. \tag{5.93}$$

Then $[\widehat{A}_c, J] = 0$, i.e., $\widehat{A}_c \in \mathcal{L}_{\mathrm{sp,s}}(H)$ and it can be represented as an element of $\mathcal{L}_s(H_{\mathrm{c}})$, i.e., it can be mapped into a prequantum observable. At the moment it is too early to say whether correspondence (5.93) has some physical meaning or it is simply a mathematical trick.

Take, e.g., the prequantum variable f_A corresponding to the **R**-linear operator (5.87). Then its quantum image given by map (5.93) is a zero operator. Thus it may happen that a nontrivial prequantum variable is mapped into the trivial quantum observable $\widehat{A}_c = 0$.

5.13.3 *On Extension of the Space of Prequantum Observables*

The space $\mathcal{V}(H|SO_m(2))$ consists of real-valued functions. This selection was done to match the formulas of QM. In the latter all observables are real valued. However, if we go beyond QM, cf. Ref. [250], and considers linear functionals of the prequantum field (see Section 3.4) then, to find correlations, we have to consider variables of the form

$$f_{u,v}(\phi) = \langle \phi, u \rangle \langle u, \phi \rangle, \tag{5.94}$$

where $u, v \in H$. Such variables are complex-valued. To incorporate these variables in our model (for a Hilbert space H of the dimension $m = 1, 2, \ldots, \infty$), we extend the functional space $\mathcal{V}(H|SO_m(2))$ and introduce a space of complex-valued functions. The only difference from $\mathcal{V}(H|SO_m(2))$ is that functions from the latter space are **C** valued. To make the theory consistent, we switch from the phase space description with $H = Q \times P$ to the complex Hilbert space description with $H_{\mathrm{c}} = Q \oplus iP$. The new functional space is denoted by the symbol $\mathcal{V}(H_{\mathrm{c}}|U_m(1))$, where elements of the group $U_m(1)$ are represented as

$$\phi \to e^{i\theta}\phi.$$

Thus, for any $f \in \mathcal{V}(H_{\mathrm{c}}|U_m(1))$,

$$f(e^{i\theta}\phi) = f(\phi) \tag{5.95}$$

Consider now $f_{u,v}(\phi)$, see (5.94). Then

$$f_{u,v}(e^{i\theta}\phi) = \langle e^{i\theta}\phi, u\rangle\langle v, e^{i\theta\phi}\rangle = \langle\phi, u\rangle\langle v, \phi\rangle = f_{u,v}(\phi).$$

To obtain the asymptotic expansion of functionals of the classical prequantum field, we need not use the U_1-gauge invariance of $f(\phi)$ (see Section 5.2). Thus we can consider a prequantum classical statistical model in which physical variable can violate (5.95). However, this model has variables which have no counterpart in the conventional quantum formalism. We remark that, in principle, there is no reason to assume that QM contains the counterparts of all possible prequantum physical variables. However, we cannot exclude a possibility that for some class of experiments, the $U(1)$-gauge invariance may play a crucial role.

Since for any J-commuting covariance operator B (in the real representation), the off-diagonal block B_{12} is skew-symmetric, we have

$$E(p_j q_i - q_j p_i) = 0, \tag{5.96}$$

where $Ep_j q_i = (B_{12}e_{pj}, e_{qi})$ and e_{qi} and e_{pj} are orthonormal bases in Q and P, respectively. Thus, for a prequantum random field, its q-modes commute in average with the p-modes. For $i = j$, we get

$$Ep_i q_i = 0. \tag{5.97}$$

Since $Ep_i = 0$, $Eq_i = 0$, this is equivalent to the equality

$$\text{cov}\,(p_i q_i) = 0. \tag{5.98}$$

Thus the coordinate and momentum modes q_i and p_i are uncorrelated. In the Gaussian case they are independent. Correlations inside a prequantum random field are very special. The physical interpretation of correlations (5.96) and (5.97) needs further elaboration.

Example 12.1. Consider now the two-dimensional configuration space, $Q = \mathbf{R}^2$, i.e., $H = Q \times P = \mathbf{R}^4; q = (q_1, q_2)$, $p = (p_1, p_2)$. Then any prequantum Gaussian probability distribution has the form (up to the normalization constant)

$$\exp\left\{-\frac{1}{2}\left[\frac{q_1^2 + p_1^2}{k_1^2} + 2\frac{q_1 p_2 - q_2 p_1}{\mu} + \frac{q_2^2 + p_2^2}{k_2^2}\right]\right\}.$$

This density can be factorized in the product of two densities:

$$g(q_1, p_2) = \exp\left\{-\frac{1}{2}\left[\frac{q_1^2}{k_1^2} + 2\frac{q_1 p_2}{\mu} + \frac{p_2^2}{k_2^2}\right]\right\}$$

and

$$w(q_2, p_1) = \exp\left\{-\frac{1}{2}\left[\frac{q_2^2}{k_2^2} - \frac{2q_2 p_1}{\mu} + \frac{p_1^2}{k_1^2}\right]\right\}.$$

Thus the coordinate q_1 and the momentum p_2 are independent of the coordinate q_2 and the momentum p_1.

What is the physical meaning of this probabilistic feature of the prequantum random (Gaussian) field?

5.14 Generalization of Quantum Mechanics

In this section we proceed by using only the real Hilbert space H, generalization to the complex case is evident, but it will make the presentation essentially more technical. We do not want to shadow the main idea of this section with technical details. The main idea is that *standard linear QM has natural multilinear generalizations* with observables given by multilinear forms. All these extensions can be considered as approximations of PCSFT of different orders.

We have created the classical statistical model inducing the quantum statistical model. The quantum description is the result of neglecting the terms of the magnitude $O(\alpha), \alpha \to 0$, in the expansion of normalized prequantum averages, $\frac{\langle f \rangle_\mu}{\alpha}$, with respect to the small parameter

$$\gamma = \alpha^{1/2}, \alpha \to 0.$$

Consider a classical statistical state which is given by a Gaussian measure μ with zero mean value, $a_\mu = 0$. In the expansion of the average $\langle f \rangle_\mu$ into a power series with respect to the parameter γ terms of the magnitude γ are absent. This expansion starts directly with $\gamma^2 = \alpha$.

Thus standard QM arises as an approximation of a special version of classical statistical mechanics on the infinite-dimensional phase space, namely, statistical mechanics of classical fields. Let us expand the classical average with respect to the small parameter and normalize it by α. Then the main term in the expansion of the normalized average $\langle f\rangle_\mu/\alpha$ is given by QM. This is the standard trace-average. This viewpoint on conventional QM implies the evident possibility to generalize this (conventional quantum) formalism by considering higher order expansions of averages $\langle f\rangle_\mu$ with respect to the small parameter $\gamma = \alpha^{1/2}$.

We recall that the momenta of a measure μ are defined by

$$a_\mu^{(k)}(z_1, \ldots, z_k) = \int_H (z_1, x)\cdots(z_k, x)d\mu(x).$$

For a Gaussian measure μ, $a_\mu = 0$ implies that all its momenta of odd orders $a_\mu^{(k)}$, $k = 2n+1$, $n = 0, 1, \ldots,$ are also equal to zero. Therefore the expansion of $\langle f\rangle_\mu$ with respect to γ does not contain terms with γ^{2n+1}. Hence this is the expansion with respect to $\alpha^n(= \gamma^{2n})$, $n = 1, 2, \ldots$

We are able to create generalizations of QM corresponding to $O(\alpha^n)$-approximations of PCSFT by neglecting the terms of the magnitude $O(\alpha^n)$, $\alpha \to 0 (n = 1, 2, \ldots)$ in the power expansion of the normalized classical average.

Since in the future consideration we shall use the expansion of $f(\phi)$ into Taylor's series and work with an arbitrary number of its terms (before we were fine with just two first terms), we prefer to operate with analytic functions, although it is possible to proceed even with smooth functions under some restrictions on the growth of derivatives. We recall that a function on the complex Hilbert space is called analytic if it can be expanded into Taylor's series

$$f(\phi) = f(0) + f'(0)(\phi) + \frac{1}{2}f''(0)(\phi, \phi) + \cdots + \frac{1}{n!}f^{(n)}(0)(\phi, \ldots, \phi) + \cdots,$$

which converges uniformly on each ball, see Refs. [128, 130, 138] for details.

As the space of classical physical variables, we choose the functional space $\mathcal{V}_{\mathrm{an}}(H)$ consisting of functions $f : H \to \mathbf{R}$ which can be extended to analytic functions on the complexification of H having the exponential growth

$$\text{there exist } C_0, C_1 \geq 0 : |f(\phi)| \leq C_0 e^{C_1\|\phi\|}, \tag{5.99}$$

and preserving the state of vacuum, $f(0) = 0$.

Let us consider the classical statistical model

$$M_{\rm cl} = (S_{\rm G}^{\alpha}(H), \mathcal{V}_{\rm an}(H)), \qquad (5.100)$$

where H is the real Hilbert space; $S_{\rm G}^{\alpha}(H)$ consists of Gaussian measures with zero mean value and dispersion α. The latter is a small parameter of the model.[a]

By taking into account that the moments of odd order $a_{\mu}^{2n+1} = 0$, $n = 0, 1, \ldots$, for $\mu \in S_{\rm G}^{\alpha}(\Omega)$, we obtain the equality

$$\langle f \rangle_{\mu} = \frac{\alpha}{2} \mathrm{Tr}\ \rho f''(0) + \sum_{k=2}^{\infty} \frac{\alpha^k}{(2k)!} \int_H f^{(2k)}(0)(y, \ldots, y) d\mu_{\rm scal}(y), \qquad (5.101)$$

where, as always, $\rho = B_{\mu_{\rm scal}} = B_{\mu}/\alpha$. Here the measure $\mu_{\rm scal}$ is obtained as $\alpha^{1/2}$-scaling of μ; so it has the unit dispersion.

We now proceed toward a new statistical model, which is a natural generalization of conventional QM. This model (in fact, a family of models) will provide an approximative description of averages with a better approximation than standard QM. In the future experimenters may first approach these generalized quantum models on the way to PCSFT.

We start with some preliminary mathematical considerations. Let A and B be two n-linear symmetric forms. We define the following generalized trace by

$$\mathrm{Tr}\ BA = \sum_{j_1, \ldots, j_n = 1}^{\infty} B(e_{j_1}, \ldots, e_{j_h}) A(e_{j_1}, \ldots, e_{j_n}). \qquad (5.102)$$

If this series converges and its sum does not depend on the choice of an orthonormal basis $\{e_j\}$ in H. We remark that

$$\frac{\langle f \rangle_{\mu}}{\alpha} = \frac{1}{2} \mathrm{Tr}\ \rho f''(0) + \sum_{k=2}^{n} \frac{\alpha^{k-1}}{2k!} \mathrm{Tr}\ a_{\mu_{\rm scal}}^{(2k)} f^{(2k)}(0) + O(\alpha^n),\ \alpha \to 0. \qquad (5.103)$$

Here we used the following result about Gaussian integrals:

Lemma 13.1. *Let A_k be a continuous k-linear form on H and let μ be a Gaussian measure with zero mean value. Then*

$$\int_H A_k(x, \ldots, x) d\mu(x) = \mathrm{Tr}\ a_{\mu}^{(k)} A_k. \qquad (5.104)$$

[a] In Sections 5.3 and 5.4 we used the spaces $S_{\rm G}^{\approx\alpha}$. Operation with random fields with dispersions which are only approximately equal to α was just a minor technicality. In this section we cannot proceed with the space $S_{\rm G}^{\approx\alpha}$,

Proof. Let $\{e_j\}_{j=1}^{\infty}$ be an orthonormal basis in H. We apply the well-known Lebesgue dominated convergence theorem. We introduce a series of functionals of the Hilbert space variable x

$$f_N(x) = \sum_{j_1,\ldots,j_k=1}^{n} A_k(e_{j_1},\ldots,e_{j_k})(e_{j_1},x)\ldots(e_{j_k},x). \qquad (5.105)$$

They are bounded by an integrable function

$$|f_N(x)| = |A_k(\sum_{j_1=1}^{N}(x,e_{j_1})e_{j_1}\ldots,\sum_{j_k=1}^{N}(x,e_{j_k})e_{j_k})| \le ||A_k||\;||x||^k. \qquad (5.106)$$

Therefore it is possible to apply the limit procedure to integrals on H

$$\int_H A_k(x,\ldots,x)d\mu(x) = \lim_{N\to\infty}\int_H f_N(x)d\mu(x)$$

$$= \sum_{j_1=1,\ldots,j_k=1}^{\infty} A_k(e_{j_1},\ldots,e_{j_k})\int_H (e_{j_1},x)\ldots(e_{j_k},x)d\mu(x) = \mathrm{Tr}\; a_\mu^{(k)}A_k. \qquad (5.107)$$

Hence, the proof is complete.

In particular, the following inequality was derived

$$|\mathrm{Tr}\; a_\mu^k A_k| \le ||A||\int_H ||x||^k d\mu(x). \qquad (5.108)$$

We state again that for a Gaussian measure (with zero mean value) integrals (5.104) are equal to zero for $k = 2l+1$. Thus $\mathrm{Tr}\; a_\mu^{(2l+1)}A_{2l+1} = 0$. It is easy to see that $2k$-linear forms (momenta of even orders) a_μ^{2k} can be expressed in terms of the covariance operator B

$$a_\mu^{(2k)} = e(k,B) = \frac{d^{2k}}{dy^{2k}}e^{-\frac{1}{2}(By,y)}|_{y=0}. \qquad (5.109)$$

In particular, $e(2,D)(z_1,z_2) = (Bz_1,z_2)$ and

$$e(4,B)(z_1,z_2,z_3,z_4) = (Bz_1,z_3)(Bz_2,z_4) + (Bz_2,z_3)(Bz_1,z_4)$$
$$+(Bz_1,z_2)(Bz_3,z_4).$$

Thus (5.103) can be rewritten as

$$\frac{\langle f\rangle_\mu}{\alpha} = \frac{1}{2}\mathrm{Tr}\;\rho f''(0) + \sum_{k=2}^{n}\frac{\alpha^{k-1}}{2k!}\mathrm{Tr}\; e(2k,\rho)f^{(2k)}(0) + O(\alpha^n),\; \alpha\to 0. \qquad (5.110)$$

This formula is the basis of *a new quantum theory* which appears as an approximation of PCSFT.[a] In this theory statistical states can still be represented by von Neumann's density operators $\rho \in \mathcal{D}(H)$, but *observables are represented by vectors* $A = (A_2, A_4, \ldots, A_{2n})$, where A_{2j} are symmetric $2j$-linear forms on the Hilbert space H. In particular, the quadratic form A_2 can be represented by a self-adjoint operator. To escape mathematical difficulties, we can assume that forms A_{2j} are continuous.

Denote the space of all such vectors A by $L_{2n,s}(H)$. We obtain the following generalization of the conventional quantum model

$$M_{\mathrm{q},2n} = (\mathcal{D}(H), L_{2n,s}(H)). \tag{5.111}$$

Here the average of an observable $A \in L_{2n,s}(H)$ with respect to a state $\rho \in \mathcal{D}(H)$ is given by

$$\langle a \rangle_\rho = \sum_{n=1}^{n} \mathrm{Tr}\, e(2k, \rho) A_{2k} \tag{5.112}$$

If one defines

$$\mathrm{Tr}\, \rho A = \sum_{k=1}^{n} \mathrm{Tr}\, e(2k, \rho) A_{2k},$$

then formula (5.112) can be written in conventional quantum mechanics (von Neumann's trace formula of nth order) as

$$\langle A \rangle_\rho = \mathrm{Tr} \rho A \tag{5.113}$$

This model is the result of the following "quantization" procedure of the classical statistical model $M_{\mathrm{cl}} = (S_{\mathrm{G}}^{\alpha}(H), \mathcal{V}_{\mathrm{an}}(H))$

$$\mu \to \rho = \frac{B_\mu}{\alpha}; \tag{5.114}$$

$$f \to A = \left(\frac{1}{2} f''(0), \frac{\alpha}{4!} f^{(4)}(0), \ldots, \frac{\alpha^{n-1}}{(2n)!} f^{(2n)}(0)\right). \tag{5.115}$$

(thus here $A_{2k} = \frac{\alpha^{k-1}}{(2k)!} f^{(2k)}(0)$). The transformation T_{2n} given by (5.114) and (5.115) maps the classical statistical model $M_{\mathrm{cl}} = (S_{\mathrm{G}}^{\alpha}(H), \mathcal{V}_{\mathrm{an}}(H))$ onto the generalized quantum model $M_{\mathrm{q},2n} = (\mathcal{D}(H), L_{2n,s}(H))$.

[a] First we created PCSFT which has QM as the first-order approximation. Then on the basis of PCSFT we created new quantum-like models. PCSFT has them as its approximations of different orders.

Theorem 13.1. *For the classical statistical model $M_{cl} = (S_G^\alpha(H), \mathcal{V}_{an}(H))$, the classical$\mapsto$quantum map T_{2n} defined by (5.114) and (5.115) is one-to-one for statistical states; it still has a huge degeneration for variables. Classical (normalized) and quantum-like averages are equal to* mod $O(\alpha^n)$, $\alpha \to 0$, *see (5.110).*

Consider, for example, the approximation of PCSFT by $M_{q,4} = (\mathcal{D}(H), L_{4,s}(H))$. Here

$$\frac{\langle f \rangle_\mu}{\alpha} = \text{Tr } \rho A_2 + \alpha \text{Tr } \rho A_4 + O(\alpha^2), \ \alpha \to 0. \tag{5.116}$$

For this generalized quantum model $M_{q,4} = (\mathcal{D}(H), L_{4,s}(H))$, the conventional QM also plays a role of approximation

$$\langle A \rangle_\rho = \langle A_2 \rangle_\rho + \alpha \langle A_4 \rangle_\rho,$$

for a generalized quantum observable $A = (A_2, A_4)$.

How can we find the form A_4? PCSFT cannot tell us anything; the conventional quantization procedure (giving the correspondence between quantities of classical mechanics of particles and quantum operators) produces only A_2.

We remark that the degree of degeneration of the map $T_{2n} : \mathcal{V}_{\text{an}}(H) \to L_{2n,s}(H)$ is decreasing for $n \to \infty$. Denote the space of polynomials of degree $2n$ containing only terms of even degrees by the symbol $P_{2n}(H)$. Thus $f \in P_{2n}(H)$ iff

$$f(x) = Q_2(x, x) + Q_4(x, x, x, x) + \cdots + Q_{2n}(x, \ldots, x),$$

where $Q_{2j} : H^{2j} \to \mathbf{R}$ is a symmetric $2j$-linear (continuous) form. The restriction of the map T_{2n} to the subspace $P_{2n}(H)$ of the space $\mathcal{V}_{\text{an}}(H)$ is one-to-one.

One can also consider a quantum model

$$M_{q,\infty} = (\mathcal{D}(H), L_{\infty,s}(H)), \tag{5.117}$$

where $L_{\infty,s}(H)$ consists of infinite sequences of $2n$-linear (continuous) symmetric forms on H

$$A = (A_2, \ldots, A_{2n}, \ldots). \tag{5.118}$$

As a simple consequence of expansion (5.110) of the classical average for a variable $f \in \mathcal{V}_{\text{an}}(H)$, we obtain

Theorem 13.2. *Averages given by the classical statistical model M_{cl} and the quantum-like model $M_{q,\infty}$ coincide.*

5.15 Coupling between the Time Scale and Dispersion of a Prequantum Random Signal

5.15.1 *Prequantum Wiener Process: Brownian Motion of Waves*

Consider the Hilbert phase space $H = Q \times P$. Let $w_s^D, s \geq 0$, be the H-valued Wiener process[a] corresponding to the trace-class (self-adjoint) operator $D \geq 0$ with $\mathrm{Tr}D = 1$. As usual, we interpret s as the time parameter. Trajectories $s \to w_s^D \in H$ of the Wiener process describe stochastic dynamics in the Hilbert space. The ordinary Wiener process taking values in $\mathbf{R}^3$ describes the Brownian motion. As we know, Einstein and Smoluchowski observed that according to the kinetic theory of fluids, the molecules of water move at random. Therefore, a small particle in water would receive a random number of impacts of random strength and from random directions in any short period of time. This random bombardment by the molecules of the fluid would make a sufficiently small particle move in the way described by Brown.

In our model the role of a particle is played by a classical field $\phi = (q, p)$. It is assumed that this field moves in a random background field. Hence a trajectory $s \to w_s^D$ can be interpreted as the result of "collisions" of the field ϕ with the random background field. The time scale s under consideration is the *prequantum time scale.* As usual, the *quantum time scale* is the time scale of quantum observations.[b] We also assume that the covariance operator D determining the Wiener process belongs to the class $\mathcal{L}_{\mathrm{sp}}(H)$. We have for any instance of prequantum time s,

$$E(\phi, w_s^D) = 0, \phi \in H, \tag{5.119}$$

$$E(\phi_1, w_s^D)(w_s^D, \phi_2) = s(D\phi_1, \phi_2), \phi_1, \phi_2 \in H. \tag{5.120}$$

[a] The reader who is not familiar with the theory of infinite-dimensional stochastic processes can either study some literature, e.g., Refs. [58, 124, 126–128, 279], or just consider the finite-dimensional phase space $H_{2n} = \mathbf{R}^{2n}$.

[b] We state again that our meaning of the "quantum time scale" has to be distinguished from a rather common use of the terminology "quantum time scale" for the Planck time scale. We shall consider the Planck time scale as one of possible candidates for the prequantum time scale.

The following time scaling law for the Wiener process is well known (see, e.g., Ref. [277]):

$$\text{Prob. law } (w^D_{\alpha s} : s \geq 0) = \text{Prob. law } (\alpha^{1/2} w^D_s : s \geq 0) \qquad (5.121)$$

for any $\alpha > 0$. Thus scaling of time $s \to t = \alpha s$ is equivalent to scaling of dispersion, multiplication of the latter by the factor α. We shall see that by (5.121) our $\alpha^{1/2}$-scaling (5.43) of $\phi \in H$ can be considered as the result of α-scaling of time.

We state again the basic postulate: the *quantum formalism arises as the result of the approximation based on time scaling.*

Let us consider a *prequantum time scale* which is essentially finer[a] than the quantum time scale. We suppose that these two time scales can be coupled by a scaling parameter α. As always, we denote the prequantum and quantum times by symbols s and t, respectively. We suppose that

$$t = \alpha s. \qquad (5.122)$$

Here α is a *dimensionless parameter.* It is assumed that

$$\alpha \ll 1. \qquad (5.123)$$

Thus the *unit interval of prequantum time* corresponds to the interval $t = \alpha$ of quantum time. We can also say that the unit interval $t = 1$ of quantum time corresponds to a huge interval $s = \frac{1}{\alpha}$ of prequantum time. Moreover, if $\alpha \to 0$, then $s = \frac{1}{\alpha} \to \infty$. At the prequantum time scale quantum processes have practically infinite duration. In this model a quantum measurement has a huge duration with respect to the prequantum time scale. In particular, in this scale the "collapse of the wave function" is a very long process. Let us consider the time scaling (5.122) for the Wiener processes w^D_s. We set

$$W^D_t = w^D_{\alpha s}.$$

Formula (5.121) implies that for any continuous function $f : H \to \mathbf{R}$ (which is integrable with respect to any Gaussian measure on H), we have

$$E f(W^D_\alpha) = E f(\alpha^{1/2} w^D_1). \qquad (5.124)$$

[a]The meaning of "essentially" will be discussed later.

We interpret W_t^D as the Wiener process with respect to quantum time t and w_s^D as the Wiener process with respect to prequantum time s. We can use either the quantum time scale and work with a random field having small dispersion α (as we did in previous sections) or the prequantum time scale and work with a random field having dispersion normalized by 1.

In our interpretation, the quantum formalism does not provide a possibility to find exactly the average $Ef(W_t^D)$ with respect to the Wiener process W_t^D. The main problem is that the interval $t = \alpha$ is negligible comparing to natural quantum scales of time. The quantum formalism provides only an approximation of the classical average $Ef(W_\alpha^D)$.

We can proceed by expanding the right-hand side of (5.124) into the Taylor series with respect to the scaling parameter $\alpha^{1/2}$ (cf. Sections 5.1 and, 5.3). We obtain the following result for the asymptotic expansion. For a sufficiently smooth function f (with some growth restrictions on the derivatives, cf. Section 5.3), we have

$$\frac{Ef(W_\alpha^D)}{\alpha} = \frac{Ef(\alpha^{1/2} w_1^D)}{\alpha} = \frac{1}{2} E(f''(0) w_1^D, w_1^D) + O(\alpha), \alpha \to 0. \tag{5.125}$$

Thus (as we have already seen in the previous sections) for nonquadratic maps $f : H \to \mathbf{R}$, the quantum statistical model gives only an approximation $< f >_D = \text{Tr}\, Df''(0)$ of the normalized classical average. The difference between statistical predictions of the quantum statistical model and PCSFT is of the magnitude α, where α is the scaling parameter for the prequantum and quantum time scales, see Eq. (5.122). What is the magnitude of the time scaling factor α?

5.15.2 *Speculation about the Prequantum Time Scale*

To estimate the basic parameter of our model α, we have to choose the quantum and prequantum time scales. There are a few different possibilities and we shall discuss one of them. We may choose the *atom time* scale in QM and the *Planck time* scale in the prequantum classical theory. We recall that Max Planck first listed his set of units and gave values for them remarkably close to those used today at a time [261] when QM had not yet been invented. He had not yet discovered the theory of black body radiation.

We assume that the prequantum time scale is based on the Planck time

$$t_{\text{prq}} = t_{\text{P}} = \sqrt{\frac{\hbar G}{c^5}} \approx 5.391 \times 10^{-44} s. \tag{5.126}$$

We remark that the Planck time t_{P} is expressed as

$$t_{\text{P}} = \frac{\hbar}{c^2 M_P}, \tag{5.127}$$

where the Planck mass is given by

$$m_P = \sqrt{\frac{\hbar c}{G}} \approx 2.176 \times 10^{-8} \text{ kg}.$$

We state again that, in contrast to the Planck time and length, the Planck mass is a *macroscopic quantity.* It is often considered as a pathological feature of the Planck system of units. We shall see that the mass scale of PCSFT has to be macroscopic. This implies that the difference between statistical predictions of quantum and prequantum models is very small.

To obtain the atom time scale, we choose the electron mass scale, $m_e \approx 9.109 \times 10^{-31}$ kg. The corresponding time scale can be set in the same way as it was done by Max Planck:

$$t_{\text{q}} = t_e = \frac{\hbar}{c^2 m_e} = 1.288 \times 10^{-21}. \tag{5.128}$$

Therefore our time scaling parameter

$$\alpha = \frac{t_{\text{prq}}}{t_q} = \frac{t_{\text{P}}}{t_e} =\approx 4.185 \times 10^{-23}. \tag{5.129}$$

We also remark that

$$\alpha = \frac{m_e}{m_P}. \tag{5.130}$$

Thus our time scaling parameter has the magnitude

$$\alpha \sim 10^{-23}.$$

For such a choice of the prequantum scale, the difference between statistical predictions of PCSFT and the quantum statistical model (given by (5.125)) is of order 10^{-23}. Of course, the previous discussion has sense only under the assumption that the Planck time

t_P really provides the correct prequantum time scale. As was already mentioned, at the moment we cannot justify this assumption. The main problem induced by such a choice is that there is a huge gap between the atomic and Planck scales. Another natural scale between the Planck and atomic scales might be found. In such a case α has to be larger. This would simplify the experimental verification of PCSFT. On the other hand, the use of the Planck scale (and hence the macroscopic mass scale) and the corresponding parameter $\alpha \equiv \alpha_e \sim 10^{-23}$ clarify why predictions of the quantum statistical model have not yet been violated, the deviation is really negligibly small. One of the reasons in favor of the Planck scale as the scale of prequantum fluctuations is that the Planck mass has a macroscopic magnitude. Let a system have the mass m. Then by choosing the corresponding time scale

$$t_m = \frac{\hbar}{c^2 m},$$

we obtain

$$\alpha = \frac{m}{m_P}.$$

Therefore, predictions of the quantum statistical model have to be violated for systems of macroscopic mass. In principle, one may expect that it would be easier to produce deviations from QM for heavy elementary particles, e.g., muons. Let us take $m = m_{\text{muon}}$. The corresponding time scale

$$\alpha_{\text{muon}} = \alpha_e \frac{m_{\text{muon}}}{m_e} \approx 207\alpha_e.$$

Thus statistical deviations for muons are essentially larger than for electrons, but they are still very small

$$\alpha_{\text{muon}}\alpha_e \sim 10^{-21}.$$

On the other hand, decreasing of the mass increases strongly the precision of quantum approximation. For electron neutrino and antineutrino,

$$\alpha_{\text{e-neutrino}} = \alpha_e \frac{m_{\text{e-neutrino}}}{m_e} < 4.31 \times 10^{-7}\alpha_e \sim 10^{-30}.$$

It is impossible to interpolate the previous consideration based on the mass scaling directly to photons. The direct interpolation would give us

$$\alpha_{\text{photon}} = 0.$$

It would imply that *the quantum statistical model is precise for photons.* However, as was already mentioned, such an interpolation is too straightforward.

Chapter 6

Supplementary Mathematical Considerations

This chapter contains rather abstract mathematical considerations. One may jump to the next chapter taking into account the main consequences of the present chapter:

(a) Results of Chapter 5 can be generalized to nonquadratic Hamiltonian functions, but it is impossible to proceed with Gaussian random fields.
(b) The background field (vacuum fluctuations) can be mathematically represented as spatial white noise, a random variable valued in the space of Schwartz distributions. Thus vacuum fluctuations are in general extremely singular.
(c) Prequantum random fields corresponding to quantum (pure and mixed) states can be obtained from the background field by using linear filters. Thus they can be considered as signals sent by the background field. In a sense, everything is done by sending the background field though linear filters (see Chapter 8 for details).
(d) By using the formalism of the rigged Hilbert space PCSFT can be easily generalized to cover unbounded quantum observables; in particular, the position and momentum observables.

Beyond Quantum
Andrei Khrennikov

ISBN 978-981-4411-73-8 (Hardcover), 978-981-4411-74-5 (eBook)
www.panstanford.com

6.1 Dispersion Preserving Dynamics with Nonquadratic Hamilton Functions

By considering nonquadratic classical variables we confront with a new interesting problem: the investigation of dynamics with nonquadratic Hamilton functions. Let us consider an arbitrary Hamilton function $\mathcal{H} : H \to \mathbf{R}$. The first important remark is that such dynamics transfers Gaussian states into Gaussian iff $\mathcal{H}$ is quadratic.

Suppose that for any $\phi \in H$, the system of Hamiltonian equations (4.2) has a unique solution, $\phi(t) \equiv U_t\phi$, $\phi(0) = \phi$. In this case there is a well-defined map (Hamiltonian flow)

$$U_t : H \to H. \tag{6.1}$$

This map induces the map U_t^* in the space of probability measures $PM(H)$ on the phase space H (see Section 4.5). As already mentioned, in the nonquadratic case the measure $U_t^*\mu, t > 0$, can be non-Gaussian even for a Gaussian measure μ. For nonquadratic Hamilton functions, we cannot restrict the classical statistical model to the model with Gaussian states. We should consider the space of statistical states consisting of all probability measures μ on H that have zero mean value and the dispersion α. Denote this class by the symbol $PM^\alpha(H)$. In principle, we have to consider the subclass $PM^\alpha_{\rm sp}(H)$ of $PM^\alpha(H)$ consisting of J-invariant measures: $J^*\mu = \mu$. But at the moment we consider arbitrary measures.[a] The appearance of non-Gaussian random fields is not surprising. In Section 3.12 we motivated the usage of Gaussian fields by a limit theorem for probability distributions corresponding to dense trains of wave pulses on the prequantum time scale. The latter is very fine comparing to the "quantum time scale," the scale of the present measurement technology.

We are interested in Hamiltonian dynamics U_t in the phase space H that induces dynamics U_t^* in $PM^\alpha(H)$. Such a dynamics preserves zero mean value and the dispersion α.

The quantum dynamics corresponding to the classical Hamiltonian dynamics with a quadratic J-invariant Hamilton functions

[a]As we have seen, for Gaussian measures, the J-invariance is equivalent to $SO(2)$-invariance. For an arbitrary μ, this is not the case.

is an example of dynamics preserving zero mean value and the dispersion. We are interested in more general dynamics with similar features. We find the dispersion of $U_t^*\mu$ for an arbitrary $\mu \in PM(H)$ having zero mean value

$$\sigma^2(U_t^*\mu) = \int_H ||U_t\phi||^2 d\mu(\phi). \tag{6.2}$$

We are interested in a Hamiltonian dynamics such that dispersions of probability measures are preserved, i.e., *dispersion preserving dynamics.*

Suppose, *U_t preserves the mean value of a measure.* By (6.2), if U_t preserves the norm on the phase space H, then U_t^* preserves the dispersion. We remark that a nonlinear norm preserving map $U : H \to H$ need not be one-to-one or surjection. Moreover, it need not be an isometry: the equality $\|U\phi\| = \|\phi\|$ for any $\phi \in H$ does not imply that $\|U\phi_1 - U\phi_2\| = \|\phi_1 - \phi_2\|$.

It is easy to find the necessary and sufficient condition for norm preservation of dynamics induced by a Hamilton function $\mathcal{H}(\phi)$. We can write the general Hamiltonian equation (4.2) in the form

$$\dot{\phi} = J\mathcal{H}'(\phi). \tag{6.3}$$

Theorem 1.1. *Let the flow U_t induced by the Hamilton function $\mathcal{H}(\phi)$ be a surjection, i.e., $U_t(H) = H$. Then it is norm preserving iff the following equality, cf. (4.38), holds:*

$$(J\mathcal{H}'(\phi), \phi) = 0, \ \phi \in H. \tag{6.4}$$

Proof. (a) Let $||U_t\phi||^2 = ||\phi||^2$ for any $\phi \in H$. By using representation (6.3) we obtain

$$0 = \frac{d}{dt}||U_t\phi||^2 = 2(J\mathcal{H}'(U_t\phi), U_t\phi).$$

Thus $(J\mathcal{H}'(U_t\phi), U_t\phi) = 0, \phi \in H$. Now we use the fact that $U_t(H) = H$ and obtain equality (6.4).

(b) Let equality (6.4) hold for any point of $\phi \in H$. Then, in particular,

$$(J\mathcal{H}'(U_t\phi), U_t\phi) = 0 \tag{6.5}$$

for any $\phi \in H$. Thus $\frac{d}{dt}||U_t\phi||^2 = 0$ and hence $||U_t\phi|| = ||\phi||, t \geq t_0, \phi \in H$.

We remark that (6.4) implies the norm preserving even in the case when U_t is not a surjection.

Denote the class of maps $f : H \to \mathbf{R}$ satisfying condition (6.4) by the symbol $W(H)$. Functions belonging to this class have some features similar to SO_2-invariant functions. However, we have not yet been able to establish a precise relation between these classes.

Corollary 1.1. *A Hamiltonian flow is norm preserving iff equality (6.5) holds.*

Equation (6.4) is a linear equation with respect to $\mathcal{H}$. It can be written in the form

$$\left(\frac{\partial \mathcal{H}}{\partial q}, p\right) = \left(\frac{\partial \mathcal{H}}{\partial p}, q\right) \tag{6.6}$$

Theorem 1.2. *Let a function $\mathcal{H}$ belong to the functional space $W(H)$. Then $\mathcal{H}''(0) \in \mathcal{L}_{\mathrm{sp},s}(H)$ (the space of J-invariant self-adjoint operators on a phase space).*

Proof. We have: $(\mathcal{H}'(\phi), J\phi) = 0$. Thus $\mathcal{H}''(\phi)J\phi + J^*\mathcal{H}'(\phi) = 0$ and hence $\mathcal{H}'''(\phi)J\phi + \mathcal{H}''(\phi)J + J^*\mathcal{H}''(\phi) = 0$. Therefore,

$$[\mathcal{H}''(0), J] = 0. \tag{6.7}$$

We remark that in general we have

$$[\mathcal{H}''(\phi), J] = -\mathcal{H}'''(\phi)J\phi. \tag{6.8}$$

We also remark that, for any map $\mathcal{H} : H \to \mathbf{R}$, we can represent its Hessian as

$$\mathcal{H}'' = \begin{pmatrix} \frac{\partial^2 \mathcal{H}}{\partial q^2} & \frac{\partial^2 \mathcal{H}}{\partial q \partial p} \\ \frac{\partial^2 \mathcal{H}}{\partial p \partial q} & \frac{\partial^2 \mathcal{H}}{\partial p^2} \end{pmatrix}$$

The condition $\mathcal{H}''(0, 0) \in \mathcal{L}_{\mathrm{sp},s}(H)$ implies that

$$\frac{\partial^2 \mathcal{H}}{\partial q^2}(0, 0) = \frac{\partial^2 \mathcal{H}}{\partial p^2}(0, 0), \quad \frac{\partial^2 \mathcal{H}}{\partial q \partial p}(0, 0) = -\frac{\partial^2 \mathcal{H}}{\partial p \partial q}(0, 0). \tag{6.9}$$

The latter equality is not be surprising even in the light of the well-known equality of mixed partial derivatives for any two times continuously differentiable map. Of course, we always have

$$\frac{\partial^2 \mathcal{H}}{\partial p_i \partial q_j} = \frac{\partial^2 \mathcal{H}}{\partial q_j \partial p_i}$$

for any i, j. Let us consider an illustrative example with the quadratic Hamilton function $\mathcal{H}(q_1, q_2, p_1, p_2) = p_1 q_2 - q_1 p_2$. Here we have

$$\frac{\partial^2 \mathcal{H}}{\partial q \partial p} = \begin{pmatrix} \frac{\partial^2 \mathcal{H}}{\partial q_1 \partial p_1} & \frac{\partial^2 \mathcal{H}}{\partial q_1 \partial p_2} \\ \frac{\partial^2 \mathcal{H}}{\partial q_2 \partial p_1} & \frac{\partial^2 \mathcal{H}}{\partial q_2 \partial p_2} \end{pmatrix} = \begin{pmatrix} 0 & -1 \\ 1 & 0 \end{pmatrix};$$

and

$$\frac{\partial^2 \mathcal{H}}{\partial p \partial q} = \begin{pmatrix} \frac{\partial^2 \mathcal{H}}{\partial p_1 \partial q_1} & \frac{\partial^2 \mathcal{H}}{\partial p_1 \partial q_2} \\ \frac{\partial^2 \mathcal{H}}{\partial p_2 \partial q_1} & \frac{\partial^2 \mathcal{H}}{\partial p_2 \partial q_2} \end{pmatrix} = \begin{pmatrix} 0 & 1 \\ -1 & 0 \end{pmatrix}$$

We remark that any polynomial of the type considered in Example 4.1 satisfies condition (6.4). Therefore each Hamilton function of such a type induces the flow $U_t(\phi)$ that preserves the norm, e.g., $\mathcal{H}(\phi) = a_1(\widehat{H}\phi, \phi) + a_2(\widehat{H}\phi, \phi)^2$, where $[\widehat{H}, J] = 0$. However, we have not yet found a general relation between the functional classes of $SO(2)$-invariant functions and functions satisfying (6.4).

On the other hand, by using condition (6.4) we can easily find Hamilton functions with flows which do not preserve the norm. Let us consider in the two-dimensional case the map $\mathcal{H}(q, p) = q^2 p$. For this map, condition (6.4) does not hold true. Therefore the Hamiltonian flow corresponding to this map does not preserve the norm.

We now investigate the conditions for preserving the average. There is given a measure μ with zero mean value: $m_\mu = 0$. We want to find a sufficient condition for preserving this value: $m_{\mu_t} = 0$ for $t \geq 0$. Let us consider the class of symmetric measures μ such that $g^*_{-1}\mu = \mu$, where

$$g_{-1}\phi = -\phi.$$

We remark that any even measure has zero mean value. (Here g^* denotes lifting to the space of measures of a map g in a phase space, see Section 4.5).

Proposition 1.1. *Let μ be a symmetric measure and let $U_t(\phi)$ be an odd Hamiltonian flow*

$$U_t(-\phi) = -U_t(\phi). \tag{6.10}$$

Then the average $m_{\mu_t} = 0$ for $t \geq 0$.

We can even prove

Proposition 1.2. *An odd Hamiltonian flow preserves the class of symmetric measures.*

Proof. We have to prove that $g_{-1}^* U_t^* \mu = U_t^* \mu$. We have

$$\int f(\phi) dg_{-1}^* U_t^* \mu(\phi) = \int f(U_t(-\phi)) d\mu(\phi) = \int f(-U_t(\phi)) d\mu(\phi)$$

$$= \int f(U_t(\phi)) d\mu(\phi) = \int f(\phi) dU_t^* \mu(\phi).$$

Proposition 1.3. *Let the Cauchy problem for Hamiltonian equations be well posed. Then the Hamiltonian flow is odd if $\mathcal{H}'$ is odd.*

Proof. (a) Let (6.10) hold. Then $\frac{dU_t}{dt}(\phi) = -\frac{dU_t}{dt}(-\phi)$. Thus

$$\mathcal{H}'(U_t(\phi)) = -\mathcal{H}'(U_t(-\phi)).$$

Hence

$$\mathcal{H}'(\phi) = -\mathcal{H}'(-\phi)) \tag{6.11}$$

for any $\phi = U_t\phi$. Since the problem is well posed, any $\phi \in H$ can be represented in this form.

(b) Let now (6.11) hold. We have $-\frac{dU_t}{dt}(-\phi) = -J\mathcal{H}'(U_t(-\phi)) = J\mathcal{H}'(-U_t(-\phi))$. The problem is well posed, hence the solution is unique. Thus (6.10) holds.

Corollary 1.2. *Let the Hamilton function $\mathcal{H}(\phi)$ be J-invariant. Then the corresponding flow preserves the averages of symmetric measures.*

Finally, we remark that any J-invariant measure is symmetric and, in particular, its average is zero.

Corollary 1.3. *Let the Hamilton function $\mathcal{H}(\phi)$ and let the measure μ be J-invariant. Then the Hamiltonian flow preserves the (zero) mean value of μ.*

Open mathematical problems:

(a) To perform comparative analysis for the space $W(H)$ and the space of $SO(2)$-invariant functions. Do they coincide?
(b) To study dispersion preserving Hamilton dynamics which are not norm preserving; first of all, to find an example of such a dynamics.

6.2 Formalism of Rigged Hilbert Space

In this section we extend PCSFT to cover QM with observables represented by unbounded operators. Instead of Hilbert spaces, we shall work with *rigged Hilbert spaces*. To proceed in the rigorous mathematical framework, we use abstract mathematical constructions; in particular, from the theory of topological linear spaces, to make our model applicable in the case of the rigged Hilbert space associated with the Schwartz space of test functions and its dual space, the Schwartz space of distributions. Hence the reader can keep in mind these concrete spaces. To simplify the presentation, we only work with *real spaces.*

We recall that a *topological linear space* E is a linear space endowed with a topology such that all algebraic operations are continuous with respect to this topology [266]. It is called *locally convex* if its topology can be determined by a family of seminorms, $\{\| \cdot \|_\alpha\}$. Here open balls (with center at zero) have the form $B_{\alpha,r} = \{x \in E : \|x\|_\alpha < r\}, r > 0$. We recall that a seminorm has all properties of a norm except nondegeneration, i.e., it can happen that $\|x\|_\alpha = 0$ for $x \neq 0$.

For example, take the space of continuous functions in $\mathbf{R}$; denote it by the symbol $C(\mathbf{R})$. Consider a family of seminorms $\|f\|_n = \max_{-n\leq x\leq n} |f(x)|$, $n = 1, 2, \ldots$ It is clear that $\|f\|_n = 0$ does not imply $f \equiv 0$. These are not norms, but only seminorms. Open balls with respect to these seminorms have the form

$$B_{n,r} = \{f \in C(\mathbf{R}) : \max_{-n\leq x\leq n} |f(x)| \leq r\}.$$

A sequence of functions $\{f_j\}$ converges in this topology to a function f if, for any n, $\max_{-n\leq x\leq n} |f_j(x) - f(x)| \to 0, j \to \infty$, i.e., if this sequence converges uniformly on each finite segment. As usual we set $f^{(0)}(x) \equiv f(x)$.

This example can be generalized. Take the space of smooth functions in **R**; denote it by the symbol $C^\infty(\mathbf{R})$. Consider a family of seminorms

$$\|f\|_{n,m} = \max_{0\le i\le m} \max_{-n\le x\le n} |f^{(i)}(x)|, \; n = 1, 2, \ldots, m = 0, 1, 2, \ldots$$

Open balls with respect to these seminorms have the form

$$B_{nm,r} = \{f \in C(\mathbf{R}) : \max_{0\le i\le m} \max_{-n\le x\le n} |f^{(i)}(x)| \le r\}.$$

A sequence of functions $\{f_j\}$ converges in this topology to the function f if the sequence of derivatives of any order converges uniformly on each finite segment.

The term *locally convex* has the following origin. Each ball is a convex set. Thus the topology of such a space has a basis of convex neighborhoods. On the other hand, any convex neighborhood in a topological linear space determines a seminorm and can be represented as the unit ball with respect to this seminorm. Hence, if there exists a basis of topology consisting of convex neighborhoods, then this topology can be determined by a system of seminorms. A locally convex space is called the *Frechet space* if it is complete and its topology is metrizable. The latter is equivalent to the countability of the family of seminorms determining the topology. Hence, e.g., the spaces $C(\mathbf{R})$ and $C^\infty(\mathbf{R})$ are metrizable. We remark that in the case of nonmetrizable topology the notion of completeness of a topological vector space is rather complicated. It cannot be described by using sequences; so-called filters have to be used. However, if a space is metrizable, then the notion of completeness is standard: any Cauchy sequence (with respect to the metric) has a limit. To be a Cauchy sequence with respect to the metric is equivalent to be such a sequence with respect to each seminorm from a countable sequence of seminorms determining the space topology.

A seminorm $\|\cdot\|$ on E is called a *Hilbert seminorm* if there exists a symmetric positively defined bilinear form $(\cdot,\cdot)$ (it may be degenerate) such that $\|x\|^2 = (x, x)$. By factorizing E with respect to the kernel of this seminorm, $L = \{x \in E : \|x\| = 0\}$, and completing the factor-space E/L, we obtain a Hilbert space H.

Let $\|\cdot\|_\alpha$ and $\|\cdot\|_\beta$ be two seminorms on E and let $\|x\|_\beta \le \|x\|_\alpha$ for any $x \in E$. Denote the kernels of these seminorms by the symbols L_α and L_β. Then $L_\alpha \subset L_\beta$. Hence $E/L_\beta \subset E/L_\alpha$. Denote the completions of E/L_α and E/L_β (Hilbert spaces) by the symbols H_α and H_β. Consider the operator of embedding $i_{\alpha\beta} : H_\beta \to H_\alpha$.

A locally convex space is called *nuclear* if its topology can be determined by a family of Hilbert seminorms and canonical operators $i_{\alpha\beta}$ acting between the corresponding Hilbert spaces are of the *Hilbert–Schmidt* type, i.e., for any basis $\{e_j\}$ in H_β, $\sum_j \|i_{\alpha\beta} e_j\|_\alpha^2 < \infty$.

Let Ω be a nuclear Frechet topological linear space and let Ω' be its dual space. Suppose that Ω is densely and continuously embedded into a Hilbert space H, so that $\Omega \subset H$. Thus the dual space H' is densely embedded into Ω'. By identifying H and H' we obtain the rigged Hilbert space

$$\Omega \subset H \subset \Omega'. \tag{6.12}$$

We introduce the form of duality between Ω and Ω' by setting for $\Psi \in \Omega$ and $\phi \in \Omega'$,

$$(\phi, \Psi) = \phi(\Psi).$$

This is the result of application of the linear continuous functional ϕ to the vector Ψ. We remark that the first argument is a variable belonging to Ω' and the second variable belongs to Ω.

In our final application we shall set $\Omega = \mathcal{S}(\mathbf{R}^3)$. This is the space of Schwartz test functions on $\mathbf{R}^3$. We state again that it consists of smooth and rapidly decreasing functions, i.e., functions that decrease at infinity more rapidly than a polynomial of any order. On this space we can introduce the following system of seminorms determining its topology:

$$\|f\|_{km} = \sum_{i,j=0}^{k,m} \int_{\mathbf{R}^3} |f^{(i)}(x)|^2 (1 + |x|^j) dx,$$

where $k, m = 0, 1, 2, \ldots$ We remark that, in fact, these are norms. Since this system of norms is countable, this space is metrizable. It is complete, so it is a Frechet space. These norms are Hilbert norms, since they correspond to scalar products. This space is a nuclear space. Here $\Omega' = \mathcal{S}'(\mathbf{R}^3)$ is the space of Schwartz distributions:

continuous linear functionals. In this case we choose $H = L_2(\mathbf{R}^3)$ and we shall consider the rigged Hilbert space

$$\mathcal{S}(\mathbf{R}^3) \subset L_2(\mathbf{R}^3) \subset \mathcal{S}'(\mathbf{R}^3). \tag{6.13}$$

The readers who are not so much interested in the general theory of topological linear spaces can consider this concrete rigged Hilbert space throughout this section.[a]

A Gaussian measure μ on Ω' is determined by its characteristic functional (Fourier transform) $\tilde{\mu}$ which is defined on Ω :

$$\tilde{\mu}(\Psi) = \int_{\Omega'} e^{(\phi,\psi)} d\mu(\phi) = e^{-\frac{1}{2}b(\Psi,\Psi)},$$

where $b : \Omega \times \Omega \to \mathbf{R}$ is a continuous positively defined quadratic form. By the well-known theorem of Minlos–Sazonov, see, e.g., [58], μ is σ-additive on Ω' and its covariance functional is equal to b. Here

$$b(\Psi_1, \Psi_2) = \int_{\Omega'} (\phi, \Psi_1)(\phi, \Psi_2) d\mu(\phi),$$

where $\Psi_1, \Psi_2 \in \Omega$. This functional defines the covariance operator $B : \Omega \to \Omega'$ by

$$(B\Psi_1, \Psi_2) = b(\Psi_1, \Psi_2).$$

This operator is self-adjoint in the following sense. The dual operator $B' : \Omega'' \to \Omega'$. Since Ω (the topological linear space under consideration) is a nuclear Frechet space, it is *reflexive.* Hence $\Omega'' = \Omega$. Thus the operator $B' : \Omega \to \Omega'$. It is meaningful to speak about self-adjoint operators (by extending the ordinary theory of self-adjoint operators in a Hilbert space). We also point to the fact that the covariance operator B is positively defined.

6.3 Quantum Pure and Mixed States from the Background Field

In this section we shall use primary notions of the theory of topological linear spaces [266]. This theory is the basic apparatus

---

[a]Of course, we can consider rigged Hilbert spaces based on other spaces of test and generalized functions. For example, we can select $\Omega = \mathcal{D}(\mathbf{R}^3)$ and $\Omega' = \mathcal{D}'(\mathbf{R}^3)$, where $\mathcal{D}(\mathbf{R}^3)$ is the space of smooth compactly supported functions. However, the space $\mathcal{S}(\mathbf{R}^3)$ is simpler, since it is metrizable, i.e., its topology can be determined by a metric.

of theory of distributions (generalized functions); its presentation in the framework of mathematical physics can be found in Refs. [130, 138]. In this theory it is a custom to use the symbol $\widehat{A}'$ for the adjoint operator. We shall use this notation in this section; the symbol $\widehat{A}^\star$ is reserved for the Hilbert space adjoint operator (in the complex Hilbert space we use the symbol $\widehat{A}^*$).

Let us consider the standard Gaussian distribution μ on H. It is defined by its covariance functional

$$b(\Psi_1, \Psi_2) = (\Psi_1, \Psi_2).$$

The corresponding covariance operator $B = I : \Omega \to \Omega'$ is the canonical embedding operator. Since the embedding $\Omega \subset H$ is continuous, $b : \Omega \times \Omega \to \mathbf{R}$ is continuous and hence the measure μ_I is σ-additive on Ω' (but not on H). Therefore the corresponding Gaussian random variable η taking values in Ω' is well defined (it is impossible to construct an H-valued process).

In the case of the rigged Hilbert space (6.13), the Gaussian random field η valued in $S'(\mathbf{R}^3)$ is nothing else than *spatial white noise.* We extend this terminology and we shall call η *white noise* even in the abstract framework. Let us consider $\sqrt{\alpha}$-scaling of white noise

$$\xi(\omega) = \sqrt{\alpha}\eta(\omega),$$

where ω is a random parameter. The $\xi(\omega)$ is the *background field* in our prequantum model (PCSFT). To simplify the consideration of Section 3.1, we have not considered $\sqrt{\alpha}$-scaling. However, to get the real physical background, we should do this. From our macroscopic viewpoint the background field consists of fluctuations (with zero average) having very small dispersion α. In the case of the rigged Hilbert space (6.13), realizations of this random field are generalized functions (Schwartz distributions).

Hence the background random field produces very singular physical fields. Of course, they are smoothed by detectors but fundamentally they are singular.

For $\Psi \in \Omega$, we introduce the projection operator $\widehat{P}_\Psi : \Omega' \to \Omega$ by setting

$$\widehat{P}_\Psi \phi = \langle \phi, \Psi \rangle \Psi,$$

i.e.,

$$\widehat{P}_\Psi = |\Psi\rangle\langle\Psi|$$

in Dirac's notation; so

$$\widehat{P}_\Psi|\phi\rangle = \langle\phi|\Psi\rangle\ |\Psi\rangle.$$

Consider now one-dimensional projections of white noise

$$\eta_\Psi(\omega) = \widehat{P}_\Psi\eta(\omega)$$

and the corresponding projections of the background field $\xi_\Psi(\omega) = \widehat{P}_\Psi\xi(\omega) = \sqrt{\alpha}\eta_\Psi(\omega)$. We have $E\xi_\Psi = 0, E\xi_\Psi^2 = \alpha||\Psi||^2$. If $||\Psi|| = 1$, then the quantum image of the probability distribution of this random variable is nothing else than the density operator corresponding to the pure state Ψ. By considering (as in Chapter 3) quadratic functionals of the prequantum field, one can proceed without $\sqrt{\alpha}$-scaling and work directly with the random field $\eta(\omega)$. However, we should remember that the physical background field has very small dispersion. The dispersion of one-dimensional projections is proportional to α.[a]

It is convenient to select Ω' as the sample space and μ_I as the probability measure, i.e., a random parameter ω can be chosen as a vector $\phi \in \Omega'$. Here $\eta_\Psi(\phi) = \widehat{P}_\Psi\phi$.

Thus pre-images (classical random fields) of pure quantum states are one-dimensional projections of the background field of the white noise type. We can say that all these prequantum random fields are produced from the background white noise. They are obtained from vacuum fluctuations by using one-dimensional linear filters.

These filters are created by nature. We have not yet elaborated any physical model of such filters; e.g., how can a pulse of electromagnetic random field, "photon," or the electronic field, "electron," be produced from the background field?

The correspondence $\Psi \to \eta_\Psi(\omega)$ can be extended from the space Ω to the Hilbert space H. If $\Psi \in H$, then $\phi \to (\phi, \Psi)$ is also well

---

[a]We state again that in the infinite-dimensional case the dispersions of $\eta(\omega)$ as well as $\xi(\omega)$ are infinite: $\int_{\Omega'} ||\phi||^2 d\mu_I(\phi) = \text{Tr}\ I = \infty$. To escape this difficulty, one can proceed with finite-dimensional Hilbert spaces (as one does in quantum information theory). However, real physical random fields are realized in $L_2(\mathbf{R}^3)$ and here the background field has the infinite dispersion. This fact has the direct relation to the infinite energy of vacuum in QFT.

defined, but is not a continuous linear functional on the space Ω'. It is defined as an element of the space of square-integrable functionals of the white noise, $L_2(\Omega', d\mu_I)$. To define this quadratic functional of white noise, we approximate $\Psi \in H$ by elements Ψ_n of Ω, $\Psi_n \to \Psi$ in H (we recall that Ω is dense in H). Then $\xi_\Psi = \lim_{n\to\infty} \xi_{\Psi_n}$ in the L_2-space.

By using this construction we can define a random field $\phi \to \sqrt{\rho}\phi$ for each density operator $\rho \in \mathcal{D}(H)$. Let $\rho e_j = p_j e_j$, where $\{e_j\}$ is the basis consisting of eigenvectors of ρ. Then

$$\sqrt{\rho}\phi = \sum_j \sqrt{p_j}\eta_{e_j}(\phi).$$

Thus any quantum state can be obtained as the image of the output of a linear filter of the background field.

At the moment we consider noncomposite systems. The treatment of composite systems (see Section 3.1) implies that in the prequantum field the background is also present as an additional additive contribution. In the case of a pure quantum state, after $1/\sqrt{\alpha}$-scaling (i.e., normalization by the standard deviation), we obtain

$$\eta_{\Psi,\text{noisy}} = \sqrt{\widehat{P}_\Psi + \epsilon I}\phi = (\sqrt{1+\epsilon} - \sqrt{\epsilon})\widehat{P}_\Psi\phi + \sqrt{\epsilon}\phi. \tag{6.14}$$

In the case of a mixed state, we have

$$\eta_{\rho,\text{noisy}} = \sqrt{\rho + \epsilon I}\phi. \tag{6.15}$$

We shall come back to this problem, i.e., realization of quantum systems by linear filters of the background field in Chapter 8, where we consider composite systems.

6.4 Classical Model for Unbounded Quantum Observables

We continue to work on the rigged Hilbert space. Let $f : \Omega' \to \mathbf{R}$. We "quantize" $f(u)$ by mapping it into $\widehat{A} = \frac{f''(0)}{2}$. For quadratic functionals $f(u) = (\widehat{A}u, u)$, $\widehat{A} \in \mathcal{L}_s(H)$, we have the precise equality of prequantum and quantum averages and we can directly use the average with respect to the canonical Gaussian random variable $\eta(\omega)$. Here

$$Ef(\eta_\Psi) = \frac{1}{2}(f''(0)\Psi, \Psi).$$

Let $f : \Omega \to \mathbf{R}$ be a smooth function. Then at any point $\Psi_0 \in \Omega$, $f''(\Psi_0) : \Omega \to \Omega'$. Therefore $f''(0)$ is in general an unbounded operator in H.

Remark 4.1 (The source of unbounded operators in quantum mechanics) In our approach the appearance of unbounded quantum observables is a consequence of the presence of classical variables $f : \Omega \to \mathbf{R}$ which cannot be extended smoothly to the Hilbert space H. If we choose $H = L_2(\mathbf{R}^3)$ and the rigged Hilbert space (6.13), then we see that unbounded operators appear in quantum mechanics, because in the prequantum classical model (PCSFT) there exist physical variables $f(\phi)$ which are well defined only for smooth and rapidly decreasing classical fields $\phi(x)$, but they cannot be smoothly extended to the space of square-integrable fields.

Moreover, in this way, i.e., starting with PCSFT, we obtain a class of linear operators, quantum observables, which is essentially larger than in the conventional quantum formalism. In general, $\widehat{A} = f''(0)/2$ maps Ω not into H, but into Ω'.

Example 4.1. Let us consider the rigged Hilbert space (6.13). We consider the map $f : \mathcal{S}(\mathbf{R}^3) \to \mathbf{R}$ determined by a fixed point $x_0 \in \mathbf{R}^3$:

$$f(\phi) = \phi^2(x_0).$$

For example, the classical field $\Psi(x) = e^{-x^2}$ is mapped into the real number $e^{-x_0^2}$. Then $(f''(0)\phi_1, \phi_2) = \phi_1(x_0)\phi_2(x_0)$. Thus,

$$\widehat{A}\phi(x) = \frac{1}{2}f''(0)\phi(x) = \delta(x - x_0)\phi(x)$$

is the operator of multiplication by the δ-function $\delta(x - x_0)$. Hence,

$$f''(0)(\mathcal{S}(\mathbf{R}^3)) \not\subset L_2(\mathbf{R}^3).$$

For any $\Psi \in \mathcal{S}(\mathbf{R}^3)$, we have

$$Ef(\eta_\Psi(\omega)) = \Psi^2(x_0) = (\widehat{A}\Psi, \Psi) \equiv \langle \widehat{A} \rangle_\Psi.$$

However, in general, for $\Psi \in L_2(\mathbf{R}^3)$, the average $\langle \widehat{A} \rangle_\Psi$ is not well defined.

Let us now consider a Gaussian measure $\mu \in S_{\mathrm{G}}^{\alpha}(H)$ which has the support on the space Ω. Hence $\mu(H \setminus \Omega) = 0$. Thus μ can be considered as a measure on Ω. For such a measure μ, its covariance operator $B_\mu : \Omega' \to \Omega$ and its Fourier transform $\tilde{\mu}$ is defined on Ω'. We denote this class of classical statistical states (Gaussian measures) by the symbol $S_{\mathrm{G}}^{\alpha}(\Omega)$. We remark that $S_{\mathrm{G}}^{\alpha}(\Omega) \subset S_{\mathrm{G}}^{\alpha}(H)$.

Let E be a complex locally convex topological linear space. We recall that the topology of E can be determined by a system of seminorms. Let $b_n : E \times \cdots \times E \to \mathbf{C}$ be a continuous n-linear symmetric form. There exists a continuous seminorm p on E such that

$$\|b_n\|_p = \sup_{p(\phi) \leq 1} |b_n(\phi, \ldots, \phi)| < \infty$$

(here p depends on b_n). Thus,

$$|b_n(\phi, \ldots, \phi)| \leq \|b_n\| \; p^n(\phi) \tag{6.16}$$

An analytic function (see, e.g., Refs. [130, 138] for details) $f : E \to \mathbf{C}$ has the exponential growth if there exists a continuous seminorm p on E such that

$$|f(\phi)| \leq a e^{bp(\phi)}, \phi \in E. \tag{6.17}$$

Here the constants and seminorm depend on $f : a \equiv a_f$, $b \equiv b_f$, $p \equiv p_f$.

Lemma 4.1. *The space of analytic functions of the exponential growth coincides with the space of analytic functions such that there exists a continuous seminorm $p = p_f$ [127, 130]:*

$$\|f^{(n)}(0)\|_p \leq c\, r^n, \;\; n = 0, 1, 2, \ldots \tag{6.18}$$

Here the constants $c = c_f$ and $r = r_f$ depend on the function f.

Proof. (a) Let f have the exponential growth. For any $\phi \in E$, we consider the function of the complex variable $z \in \mathbf{C} : g_\phi(z) = f(z\phi)$. We have $|g_\phi^{(n)}(0)| \leq n! R^{-n} \sup_{0 \leq \theta \leq 2\pi} |f(Re^{i\theta}\phi)| \leq a_f n! R^{-n} e^{b_f R p(\phi)}$. We obtain

$$\|f^{(n)}(0)\|_p \leq a'_f e^{-n} n^{1/2} e^{b_f n}.$$

Thus the derivatives of f satisfy inequalities (6.18) with $r_f = e^{b_f}$.

(b) Let now the derivatives of f satisfy inequalities (6.18) for some continuous seminorm p. Then by inequalities (6.16) we have

$$|f(\phi)| \leq \sum_{n=0}^{\infty} \|f^{(n)}(0)\|_p p^n(\phi)/n! \leq c_f e^{r_f p(\phi)}.$$

Thus f has the exponential growth with $b_f = r_f$ and the same continuous seminorm p as in (6.18).

We denote by $\mathcal{V}_{\text{an}}(\Omega)$ the class of functions $f : \Omega \to \mathbf{R}$, $f(0) = 0$, which can be analytically continued on the complexification of Ω to functions of the exponential growth.

Lemma 4.2. *Let* $\mu \in S^{\alpha}_{\mathrm{G}}(\Omega)$. *Then for any function* $f \in \mathcal{V}_{an}(\Omega)$, *the following asymptotic equality holds:*

$$\langle f\rangle_\mu \equiv \int_\Omega f(\phi)d\mu(\phi) = \frac{\alpha}{2}\int_\Omega (f''(0)\phi, \phi)d\mu_\rho(\phi) + O(\alpha^2), \ \alpha \to 0, \tag{6.19}$$

where $\rho = B_\mu/\alpha$. *Here*

$$O(\alpha^2) = \alpha^2 R(\alpha, f, \mu), \tag{6.20}$$

where

$$|R(\alpha, f, \mu)| \leq c_f \int_\Omega e^{r_f p(\phi)} d\mu_\rho(\phi). \tag{6.21}$$

The seminorm p is determined by inequality (6.17).

We remark that for a measure $\mu \in S^\alpha(\Omega)$, its covariance operator $B_\mu : \Omega' \to \Omega$, and for a function $f \in \mathcal{V}_{\text{an}}(\Omega)$, its second derivative $\widehat{A} = \frac{f''(0)}{2} : \Omega \to \Omega'$, so $\widehat{C} = B_\mu \widehat{A} : \Omega \to \Omega$. In general, this operator cannot be extended to a continuous operator in H. We would like to obtain, for a linear continuous operator $\widehat{A} : \Omega \to \Omega'$, an analogue of the formula

$$\int_\Omega (\widehat{A}\Psi, \Psi)d\mu(\Psi) = \text{Tr } B_\mu \widehat{A}. \tag{6.22}$$

The main mathematical problem is that in general the operator $\widehat{C} = B_\mu \widehat{A}$ is not even continuous in H, so it is not a trace-class operator in the Hilbert space H. Nevertheless, we can introduce a notion of trace even in such a framework. However, it will be not simply the Hilbert space trace; we use the notion of trace in a topological linear space.

We recall that systems of vectors $\{e_j\}_{j=1}^{\infty}, e_j \in \Omega$, and $\{e'_j\}_{j=1}^{\infty}, e'_j \in \Omega'$, are called *bi-orthogonal topological bases* in Ω and Ω' if

$$(e'_j, e_i) = \delta_{ij}, \text{ and } \Psi = \sum_{j=1}^{\infty} (e'_j, \Psi) e_j, \Psi \in \Omega,$$

$$\phi = \sum_{j=1}^{\infty} (\phi, e_j) e'_j, \phi \in \Omega',$$

where the series converge in Ω and Ω', respectively.

Definition 4.1. *A linear continuous operator $\widehat{C} : \Omega \to \Omega$ is called a trace-class operator if for any pair of bi-orthogonal topological bases, the series*

$$Tr\, \widehat{C} = \sum_{j=1}^{\infty} (e'_j, \widehat{C} e_j) \qquad (6.23)$$

converges and its sum does not depend on bases.

Lemma 4.3. *Let μ be a Gaussian measure on Ω and let $\widehat{A} : \Omega \to \Omega'$ be a continuous operator. Then the operator $\widehat{C} = B_\mu \widehat{A}$ belongs to the trace-class and equality (6.22) holds.*

Proof. We remark that for any $\Psi \in \Omega, \phi \in \Omega'$, we have $(\phi, \Psi) = \sum_{j=1}^{\infty} (\phi, e_j)(e'_j, \Psi)$. We have $(\widehat{A}\Psi, \Psi) = \sum_{i,j=1}^{\infty} (\widehat{A} e_i, e_j)$ $(e'_i, \Psi)(e'_j, \Psi)$. Thus,

$$\int_\Omega (\widehat{A}\Psi, \Psi) d\mu(\Psi) = \sum_{i,j=1}^{\infty} (\widehat{A} e_i, e_j)(e'_j, B_\mu e'_i).$$

Therefore,

$$\int_\Omega (\widehat{A}\Psi, \Psi) d\mu(\Psi) = \sum_{i=1}^{\infty} (\widehat{A} e_i, B_\mu e'_i) = \sum_{i=1}^{\infty} (e'_i, B_\mu \widehat{A} e_i)$$

(here we used the fact that $B'_\mu = B_\mu$).

As a consequence of Lemmas 4.2 and 4.3, we obtain:

Theorem 4.1. *Let $\mu \in S_G^\alpha(\Omega)$. Then, for any function $f \in \mathcal{V}_{an}(\Omega)$, the following asymptotic equality holds:*

$$\frac{\langle f \rangle_\mu}{\alpha} \equiv \frac{1}{\alpha} \int_\Omega f(\phi) d\mu(\phi) = Tr\, \rho f''(0)/2 + O(\alpha),\ \alpha \to 0, \quad (6.24)$$

where $\rho = B_\mu / \alpha$.

PCSFT motivates to extend the set of quantum observables and consider all continuous operators $\widehat{A} : \Omega \to \Omega'$. The operators should be self-adjoint in the ordinary sense: $\widehat{A}' = \widehat{A}$. Here $\widehat{A}' : \Omega'' \to \Omega'$, but $\Omega'' \equiv \Omega$, since Ω is a nuclear Frechet space and hence it is reflexive (i.e., the dual space of the dual space coincides with the original space). Denote the set of such operators by the symbol $\mathcal{L}_s(\Omega, \Omega')$. Denote the set of covariance operators of Gaussian measures belonging to the space $S^1_{\mathrm{G}}(\Omega)$ by the symbol $\mathcal{D}(\Omega', \Omega)$.[a]

Definition 4.2. *A statistical quantum model corresponding to a rigged Hilbert space $\mathcal{T}$ given by (6.12) is the pair $N_q(\mathcal{T}) = (\mathcal{D}(\Omega', \Omega), \mathcal{L}_s(\Omega, \Omega'))$. A generalized density operator $\rho \in \mathcal{D}(\Omega', \Omega)$ represents a statistical state; a linear operator $\widehat{A} \in \mathcal{L}_s(\Omega, \Omega')$ represents a quantum observable. The average is given by the following generalization of the von Neumann trace formula $\langle \widehat{A} \rangle_\rho = \mathrm{Tr}\, \rho \widehat{A}$, where the trace is taken in the sense of definition (6.23).*

We choose the state space Ω as nuclear Frechet space. For a rigged Hilbert space $\mathcal{T}$ given by (6.12), we consider the classical statistical model $M(\mathcal{T}) = (S^\alpha_{\mathrm{G}}(\Omega), \mathcal{V}_{\mathrm{an}}(\Omega))$. Here, as always, $\langle f \rangle_\mu = \int_\Omega f(\phi) d\mu(\phi)$. As usual, we can map this classical model onto the generalized quantum model and classical and quantum averages are coupled by the asymptotic equality (6.24).[b]

Example 4.2. The position operators $\widehat{x}_j$, $j = 1, 2, 3$ can be obtained as $\widehat{x}_j = \frac{1}{2} f''_{x_j}(0)$, where

$$f_{x_j}(\phi) = \int_{\mathbf{R}^3} x_j \phi^2(x) dx.$$

Hence $\widehat{x}_j : \mathcal{S}(\mathbf{R}^3) \to \mathcal{S}'(\mathbf{R}^3)$ is continuous. Thus, for any measure $\mu \in S^\alpha_{\mathrm{G}}(\mathcal{S}(\mathbf{R}^3))$, we have

$$\frac{\langle f_{x_j} \rangle_\mu}{\alpha} \equiv \frac{1}{\alpha} \int_{\mathcal{S}(\mathbf{R}^3)} \int_{\mathbf{R}^3} x \phi^2(x) dx d\mu(\phi) = \mathrm{Tr}\, \rho \widehat{x}_j,$$

$\rho = B_\mu / \alpha$ (here the trace of the composition $\rho \widehat{x}_j$ is well defined).

---

[a] Of course, it would be better to present a direct description of this space, i.e., without coupling to Gaussian measures. And, in principle, it is possible. However, it will need a new portion of measure theory on topological linear spaces. We proceed by considering the space of covariance operators of a special class of Gaussian measures.

[b] In the same way as in Section 5.3, we can consider a more general space of classical statistical states, namely, $S^{\approx\alpha}_{\mathrm{G}}(\Omega)$.

Example 4.3. Let x_0 be a fixed point in $\mathbf{R}^3$. Let now $\widehat{A}\phi(x) = \delta(x - x_0)\phi(x)$, $\phi \in \mathcal{S}(\mathbf{R}^3)$. This operator does not belong to the domain of the conventional quantum formalism. It cannot be represented as an unbounded operator in $H = L_2(\mathbf{R}^3)$ with a dense domain of definition. Nevertheless,

$$\frac{1}{\alpha}\int_{\mathcal{S}(\mathbf{R}^3)} (A\phi, \phi)d\mu(\phi) = \frac{1}{\alpha}\int_{\mathcal{S}(\mathbf{R}^3)} \phi^2(x_0)d\mu(\phi) = \mathrm{Tr}\ \rho\widehat{A}$$

and the trace of the composition $\rho\widehat{A}$ is well defined.

In this section we worked with real rigged Hilbert spaces. It is easy to generalize this theory to the case of complex rigged Hilbert spaces. In this case we consider the phase space $\mathcal{S}(\mathbf{R}^3) \times \mathcal{S}(\mathbf{R}^3)$ and its complex realization, the space of complex-valued Schwartz test functions $\mathcal{S}(\mathbf{R}^3; \mathbf{C})$.

Example 4.4. The momentum operators $\widehat{p}_j$, $j = 1, 2, 3$, can be obtained as $\widehat{p}_j = \frac{1}{2} f''_{p_j}(0)$, where

$$f_{p_j}(\phi) = -i\int_{\mathbf{R}^3} \frac{\partial\phi}{\partial x_j}(x)\overline{\phi(x)}dx.$$

The operator $\widehat{p}_j : \mathcal{S}(\mathbf{R}^3) \to \mathcal{S}'(\mathbf{R}^3)$ is continuous. Thus,

$$\frac{\langle f_{p_j}\rangle_\mu}{\alpha} \equiv \frac{-i}{\alpha}\int_{\mathcal{S}(\mathbf{R}^3;\mathbf{C})}\int \frac{\partial\phi}{\partial x_j}(x)\overline{\phi(x)}dx d\mu(\phi) = \mathrm{Tr}\ \rho\ \widehat{p}_j.$$

Here $\rho\widehat{p}_j : \mathcal{S}(\mathbf{R}^3; \mathbf{C}) \to \mathcal{S}(\mathbf{R}^3; \mathbf{C})$ is the trace-class operator. Similar considerations can be used for angular momentum operators.

Chapter 7

Mathematical Presentation for Composite Systems

In this chapter we turn to formulas for Gaussian correlations used in Section 3.2, and present them in the rigorous mathematical framework. Moreover, we proceed without the condition that quantum observables are represented by real operators, i.e., satisfying equality (3.14). Now we are able to work with arbitrary (bounded) self-adjoint operators. Basic formulas coupling quantum and prequantum averages (equalities (3.15) and (3.20)) become more complicated.

Then we generalize our model to describe quantum correlations with respect to mixed states. Finally, we proceed to the correlation for a triparticle quantum system. The latter case is essentially more complicated than the biparticle case.

7.1 Derivation of Basic Formulas

Let Q and P be two copies of a real Hilbert space. Let us consider their Cartesian product $H = Q \times P$, the phase space, endowed with the symplectic operator $J = \begin{pmatrix} 0 & 1 \\ -1 & 0 \end{pmatrix}$. As in Section 5.4, we

Beyond Quantum
Andrei Khrennikov

ISBN 978-981-4411-73-8 (Hardcover), 978-981-4411-74-5 (eBook)
www.panstanford.com

consider the class of Gaussian measures (with zero mean value) which are invariant with respect to the action of the group $SO_\infty(2)$. Denote this class $S_G(H|SO_\infty(2))$. As we have seen in Chapter 5, $\mu \in S_G(H|SO_\infty(2))$ if and only if its covariance operator commutes with the symplectic operator.

As always, we consider the complex structure on $H = Q \times P$; the corresponding complex Hilbert space will be denoted by the same symbol, $H = Q \oplus iP$. The complex scalar product is denoted by the symbol $\langle \cdot, \cdot \rangle$. We shall also use the operation of complex conjugation $*$ in the complex Hilbert space H, for $\phi = \phi_1 + i\phi_2$, $\phi_1 \in Q$, $\phi_2 \in P$, we set $*(\phi) \equiv \overline{\phi} = \phi_1 - i\phi_2$. We remark that

$$\langle \overline{u}, \overline{v} \rangle = \langle v, u \rangle, \; u, v \in H. \tag{7.1}$$

We work with the complex covariance operator of a measure μ on the complex Hilbert space H, see Eq. (4.53):

$$\langle Dy_1, y_2 \rangle = \int_H \langle y_1, \phi \rangle \langle \phi, y_2 \rangle d\mu(\phi).$$

We also consider the complex Fourier transform of μ

$$\tilde{\mu}(y) = \int_H exp\{i(\langle y, \phi \rangle + \langle \phi, y \rangle)\} d\mu(\phi). \tag{7.2}$$

Any Gaussian measure of the class $S_G(H|SO_\infty(2))$ is determined by its complex Fourier transform,

$$\tilde{\mu}(y) = exp\{-\langle Dy, y \rangle\}. \tag{7.3}$$

Lemma 1.1. For any measure $\mu \in S_G(H|SO_\infty(2))$, the following representation takes place:

$$\int_H \langle \xi_1, \phi \rangle \langle \eta_1, \phi \rangle \langle \phi, \xi_2 \rangle \langle \phi, \eta_2 \rangle d\mu(\phi) = \langle D\xi_1, \eta_2 \rangle \langle D\eta_1, \xi_2 \rangle + \langle D\xi_1, \xi_2 \rangle \langle D\eta_1, \eta_2 \rangle.$$

To prove this formula, one differentiates the Fourier transform (7.2)–(7.3) four times.

Let μ be a measure on the Cartesian product $H_1 \times H_2$ of two Hilbert spaces. Then its covariance operator has the block structure

$$D = \begin{pmatrix} D_{11} & D_{12} \\ D_{21} & D_{22} \end{pmatrix}, \tag{7.4}$$

where $D_{ii} : H_i \to H_i$ and $D_{ij} : H_j \to H_i$. The operator D is self-adjoint. Hence $D_{ii}^* = D_{ii}$, and $D_{12}^* = D_{21}$.

Let H be a complex Hilbert space and let $\widehat{A} \in \mathcal{L}(H)$. We consider its quadratic form $\phi \to f_A(\phi) = \langle \widehat{A}\phi, \phi\rangle$.

Theorem 1.1. *Let $\mu \in S_G(H_1 \times H_2|SO_\infty(2))$ with the (complex) covariance operator D and let $\widehat{A}_i \in \mathcal{L}(H_i)$, $i = 1, 2$. Then*

$$\int_{H_1\times H_2} f_{A_1}(\phi_1) f_{A_2}(\phi_2) d\mu(\phi_1, \phi_2) = \mathrm{Tr} D_{11}\widehat{A}_1\, \mathrm{Tr} D_{22}\widehat{A}_2 + \mathrm{Tr} D_{12}\widehat{A}_2 D_{21}\widehat{A}_1. \quad (7.5)$$

This theorem is a consequence of the following general result:

Lemma 1.2. *Let $\mu \in S_G(H|SO_\infty(2))$ with the (complex) covariance operator D and let $\widehat{A}_i \in \mathcal{L}(H)$, $i = 1, 2$. Then*

$$\int_H f_{A_1}(\phi) f_{A_2}(\phi) d\mu(\phi) = \mathrm{Tr} D\widehat{A}_1 \mathrm{Tr} D\widehat{A}_2 + \mathrm{Tr} D\widehat{A}_2 D\widehat{A}_1. \quad (7.6)$$

Proof. By Lemma 1.1 the integral can be represented as

$$I = \sum_{i_1 j_1}\sum_{i_2 j_2} \langle \widehat{A}_1 e_{i_1}, e_{j_1}\rangle \langle \widehat{A}_2 e_{i_2}, e_{j_2}\rangle \times [\langle De_{j_1}, e_{i_2}\rangle\langle De_{j_2}, e_{i_1}\rangle + \langle De_{j_1}, e_{i_1}\rangle\langle De_{j_2}, e_{i_2}\rangle] = I_1 + I_2, \quad (7.7)$$

where $\{e_i\}$ is an orthonormal basis in H. Here

$$I_1 = \sum_{i_1 i_2}\sum_{j_1} \langle \widehat{A}_1 e_{i_1}, e_{j_1}\rangle\langle e_{j_1}, De_{i_2}\rangle \sum_{j_2}\langle \widehat{A}_2 e_{i_2}, e_{j_2}\rangle\langle e_{j_2}, De_{i_1}\rangle$$
$$= \sum_{i_1 i_2} \langle D\widehat{A}_1 e_{i_1}, e_{i_2}\rangle\langle e_{i_2}, \widehat{A}_2^* De_{i_1}\rangle = \sum_i \langle D\widehat{A}_2 D\widehat{A}_1 e_{i_1}, e_{i_1}\rangle$$
$$= \mathrm{Tr} D\widehat{A}_2 D\widehat{A}_1.$$
$$I_2 = \sum_{i_1 i_2}\sum_{j_2} \langle \widehat{A}_1 e_{i_1} e_{j_1}\rangle\langle e_{j_1}, De_{i_1}\rangle \sum_{j_2}\langle \widehat{A}_2 e_{i_2} e_{j_2}\rangle\langle e_{j_2}, De_{i_2}\rangle$$
$$= \sum_{i_1 i_2} \langle \widehat{A}_1 e_{i_1}, De_{i_1}\rangle\langle \widehat{A}_2 e_{i_2}, De_{i_2}\rangle = \mathrm{Tr} D\widehat{A}_1 \mathrm{Tr} D_{A_2}.$$

Let H_1 and H_2 be two complex Hilbert spaces and let $D_{21} \in \mathcal{L}(H_1, H_2)$, $D_{12} \in \mathcal{L}(H_2, H_1)$.[a] Then $D_{21} \otimes D_{12} \in \mathcal{L}(H_1 \otimes H_2, H_2 \otimes H_1)$. Let us consider the permutation operator $\sigma : H_2 \otimes H_1 \to H_1 \otimes H_2$, $\sigma(\phi_2 \otimes \phi_1) = \phi_1 \otimes \phi_2$. We remark that $\sigma \in \mathcal{L}(H_2 \otimes H_1, H_1 \otimes H_2)$.

[a]As always, $\mathcal{L}(H_i, H_j)$ denotes the space of bounded linear operators from H_i to H_j.

Proposition 1.1. *Let the conditions of Theorem 1.1 hold. Then*

$$\int_{H_1\times H_2} f_{A_1}(\phi_1) f_{A_2}(\phi_2) d\mu(\phi_1, \phi_2) = \mathrm{Tr}(D_{11} \otimes D_{22} + \sigma(D_{21} \otimes D_{12}))$$
$$\widehat{A}_1 \otimes \widehat{A}_2. \tag{7.8}$$

Proof. It is sufficient to prove that

$$\mathrm{Tr}\sigma(D_{21} \otimes D_{12})\widehat{A}_1 \otimes \widehat{A}_2 = \mathrm{Tr} D_{12}\widehat{A}_2 D_{21}\widehat{A}_1.$$

We have

$$\begin{aligned}\mathrm{Tr}\sigma(D_{21} \otimes D_{12})\widehat{A}_1 \otimes A_2 &= \sum_{ij} \langle \sigma(D_{21}\otimes D_{12})\widehat{A}_1 \otimes \widehat{A}_2 e_i \otimes f_j, e_i \otimes f_j\rangle \\ &= \sum_{ij} \langle D_{12}\widehat{A}_2 f_j \otimes D_{21}\widehat{A}_1 e_i, e_i \otimes f_j\rangle \\ &= \sum_{ij} \langle D_{12}\widehat{A}_2 f_j, e_i\rangle\langle D_{21}\widehat{A}_1 e_i, f_j\rangle.\end{aligned}$$

On the other hand,

$$\mathrm{Tr}\mathrm{D}_{12}\widehat{\mathrm{A}}_2\mathrm{D}_{21}\widehat{\mathrm{A}}_1 =$$
$$\sum_i \langle D_{12}\widehat{A}_2 D_{21}\widehat{A}_1 e_i, e_i\rangle \sum_i \langle D_{21}\widehat{A}_1 e_i, A_2^* D_{21} e_i\rangle$$
$$\sum_{ij} \langle D_{21}\widehat{A}_1 e_i, f_j\rangle\langle f_j, A_2^* D_{21} e_i\rangle.$$

7.2 Vector and Operator Realizations of the Tensor Product

In quantum theory, a pure state of a composite system is represented by a normalized *vector* belonging to the tensor product $H_1 \otimes H_2$. On the other hand, in functional analysis it is common to use elements of $H_1 \otimes H_2$ as operators. The standard construction provides the realization of any vector $\Psi \in H_1 \otimes H_2$ by a linear operator from $H_2^* \to H_1$, where H_2^* is the dual space to H_2. For a vector $Z = u \otimes v$, one puts $\widehat{Z} x^* = x^*(v)u$ for $x^* \in H_2^*$. This correspondence $Z \to \widehat{Z}$ is extended to an isomorphism of $H_1 \otimes H_2$ with the space of Hilbert–Schmidt operators $\mathcal{HS}(H_2^*, H_1)$ or (equivalently) with the space of anti-linear HS-operators from H_2 to H_1. The main complication in the coming consideration is related to our need to represent vectors from the tensor product by *linear* operators from H_2 to H_1. We recall the definition of the HS-norm corresponding to the trace scalar product in $\mathcal{HS}(H_2, H_1)$. Take an arbitrary orthonormal basis $\{f_k\}$ in H_2. Then

$$\langle \widehat{L_1}, \widehat{L_2} \rangle = \sum_k \langle \widehat{L_1} f_k, \widehat{L_2} f_k \rangle = \mathrm{Tr} \widehat{L_2}^* \widehat{L_1}, \ \ \|\widehat{L}\|_2^2 = \mathrm{Tr} \widehat{L}^* \widehat{L}$$

(index 2 is typically used for the Hilbert–Schmidt norm).

Let $\{e_j\}$ and $\{f_j\}$ be two orthonormal bases in H_1 and H_2, respectively. Then

$$\Psi = \sum_{ij} \Psi_{ij} e_i \otimes f_j, \ \Psi_{ij} \in \mathbf{C},$$

and $||\Psi||^2 = \sum_{ij} |\Psi_{ij}|^2$. We remark that for an orthonormal basis, say $\{f_j\}$, the system of complex conjugate vectors $\{\bar{f}_j\}$ is also an orthonormal basis. Indeed, by (7.1) we get $\langle \bar{f}_j, \bar{f}_i \rangle = \langle f_i, f_j \rangle = \delta_{ij}$.

For $\phi \in H_2$, we set

$$\widehat{\Psi} \phi = \sum_{ij} \Psi_{ij} \langle \phi_2, \bar{f}_j \rangle e_i. \tag{7.9}$$

We emphasize that, for the vector Ψ, the expansion with respect to the basis $\{e_i \otimes f_j\}$ was used. In contrast to this, in the expansion of the operator $\widehat{\Psi}$ the basis $\{e_i \otimes \bar{f}_j\}$ was used. If the basis is real, then the definition is essentially simplified:

$$\widehat{\Psi} \phi = \sum_{ij} \Psi_{ij} \langle \phi_2, f_j \rangle e_i. \tag{7.10}$$

Take now a factorizable vector $\Psi = u \otimes v$. It defines the rank-one operator

$$\widehat{\Psi} \phi = \langle \phi, \bar{v} \rangle u = \langle v, \bar{\phi} \rangle u.$$

To play with the definition, take $\Psi = cu \otimes v = (cu) \otimes v = u \otimes (cv)$, where $c \in \mathbf{C}$. On one hand we have $\widehat{\Psi} \phi = c \langle \phi, \bar{v} \rangle u$, on the other hand $\widehat{\Psi} \phi = \langle \phi, \overline{cv} \rangle u = \langle \phi, \bar{c} \bar{v} \rangle u = c \langle \phi, \bar{v} \rangle u$. Thus our definition is consistent with scaling by a complex constant.

Consider now very special but at the same time very important case $H_i = L_2(\mathbf{R}^{n_i})$, $i = 1, 2$. Here our definition gives the following representation (a special case of (7.10)), see also (3.13):

$$\widehat{\Psi} \phi(x) = \int \Psi(x, y) \phi(y) dy \tag{7.11}$$

We proceed formally and use the real basis $\{e_x \otimes f_y\}$, where $e_x(t) = \delta(t - x)$ and $f_y(s) = \delta(s - y)$.

Finally, we recall a result of the theory of integral operators. The operator given by (7.11) is of the HS-type in L_2-spaces if and only if its kernel $\Psi(x, y)$ is square-integrable. Thus the condition $\Psi \in L_2(\mathbf{R}^{n_1} \times \mathbf{R}^{n_2})$ for a function Ψ is equivalent to the HS-condition for the operator $\widehat{\Psi}$. In the abstract form this fact will be formulated in the following lemma.

Lemma 2.1. *Each vector $\Psi \in H_1 \otimes H_2$ determines (uniquely) an operator $\widehat{\Psi} \in \mathcal{L}(H_2, H_1)$ and, moreover, $\widehat{\Psi} \in \mathcal{HS}(H_1, H_2)$ and $||\widehat{\Psi}||_2 = ||\Psi||$ (the norm of the vector coincides with the Hilbert–Schmidt norm of the corresponding operator).*

Proof. We start with the proof of the correctness of Definition (7.9), i.e., we prove that it does not depend on the choice of orthonormal bases. Let $\{e_j'\}$ and $\{f_j'\}$ be two orthonormal bases in H_1 and H_2, respectively, which are in general different from the bases $\{e_j\}$ and $\{f_j\}$. Consider the expansion $\Psi = \sum_{ij} \Psi_{ij}' e_i' \otimes f_j'$. We remark that $\bar{f}_j' = \sum_k \langle \bar{f}_j', \bar{f}_k \rangle \bar{f}_k = \sum_k \langle f_k, f_j' \rangle \bar{f}_k$, which can also be obtained by complex conjugation from the expansion $f_j' = \sum_k \langle f_j', f_k \rangle f_k$. In this basis the vector Ψ defines the operator

$$\widehat{\Psi}'\phi = \sum_{ij} \Psi_{ij}' \langle \phi, \bar{f}_j' \rangle e_i'. \tag{7.12}$$

We have

$$\widehat{\Psi}'\phi = \sum_{ij} \sum_{nm} \Psi_{nm} \langle e_n, e_i' \rangle \langle f_m, f_j' \rangle \langle \phi, \sum_k \langle \bar{f}_j', \bar{f}_k \rangle \bar{f}_k \rangle \sum_p \langle e_i', e_p \rangle e_p$$
$$= \sum_{nm} \Psi_{nm} \sum_{kp} \langle \phi, \bar{f}_k \rangle \sum_i \langle e_n, e_i' \rangle \langle e_i', e_p \rangle \sum_j \langle \bar{f}_k, \bar{f}_j' \rangle \langle f_m, f_j' \rangle e_p.$$

We remark

$$\sum_j \langle \bar{f}_k, \bar{f}_j' \rangle \langle f_m, f_j' \rangle = \sum_j \langle f_m, f_j' \rangle \langle f_j', f_k \rangle = \delta_{km}$$

Thus, $\widehat{\Psi}'\phi = \sum_{nm} \Psi_{nm} \langle \phi, \bar{f}_m \rangle e_n$. It coincides with (7.9). We now apply the Cauchy–Bunyakovsky inequality to expression (7.9) and obtain $\|\widehat{\Psi}\phi\|^2 = \sum_i |\sum_j \Psi_{ij} \langle \phi, \bar{f}_j \rangle|^2 \leq \sum_i \sum_j |\Psi_{ij}|^2 \sum_j |\langle \phi, \bar{f}_j \rangle|^2 =$

$\|\Psi\|^2\|\phi\|^2$. Thus this operator belongs to the space $\mathcal{L}(H_2, H_1)$. Finally, we show that it belongs even to the space $\mathcal{HS}(H_2, H_1)$:

$$\|\widehat{\Psi}\|_2^2 = \text{Tr}\widehat{\Psi}^*\widehat{\Psi} = \sum_k \langle \widehat{\Psi}\bar{f}_k, \widehat{\Psi}\bar{f}_k\rangle = \sum_k \langle \sum_{i_1} \Psi_{i_1 k} e_{i_1}, \sum_{i_1} \Psi_{i_2 k} e_{i_2}\rangle$$

$$= \sum_{ki} \Psi_{ik}\bar{\Psi}_{ik}.$$

Moreover, it is easy to show that any operator $\widehat{L} \in \mathcal{HS}(H_2, H_1)$ can be represented as Eq. (7.9) for some vector $\Psi \in H_1 \otimes H_2$. Combining this with Lemma 3, we get:

Corollary 2.1. The equality (7.9) establishes the isomorphism of Hilbert spaces $H_1 \otimes H_2$ and $\mathcal{HS}(H_2, H_1)$.

We now find the adjoint operator. We have

$$\langle \widehat{\Psi} y, x\rangle = \sum_{nm} \Psi_{nm}\langle y, \bar{f}_m\rangle\langle e_n, x\rangle = \langle y, \sum_{nm} \bar{\Psi}_{nm}\langle x, e_n\rangle \bar{f}_m\rangle.$$

Thus,

$$\widehat{\Psi}^* x = \sum_{nm} \bar{\Psi}_{nm}\langle x, e_n\rangle \bar{f}_m.$$

7.3 Operation of the Complex Conjugation in the Space of Self-Adjoint Operators

Let $\widehat{A} \in \mathcal{L}_s(W)$, where W is a Hilbert space. We define the "complex conjugate operator" $\widehat{\bar{A}}$ by its bilinear form:

$$\langle \widehat{\bar{A}} u, v\rangle = \langle \bar{v}, \widehat{A}\bar{u}\rangle. \tag{7.13}$$

Let $\{f_j\}$ be an orthonormal basis in W. We find the matrix of the operator $\widehat{B} \equiv \widehat{\bar{A}}$ with respect to this basis: $b_{ij} = \langle \widehat{\bar{A}} f_i, f_j\rangle = \langle \bar{f}_j, \widehat{A}\bar{f}_i\rangle$. In the special case of the real basis, i.e., $\bar{f}_j = f_j$, we have

$$b_{ij} = \overline{a_{ij}} = a_{ji}.$$

Thus its matrix is given by the transposition of the matrix of $\widehat{A}$.

If $\widehat{\bar{A}} = \widehat{A}$, then the operator is called *real*. Only such operators were used in Chapter 3.

We remark that the operator $\widehat{\bar{A}}$ is self-adjoint (we state again that in this consideration the operator $\widehat{A}$ is self-adjoint):

$$\langle \widehat{\bar{A}} u, v\rangle = \langle \widehat{A}\bar{v}, \bar{u}\rangle = \overline{\langle \bar{u}, \widehat{A}\bar{v}\rangle} = \overline{\langle \widehat{\bar{A}} v, u\rangle} = \langle u, \widehat{\bar{A}} v\rangle. \tag{7.14}$$

For a positive operator $\widehat{N}$, the operator $\widehat{\bar{N}}$ is also positive, since

$$\langle \widehat{\bar{N}}u, u\rangle = \langle \bar{u}, \widehat{N}\bar{u}\rangle = \langle \widehat{N}\bar{u}, \bar{u}\rangle. \tag{7.15}$$

Consider the quadratic form of the complex conjugate operator $\widehat{\bar{A}}$ of a self-adjoint operator $\widehat{A}$

$$f_{\bar{A}}(\phi) = \langle \widehat{\bar{A}}\phi, \phi\rangle = \langle \bar{\phi}, \widehat{A}\bar{\phi}\rangle = f_A(\bar{\phi}). \tag{7.16}$$

Consider the group $\{e, *\}$, where $*$ is the operation of complex conjugation in a complex Hilbert space W. It induces the action in the space of real-valued functions on $W : f \to \bar{f}$, where $\bar{f}(\phi) = f(\bar{\phi})$. (We hope that the symbol $\bar{f}$ will not be misleading. Only real-valued functions are considered. Thus it cannot be mixed with the operation of complex conjugation in the range of values.)

For $\widehat{A}_i \in \mathcal{L}_s(H_i, H_i)$, $i = 1, 2$, Theorem 1.1 implies

$$\int_{H_1 \times H_2} f_{A_1}(\phi_1)\bar{f}_{A_2}(\phi_2)d\mu(\phi_1, \phi_2) = \mathrm{Tr}D_{11}\widehat{A}_1\ \mathrm{Tr}D_{22}\widehat{\bar{A}_2}$$
$$+\mathrm{Tr}D_{12}\widehat{\bar{A}_2}D_{21}\widehat{A}_1.$$

7.4 The Basic Operator Equality for Arbitrary (Bounded) Self-Adjoint Operators

Lemma 4.1. Let $\Psi \in H_1 \otimes H_2$. Then for any pair of operators $\widehat{A}_j \in \mathcal{L}_s(H_j)$, $j = 1, 2$,

$$\mathrm{Tr}\widehat{\Psi}\widehat{\bar{A}}_2\widehat{\Psi}^*\widehat{A}_1 = \langle \widehat{A}_1 \otimes \widehat{A}_2\rangle_\Psi \equiv \langle \widehat{A}_1 \otimes \widehat{A}_2\Psi, \Psi\rangle. \tag{7.17}$$

Proof. We have

$$\mathrm{Tr}\widehat{\Psi}\widehat{\bar{A}}_2\widehat{\Psi}^*\widehat{A}_1 = \sum_k \langle \widehat{\Psi}\widehat{\bar{A}}_2\widehat{\Psi}^*\widehat{A}_1 e_k, e_k\rangle = \sum_k \sum_{ij} \Psi_{ij}\langle \widehat{\bar{A}}_2\widehat{\Psi}^*\widehat{A}_1 e_k, \bar{f}_j\rangle\delta_{ik}$$

$$= \sum_{ij} \Psi_{ij}\langle \widehat{\bar{A}}_2\widehat{\Psi}^*\widehat{A}_1 e_i, \bar{f}_j\rangle = \sum_{i_1 j_1}\sum_{i_2 j_2} \Psi_{i_1 j_1}\bar{\Psi}_{i_2 j_2}\langle \widehat{A}_1 e_{i_1}, e_{i_2}\rangle\langle \bar{f}_{j_2}, \widehat{\bar{A}}_2\bar{f}_{j_1}\rangle.$$

By (7.14) and (7.13) we obtain $\langle \bar{f}_{j_2}, \widehat{\bar{A}}_2\bar{f}_{j_1}\rangle = \langle \widehat{\bar{A}}_2\bar{f}_{j_2}, \bar{f}_{j_1}\rangle = \langle f_{j_1}, \widehat{A}_2 f_{j_2}\rangle = \langle \widehat{A}_2 f_{j_1}, f_{j_2}\rangle$. Thus,

$$\mathrm{Tr}\widehat{\Psi}\widehat{\bar{A}}_2\widehat{\Psi}^*\widehat{A}_1 = \sum_{i_1 j_1}\sum_{i_2 j_2} \Psi_{i_1 j_1}\bar{\Psi}_{i_2 j_2}\langle \widehat{A}_1 e_{i_1}, e_{i_2}\rangle\langle \widehat{A}_2 f_{j_1}, f_{j_2}\rangle.$$

On the other hand, we have

$$\langle \widehat{A}_1 \otimes \widehat{A}_2 \Psi, \Psi \rangle = \sum_{i_1 j_1} \sum_{i_2 j_2} \Psi_{i_1 j_1} \bar{\Psi}_{i_2 j_2} \langle \widehat{A}_1 e_{i_1}, e_{i_2} \rangle \langle \widehat{A}_2 f_{j_1}, f_{j_2} \rangle$$
$$= \mathrm{Tr} \widehat{\Psi} \widehat{\bar{A}}_2 \widehat{\Psi}^* \widehat{A}_1 .$$

In Section 3.2.2, a particular form of this equality (3.15) was used for a special class of self-adjoint operators, namely, real operators, see (3.14).

If Ψ is normalized by 1, then the right-hand side of equality (7.17) is nothing else than the average of the observable $\widehat{C} = \widehat{A}_1 \otimes \widehat{A}_2$ describing the correlations between measurement of observables $\widehat{A}_1$ and $\widehat{A}_2$ on the subsystems S_1 and S_2 of a composite system $S = (S_1, S_2)$ which is prepared in the state Ψ. On the other hand, the left-hand side of equality (7.17) has the form of the second term on the right-hand side of formula (7.5) giving Guassian integral of the product of two quadratic forms corresponding to the operators $\widehat{A}_1$ and $\widehat{A}_2$. These mathematical coincidences provide a possibility to couple quantum correlations with classical Gaussian correlations by selecting the covariance operator of the prequantum Gaussian distribution (corresponding to the quantum state Ψ) in the right way.

7.5 Operator Representation of Reduced Density Operators

We now present equalities (3.22) and (3.23) in the rigorous mathematical framework. Since we consider not only real, but arbitrary self-adjoint operators, there are some modifications.

Lemma 5.1. *For any vector $\Psi \in H_1 \otimes H_2$, the following equality holds:*

$$Tr_{H_2} \rho_\Psi = \widehat{\Psi} \widehat{\Psi}^* . \tag{7.18}$$

where $\rho_\Psi = \Psi \otimes \Psi$.

Proof. The operator $\widehat{\Psi}\widehat{\Psi}^*$ acts on a vector $x \in H_1$ as

$$\begin{aligned}
\widehat{\Psi}\widehat{\Psi}^* x &= \sum_{nm} \bar{\Psi}_{nm}\langle x, e_n\rangle \widehat{\Psi}\bar{f}_m \\
&= \sum_{nm}\sum_{kl} \Psi_{kl}\bar{\Psi}_{nm}\langle \bar{f}_m, \bar{f}_l\rangle\langle x, e_n\rangle e_k \\
&= \sum_{nk}\left[\sum_m \Psi_{km}\bar{\Psi}_{nm}\right]\langle x, e_n\rangle e_k.
\end{aligned}$$

Thus its bilinear form is given by

$$\langle \widehat{\Psi}\widehat{\Psi}^* x, y\rangle = \sum_{nk}\left[\sum_m \Psi_{km}\bar{\Psi}_{nm}\right]\langle x, e_n\rangle\langle e_k, y\rangle.$$

Take an orthonormal basis $\{f_m\}$ in H_2. The bilinear form of the operator $\mathrm{Tr}_{H_2}\rho_\Psi$ is given by

$$\begin{aligned}
\langle \mathrm{Tr}_{H_2}\rho_\Psi x, y\rangle &= \sum_m \langle \mathrm{Tr}_{H_2}\rho_\Psi x \otimes f_m, y \otimes f_m\rangle \\
&= \sum_m\sum_{kl}\langle x, e_k\rangle\langle e_l, y\rangle\langle e_k \otimes f_m, \Psi\rangle\langle \Psi, e_l \otimes f_m\rangle \\
&= \sum_{kl}\left[\sum_m \Psi_{lm}\bar{\Psi}_{km}\right]\langle x, e_k\rangle\langle e_l, y\rangle.
\end{aligned}$$

By setting $k \to n, l \to k$ we obtain the coincidence of these two bilinear forms and hence the coincidence of the operators.

Hence the use of arbitrary self-adjoint operators did not modify equality (3.22).

If the vector Ψ is normalized, then ρ_Ψ is the corresponding density operator and $\rho_\Psi^{(1)} = \mathrm{Tr}_{H_2}\rho_\Psi$ is the H_1-reduced density operator. By (7.18) we obtain

$$\rho_\Psi^{(1)} = \widehat{\Psi}\widehat{\Psi}^*. \tag{7.19}$$

Unfortunately, for the H_1-reduced density operator $\rho_\Psi^{(2)} = \mathrm{Tr}_{H_1}\rho_\Psi$, a similar statement is not true; in general,

$$\rho_\Psi^{(2)} \neq \widehat{\Psi}^*\widehat{\Psi}. \tag{7.20}$$

Thus, in general, equality (3.23) does not hold true. It has to be modified. Let us find the bilinear form of the operator $\widehat{\Psi}^*\widehat{\Psi} : H_2 \to$

H_2 :

$$\widehat{\Psi}^* \widehat{\Psi} y = \sum_{nm} \Psi_{nm} \langle y, \bar{f}_m \rangle \widehat{\Psi}^* e_n$$
$$= \sum_{nm,ij} \Psi_{nm} \bar{\Psi}_{ij} \langle y, \bar{f}_m \rangle \langle e_n, e_i \rangle \bar{f}_j$$
$$= \sum_{mj} \left[\sum_i \Psi_{im} \bar{\Psi}_{ij} \right] \langle y, \bar{f}_m \rangle \bar{f}_j.$$

$$\langle \widehat{\Psi}^* \widehat{\Psi} y, u \rangle = \sum_{mj} \left[\sum_i \Psi_{im} \bar{\Psi}_{ij} \right] \langle y, \bar{f}_m \rangle \langle \bar{f}_j, u \rangle. \tag{7.21}$$

However, for any orthonormal basis $\{e_m\}$ in H_1, we obtain

$$\langle \mathrm{Tr}_{H_1} \rho_\Psi y, u \rangle = \sum_m \langle \rho_\Psi e_m \otimes y, e_m \otimes u \rangle$$
$$= \sum_{mjl} \langle e_m \otimes f_j, \Psi \rangle \langle \Psi, e_m \otimes f_l \rangle \langle y, f_j \rangle \langle f_l, u \rangle$$
$$= \sum_{jl} \left[\sum_m \Psi_{ml} \bar{\Psi}_{mj} \right] \langle y, f_j \rangle \langle f_l, u \rangle. \tag{7.22}$$

By setting $m \to i,\ j \to m,\ l \to j$ in (7.22) we obtain

$$\langle \mathrm{Tr}_{H_1} \rho_\Psi y, u \rangle = \sum_{mj} \left[\sum_i \Psi_{ij} \bar{\Psi}_{im} \right] \langle y, f_m \rangle \langle f_j, u \rangle. \tag{7.23}$$

Comparing (7.21) and (7.23) we see that in general they do not coincide. To show this, let us take the real basis $\{f_m\}$. In this case (7.21) has the form

$$\langle \widehat{\Psi}^* \widehat{\Psi} y, u \rangle = \sum_{mj} \left[\sum_i \Psi_{im} \bar{\Psi}_{ij} \right] \langle y, f_m \rangle \langle f_j, u \rangle. \tag{7.24}$$

Thus in this special basis the matrix elements of operators $\widehat{\Psi}^* \widehat{\Psi}$ and $\mathrm{Tr}_{H_2} \rho_\Psi$ are coupled via the complex conjugation. In general, we have

Lemma 5.1* *For any vector $\Psi \in H_1 \otimes H_2$, the following equality holds for the operator $\widehat{T} = Tr_{H_1} \rho_\Psi$:*

$$\widehat{\bar{T}} = \widehat{\Psi}^* \widehat{\Psi}. \tag{7.25}$$

Proof. We now present a proof which is not based on matrix elements. In particular, it illustrates well the features of complex conjugate operators. For $\widehat{T}$ defined in the formulation, we obtain

$$\langle \widehat{\bar{T}} y,\, u\rangle = \langle \bar{u},\, \widehat{T}\bar{y}\rangle = \overline{\langle \widehat{T}\bar{y},\, \bar{u}\rangle}.$$

By (7.22) we get

$$\langle \widehat{\bar{T}} y,\, u\rangle = \sum_{mj}\left[\sum_i \bar{\Psi}_{ij}\Psi_{im}\right]\langle f_m,\, \bar{y}\rangle\langle \bar{u},\, f_j\rangle.$$

By (7.1) $\langle \bar{u},\, v\rangle = \langle \bar{v},\, u\rangle$ and hence $\langle \bar{u},\, f_j\rangle = \langle \bar{f}_j,\, u\rangle$ and $\langle u,\, \bar{v}\rangle = \langle v,\, \bar{u}\rangle$ and hence $\langle f_m,\, \bar{y}\rangle = \langle y,\, \bar{f}_m\rangle$. Thus,

$$\langle \widehat{\bar{T}} y,\, u\rangle = \sum_{mj}\left[\sum_i \bar{\Psi}_{ij}\Psi_{im}\right]\langle y,\, \bar{f}_m\rangle\langle \bar{f}_j,\, u\rangle = \langle \widehat{\Psi}^*\widehat{\Psi}y,\, u\rangle.$$

In particular, if Ψ is normalized (pure quantum state), then we obtain

$$\overline{\rho_\Psi^{(2)}} = \widehat{\Psi}^*\widehat{\Psi}. \tag{7.26}$$

Lemma 5.2. *Let ρ be a density operator in a Hilbert space W. Then, for any $\widehat{A} \in \mathcal{L}_s(W)$, the following equality holds:*

$$\mathrm{Tr}\, \bar{\rho}\widehat{\bar{A}} = \mathrm{Tr}\, \rho\widehat{A}. \tag{7.27}$$

Proof. Let $\{e_k\}$ be an orthonormal basis in W. Then

$$\begin{aligned}
\mathrm{Tr}\, \bar{\rho}\widehat{\bar{A}} &= \sum_k \langle \bar{\rho}\widehat{\bar{A}}e_k,\, e_k\rangle = \sum_k \overline{\langle \bar{e}_k,\, \rho\widehat{\bar{A}}e_k\rangle} \\
&= \sum_k \overline{\langle \rho\bar{e}_k,\, \widehat{\bar{A}}e_k\rangle} = \sum_k \langle \widehat{\bar{A}}e_k,\, \overline{\rho\bar{e}_k}\rangle = \sum_k \langle \rho\bar{e}_k,\, \widehat{A}\bar{e}_k\rangle \\
&= \sum_k \langle \widehat{A}\rho\bar{e}_k,\, \bar{e}_k\rangle.
\end{aligned}$$

Since the trace does not depend on the choice of a basis, we can select the real basis. Hence it was proved that

$$\mathrm{Tr}\, \bar{\rho}\widehat{\bar{A}} = \mathrm{Tr}\, \widehat{A}\rho.$$

Finally, we will prove that

$$\mathrm{Tr}\, \widehat{A}\rho = \mathrm{Tr}\, \rho\widehat{A}.$$

Take the basis consisting of the eigenvectors of the density operator: $\rho = \sum_k p_k e_k \otimes e_k$. Then

$$\mathrm{Tr}\, \rho\widehat{A} = \sum_k \langle \rho\widehat{A}e_k,\, e_k\rangle = \sum_k p_k\langle \widehat{A}e_k,\, e_k\rangle = \sum_k \langle \widehat{A}\rho e_k,\, e_k\rangle.$$

Corollary 5.1. *Let $\Psi \in H_1 \otimes H_2$ be normalized (pure quantum state). Then*

$$\mathrm{Tr}\, \widehat{\Psi}^*\widehat{\Psi}\widehat{\bar{A}} = \mathrm{Tr}\, \rho_\Psi^{(2)}\widehat{A}. \tag{7.28}$$

7.6 Classical Random Field Representation of Quantum Correlations

Consider a composite quantum system $S = (S_1, S_2)$. Here S_j has the state space H_j, a complex Hilbert space. Let $\phi_1(\omega)$ and $\phi_2(\omega)$ be two Gaussian random fields in the Hilbert spaces H_1 and H_2, respectively. Consider the Cartesian product of these Hilbert spaces, $H_1 \times H_2$, and the vector Gaussian random field $\phi(\omega) = (\phi_1(\omega), \phi_2(\omega)) \in H_1 \times H_2$. In the case under consideration, its covariance operator has the block structure given by (7.4). Set

$$\langle f_{A_1}, f_{A_2} \rangle \equiv E f_{A_1} \bar{f}_{A_2} = \int_{H_1 \times H_2} f_{A_1}(\phi_1) \bar{f}_{A_2}(\phi_2) d\mu(\phi_1, \phi_2)$$

$$= \int_{H_1 \times H_2} f_{A_1}(\phi_1) f_{A_2}(\overline{\phi}_2) d\mu(\phi_1, \phi_2)$$

Set also

$$\mathrm{cov}\,(f_{A_1}, f_{A_2}) = \langle f_{A_1}, f_{A_2} \rangle - \langle f_{A_1} \rangle \langle \bar{f}_{A_2} \rangle$$

Equalities (7.17) and (7.17) imply

Proposition 6.1. *Let $\widehat{A}_i \in \mathcal{L}_s(H_i)$, $i = 1, 2$ and let $\Psi \in H_1 \otimes H_2$ with the unit norm. Then, for any Gaussian random field $\phi(\omega)$ in $H_1 \times H_2$ with the covariance matrix D such that the nondiagonal block*

$$D_{12} = \widehat{\Psi}, \tag{7.29}$$

the following equality takes place

$$\mathrm{cov}\,(f_{A_1}, f_{A_2}) = (\widehat{A}_1 \otimes \widehat{A}_2 \Psi, \Psi) \equiv \langle \widehat{A}_1 \otimes \widehat{A}_2 \rangle_\Psi. \tag{7.30}$$

This equality establishes a coupling between the quantum and classical correlations. In Chapter 2 a particular form of this equality (3.21) was used for a special class of self-adjoint operators, namely, real operators, see (3.14).

In the next section we shall unify classical descriptions for a single system and a composite system.

7.6.1 *Making Consistent PCSFT Models for Ensembles of Noncomposite and Composite Systems*

The operators D_{ii} are responsible for averages of functionals depending only on one of the components of the vector random field $\phi(\omega)$. In particular, $E f_{A_1}(\phi_1)(\omega)) = \mathrm{Tr} D_{11} \widehat{A}_1$ and $E \bar{f}_{A_2}(\phi_2)(\omega)) = \mathrm{Tr} D_{22} \widehat{\bar{A}}_2$. We will construct a random field such that these "marginal averages" will match those given by QM. For the latter, we have

$$\langle \widehat{A}_1 \rangle_\Psi = (\widehat{A}_1 \otimes I \Psi, \Psi) = \mathrm{Tr} \rho_\Psi^{(1)} \widehat{A}_1, \ \langle \widehat{A}_2 \rangle_\Psi = (I \otimes \widehat{A}_2 \Psi, \Psi) = \mathrm{Tr} \rho_\Psi^{(2)} \widehat{A}_2,$$

where I denotes the unit operator. By equality (7.19) the first average can be written as $\langle \widehat{A}_1 \rangle_\Psi = \mathrm{Tr}(\widehat{\Psi} \widehat{\Psi}^*) \widehat{A}_1$. By equality (7.27) the second average can be written as $\langle \widehat{A}_2 \rangle_\Psi = \mathrm{Tr} \overline{\rho_\Psi^{(2)}} \widehat{\bar{A}}_2$ and, finally, by (7.26), $\langle \widehat{A}_2 \rangle_\Psi = \mathrm{Tr}(\widehat{\Psi}^* \widehat{\Psi}) \widehat{\bar{A}}_2$, Thus it would be natural to take

$$D = \begin{pmatrix} \widehat{\Psi} \widehat{\Psi}^* & \widehat{\Psi} \\ \widehat{\Psi}^* & \widehat{\Psi}^* \widehat{\Psi} \end{pmatrix}.$$

Its off-diagonal block reproduces correct quantum correlations between the systems S_1 and S_2, and its diagonal blocks produce correct quantum averages for the systems S_1 and S_2. However, in general (i.e., for an arbitrary pure state Ψ) this operator is not positively defined.

Proposition 6.2. *For any normalized vector $\Psi \in H_1 \otimes H_2$ and any $\epsilon \geq 1/4$, the operator*

$$D = \begin{pmatrix} (\widehat{\Psi} \widehat{\Psi}^* + \epsilon I) & \widehat{\Psi} \\ \widehat{\Psi}^* & (\widehat{\Psi}^* \widehat{\Psi} + \epsilon I) \end{pmatrix} \tag{7.31}$$

is positively defined.

Proof. For any vector $\phi = (\phi_1, \phi_2) \in \Psi \in H_1 \times H_2$, we have

$$(\tilde{D}_\Psi \phi, \phi) = ||\widehat{\Psi}^* \phi_1||^2 + \frac{||\phi_1||^2}{4} + (\widehat{\Psi} \phi_2, \phi_1) + (\widehat{\Psi}^* \phi_1, \phi_2) + ||\widehat{\Psi} \phi_2||^2$$
$$+ \frac{||\phi_2||^2}{4} \geq \left(||\Psi \phi_1||^2 - ||\Psi^* \phi_1|| ||\phi_2|| + \frac{||\phi_2||^2}{4} \right)$$
$$+ \left(||\widehat{\Psi} \phi_2||^2 - ||\phi_1|| ||\widehat{\Psi} \phi_2|| + \frac{||\phi_1||^2}{4} \right) \geq 0.$$

Thus the operator D is positively defined.

7.7 Infinite-Dimensional Case

In the infinite-dimensional case the Gaussian distribution with the covariance operator given by (7.31) is not σ-additive on the Hilbert space $\mathbf{H} = H_1 \times H_2$. To make it σ-additive, one has to consider a *rigged Hilbert space* $\Omega \subset \mathbf{H} \subset \Omega'$, where Ω is a nuclear topological linear space (see Section 6.2).

Proposition 7.1. *For any normalized vector $\Psi \in H_1 \otimes H_2$, the operator D given by (7.31) determines a σ-additive Gaussian distribution $\mu \equiv \mu_D$ on the space Ω' or equivalently a Gaussian random field $\phi(\omega)$ valued in Ω'. Any trace-class operator $\widehat{A} \in \mathcal{L}_s(\mathbf{H})$ determines a measurable quadratic form f_A which is defined, a.e. on Ω'; for any pair of trace-class operators $\widehat{A}_i \in \mathcal{L}_s(H_i)$, $i = 1, 2$, the measurable quadratic forms f_{A_i} satisfy equalities (3.29), (3.29), and (7.30).*

Proof. Take $g \in \mathbf{H}$ and consider the map $\omega \to \langle g, \omega\rangle$, $\omega \in \mathbf{H}$. It can be continued to a measurable linear functional on Ω'. belonging to the space of square-integrable functions $L_2(\Omega', \mu)$. Although this fact is well known, we present the proof for completeness of the presentation. We consider an orthonormal basis in $\mathbf{H}$ consisting of elements of Ω : $\{g_j\}$. It determines a sequence of Gaussian random variables $g_j(\omega) = \langle g_j, \omega\rangle$, $\omega \in \Omega'$. We set $\langle g, \omega\rangle = \sum_{j=1}^{\infty}\langle g, g_j\rangle\langle g_j, \omega\rangle$. We prove that this series converges in $L_2(\Omega', \mu)$. Set $S_N(\omega) = \sum_{j=1}^{N}\langle g, g_j\rangle\langle g_j, \omega\rangle$. We have

$$E\,S_N^2(\omega) = \sum_{j,k=1}^{N}\langle g, g_j\rangle\langle g_k, g\rangle E\,\langle g_j, \omega\rangle\langle \omega, g_k\rangle$$

$$= \sum_{j,k=1}^{N}\langle g, g_j\rangle\langle g_k, g\rangle\langle Dg_j, g_k\rangle = \langle DL_N g, L_N g\rangle,$$

where L_N is the orthonormal projector on the subspace generated by the first N basis vectors. We have: a) $E\,S_N^2(\omega) \leq \|D\|\|g\|^2$; b) $S_N^2(\omega) \leq S_{N+1}^2(\omega)$. Thus by Beppo Levi's theorem we obtain that $S^2(\omega) = \lim_{N\to\infty} S_N^2(\omega)$, a.e., is an integrable function. This implies that $S(\omega) = \lim_{N\to\infty} S_N(\omega)$ in the L_2-space. We set $\langle g, \omega\rangle \equiv S(\omega)$.

Now take a self-adjoint trace-class operator $\widehat{A}$. We define the corresponding measurable quadratic form by setting

$$f_A(\omega) \equiv \sum_{j,i=1}^{\infty} \langle \widehat{A} e_j, e_i \rangle \langle e_j, \omega \rangle \langle \omega, e_i \rangle, \tag{7.32}$$

where $\{e_j\}$ is an orthonormal basis in $\mathbf{H}$. To simplify the presentation, we take the basis consisting of eigenvectors of $\widehat{A}$ with the corresponding eigenvalues a_j. Here $f_A(\omega) \equiv \sum_{j=1}^{\infty} a_j |\langle e_j, \omega \rangle|^2$.

Denote the orthonormal projector on the subspace generated by the first N eigenvectors by the symbol P_N. Set $\widehat{A_N} = P_N \widehat{A}$. We will also consider the absolute values $|\widehat{A}|$ and $|\widehat{A}_N|$ of the operators $\widehat{A}$ and $\widehat{A_N}$. Then $f_{|A_N|}(\omega) = \sum_{j=1}^{N} |a_j| |\langle e_j, \omega \rangle|^2$. We have

$$\begin{aligned} E f^2_{|A_N|}(\omega) &= (\mathrm{Tr} D |\widehat{A_N}|)^2 + 2\mathrm{Tr}(D|\widehat{A_N}|)^2 \\ &= \left(\sum_{j=1}^{N} |a_j| \langle D e_j, e_j \rangle \right)^2 + \sum_{j=1}^{N} |a_j| \langle D e_j, P_N \widehat{A} D e_j \rangle \\ &\leq (\|D\| \mathrm{Tr}|\widehat{A}|)^2 + \|D\|^2 \|\widehat{A}\| \mathrm{Tr}|\widehat{A}|. \end{aligned}$$

We now again apply Beppo Levi's theorem to the sequence $\{f^2_{|A_N|}(\omega)\}$. It converges to an integrable function. Hence the series $\sum_{j=1}^{\infty} a_j |\langle e_j, \omega \rangle|^2$ converges in the L_2-space and it determines (a.e.) a measurable quadratic functional $f_A(\omega)$. We now consider equalities (3.29), (3.30), and (7.30). We prove (3.29). Consider the measurable quadratic functional $f_{A_1}(\omega_1) = \sum_{j=1}^{\infty} a_{1j} |\langle e_{1j}, \omega_1 \rangle|^2$, where $\widehat{A}_1 e_{1j} = a_{1j} e_{1j}$. Then

$$\begin{aligned} E f_{A_1}(\omega_1) &= \sum_{j=1}^{\infty} a_{1j} E |\langle e_{1j}, \omega_1 \rangle|^2 \\ &= \sum_{j=1}^{\infty} a_{1j} \langle D e_{1j}, e_{1j} \rangle = \langle \widehat{A}_1 \rangle_\Psi + \epsilon \mathrm{Tr} \widehat{A}_1. \end{aligned}$$

By Proposition 7.1 quantum observables are represented by measurable quadratic Gaussian functionals.

Example 7.1. (Non-Hilbert case). For example, $\mathbf{H} = L_2(\mathbf{R}^k)$ is the space of square-integrable functions, $\Omega = \mathcal{S}(\mathbf{R}^k)$ is the space of Schwartz test functions, and $\Omega' = \mathcal{S}'(\mathbf{R}^k)$ is the space of Schwartz distributions. We remark that the space of Schwartz test functions $\mathcal{S}(\mathbf{R}^k)$ is a nuclear space. In this case, the formula for the Gaussian integral of a quadratic form can be written as

$$\int_{\mathcal{S}'(\mathbf{R}^k)} f_A(\phi) d\mu(\phi) = \text{Tr } D\widehat{A}. \quad (7.33)$$

We remark that the function $\phi \to f_A(\phi)$ from $\mathcal{S}'(\mathbf{R}^k)$ to $\mathbf{R}$ is a measurable quadratic form on the space of Schwartz distributions. As usual, correlations are represented in the form

$$\int_{\mathcal{S}'(\mathbf{R}^3)\times\mathcal{S}'(\mathbf{R}^3)} f_{A_1}(\phi_1) f_{A_2}(\phi_2) d\mu_D(\phi_1, \phi_2)$$
$$= \text{Tr}_{L_2(\mathbf{R}^3)} D\widehat{A}_1 \text{Tr}_{L_2(\mathbf{R}^3)} D\widehat{A}_2 + \text{Tr}_{L_2(\mathbf{R}^3)} D\widehat{A}_2 D\widehat{A}_1. \quad (7.34)$$

7.8 Correlations in Triparticle Systems

Consider a triparticle system $S = (S_1, S_2, S_3)$, where the subsystem S_j has the state space H_j, $j = 1, 2, 3$. Take a pure state Ψ, $\Psi \in H = H_1 \otimes H_2 \otimes H_3$, $||\Psi|| = 1$. As always, we work with the corresponding density operator $\rho \equiv \rho_\Psi = \Psi \otimes \Psi (= |\Psi\rangle\langle\Psi|)$ and its partial trace operators $\rho^{(ij)} = \text{Tr}_{H_k} \rho$, where $i \neq i \neq k$.

To simplify our consideration, we proceed in the case of a *real Hilbert space.*

Set

$$H_{12} = H_2 \otimes H_2.$$

Then

$$H = H_{12} \otimes H_3.$$

This corresponds to the representation of a composite system $S = (S_1, S_2, S_3)$ consisting of three subsystems as a composite system $S = (S_{12}, S_3)$, where $S_{12} = (S_1, S_2)$ by itself consists of two subsystems.

Now we apply to this case the previously developed theory coupling quantum and classical bi-correlations. The pure state (wave function) Ψ can be realized as the operator $\widehat{\Psi}_3 : H_3 \to H_{12}$. It determines the covariance operator

$$D^{(3)} = \begin{pmatrix} \widehat{\Psi}_3 \widehat{\Psi}_3^* + \epsilon I & \widehat{\Psi}_3 \\ \widehat{\Psi}_3^* & \widehat{\Psi}_3 \widehat{\Psi}_3 + \epsilon I \end{pmatrix}, \quad (7.35)$$

where $D^{(3)} : H_{12} \times H_3 \to H_{12} \times H_3$. (We remark that $(\widehat{\Psi}_3 y_3, y_1 \otimes y_2) = (\Psi, y_1 \otimes y_2 \otimes y_3), y_j \in H_j, j = 1, 2, 3$.) Consider the corresponding Gaussian random vector

$$\omega \to (\phi_{12}(\omega), \phi_3(\omega)) \in H_{12} \times H_3.$$

Denote by $(\Omega_3, \mathcal{F}_3, P_3)$ the corresponding Kolmogorov probability space.

Now we consider the component ϕ_{12} of this vector. For each $\omega \in \Omega_3$, $\phi_{12}(\omega) \in H_{12}$. Thus it determines the covariance operator

$$D_{12}(\omega) = \begin{pmatrix} \widehat{\phi}_{12}(\omega)\widehat{\phi}_{12}^*(\omega) + \epsilon I & \widehat{\phi}_{12}(\omega) \\ \widehat{\phi}_{12}^*(\omega) & \widehat{\phi}_{12}^*\widehat{\phi}_{12}(\omega) + \epsilon I \end{pmatrix}. \tag{7.36}$$

For each $\omega \in \Omega_3$, take the corresponding Gaussian random vector

$$\tilde{\omega} \to (\phi_1(\tilde{\omega}; \omega), \phi_2(\tilde{\omega}; \omega)) \in H_1 \times H_2.$$

For each ω, we introduce its own Kolmogorov probability space

$$\mathcal{P}_{12}(\omega) = (\Omega_{12}(\omega), \mathcal{F}_{12}(\omega), P_{12}(\omega)).$$

Remark 8.1. The set Ω_{12} (of random parameters $\tilde{\omega}$) and the σ-algebra $\mathcal{F}_{12}$ can be chosen independently of ω. We can take

$$\Omega_{12} = H_1 \times H_2$$

and $\mathcal{F}_{12}$ as the σ-algebra of its Borel subsets. We also remark that we can take

$$\Omega_3 = H_{12} \times H_3.$$

We now define the Kolmogorov space with the set

$$\Omega_{12,3} = \Omega_{12} \times \Omega_3$$

(random parameters $\xi = (\tilde{\omega}, \omega)$) and the probability measure

$$P_{12,3}(O_{12} \times O_3) = \int_{O_{12} \times O_3} dP(\xi) = \int_{O_3} dP_3(\omega) \int_{O_{12}} dP_{12}(\tilde{\omega}; \omega),$$

where $O_{12} \in \mathcal{F}_{12}$ and $O_3 \in \mathcal{F}_3$. Thus formally,

$$dP_{12,3}(\xi) = dP_3(\omega) dP_{12}(\tilde{\omega}; \omega).$$

We also define the random vector

$$\xi \to (\phi_1(\xi), \phi_2(\xi), \phi_3(\xi)) \in H_1 \times H_2 \times H_3,$$

where

$$\phi_3(\xi) = \phi_3(\omega),\ \phi_i(\xi) = \phi_i(\tilde{\omega}; \omega),\ i = 1, 2.$$

Denote its probability distribution (on the space $H_1 \times H_2 \times H_3$) by the symbol $\mu_{12,3}$. We find the "cross-correlation block" of its third momentum ($y_j \in H_j$, $j = 1, 2, 3$)

$$Q_{123}(y_1, y_2, y_3) = E(y_1, \phi(\xi))(y_2, \phi(\xi))(y_3, \phi(\xi))$$
$$= \int_{H_1 \times H_2 \times H_3} d\mu_{12,3}(x_1, x_2, x_3)(y_1, x_1)(y_2, x_2)(y_3, x_3)$$
$$\int_{\Omega_3} dP_3(\omega) \int_{\Omega_{12}} dP_{12}(\tilde{\omega}; \omega)(y_1, \phi_1(\tilde{\omega}; \omega))(y_2, \phi_2(\tilde{\omega}; \omega))(y_3, \phi_3(\omega))$$
$$= \int_{\Omega_3} dP_3(\omega)(y_3, \phi_3(\omega))(\widehat{\phi}_{12}(\omega)y_2, y_1)$$
$$= \int_{\Omega_3} dP_3(\omega)(y_3, \phi_3(\omega))(\phi_{12}(\omega), y_1 \otimes y_2)$$
$$= (\widehat{\Psi}_3 y_3, y_1 \otimes y_2) = (\Psi, y_1 \otimes y_2 \otimes y_3).$$

Thus this block coincides with the wave function of the triparticle system

$$Q_{123} = \Psi. \tag{7.37}$$

This equality induces the PCSFT interpretation of the wave function of the triparticle system. It represents correlations between degrees of freedom of particles.

We now find triple correlations of classical variables corresponding to three quantum observables, $\widehat{A}_j \in \mathcal{L}(H_j)$, $j = 1, 2, 3$. They are given by quadratic forms of prequantum fields. For simplicity, we set $\epsilon = 1$ in the covariation operators given by (7.35) and (7.36). We have

$$E f_{A_1}(\phi_1) f_{A_2}(\phi_2) f_{A_3}(\phi_3)$$
$$\equiv \int_{H_1 \times H_2 \times H_3} d\mu_{12,3}(x_1, x_2, x_3) f_{A_1}(x_1) f_{A_2}(x_2) f_{A_2}(x_3)$$
$$= \int dP_3(\omega) f_{A_3}(\phi_3(\omega)) \int dP_{12}(\tilde{\omega}) f_{A_1}(\phi_1(\tilde{\omega}; \omega)) f_{A_2}(\phi_2(\tilde{\omega}; \omega))$$
$$= \int dP_3(\omega) f_{A_3}(\phi_3(\omega)) G_{A_1 A_2}(\omega),$$

where

$$G_{A_1A_2}(\omega) = (\mathrm{Tr}(\widehat{\phi}_{12}(\omega)\widehat{\phi}^*_{12}(\omega) + I)\widehat{A}_1)(\mathrm{Tr}(\widehat{\phi}^*_{12}(\omega)\widehat{\phi}_{12}(\omega) + I)\widehat{A}_2)$$
$$+2(\widehat{A}_1 \otimes \widehat{A}_2\phi_{12}(\omega), \phi_{12}(\omega)).$$

First we find the most important term

$$\int dP_3(\omega) f_{A_3}(\phi_3(\omega)) f_{A_1\otimes A_2}(\phi_{12}(\omega))$$
$$= (\mathrm{Tr}(\widehat{\Psi}_3^*\widehat{\Psi}_3 + I)\widehat{A}_1 \otimes \widehat{A}_2)(\mathrm{Tr}(\widehat{\Psi}_3\widehat{\Psi}_3^* + I)\widehat{A}_3)$$
$$+2(\widehat{A}_1 \otimes \widehat{A}_2 \otimes \widehat{A}_3\Psi, \Psi)$$
$$= (\mathrm{Tr}\rho_\Psi^{(12)}\widehat{A}_1 \otimes \widehat{A}_2 + \mathrm{Tr}\widehat{A}_1\mathrm{Tr}\widehat{A}_2)(\mathrm{Tr}\rho_\Psi^{(3)}\widehat{A}_3 + \mathrm{Tr}\widehat{A}_3)$$
$$+2\mathrm{Tr}\rho_\Psi\widehat{A}_1 \otimes \widehat{A}_2 \otimes \widehat{A}_3$$
$$= (\langle \widehat{A}_1 \otimes A_2\rangle_\Psi + \mathrm{Tr}\widehat{A}_1\mathrm{Tr}\widehat{A}_2)(\langle \widehat{A}_3\rangle_\Psi + \mathrm{Tr}\widehat{A}_3) + \langle \widehat{A}_1 \otimes \widehat{A}_2 \otimes \widehat{A}_3\rangle_\Psi.$$

We remark that

$$\mathrm{Tr}(\widehat{\phi}_{12}(\omega)\widehat{\phi}^*_{12}(\omega) + I)\widehat{A}_1 = E_{\tilde{\omega}} f_{A_1}(\phi_1(\tilde{\omega}; \omega)), \tag{7.38}$$

$$\mathrm{Tr}(\widehat{\phi}^*_{12}(\omega)\widehat{\phi}_{12}(\omega) + I)\widehat{A}_2 = E_{\tilde{\omega}} f_{A_2}(\phi_2(\tilde{\omega}; \omega)), \tag{7.39}$$

where $E_{\tilde{\omega}}$ denotes the average with respect to $\tilde{\omega}$. Thus we obtain

$$E f_{A_1} f_{A_2} f_{A_3} - E_\omega f_{A_3}(\phi_3(\omega)) E_{\tilde{\omega}} f_{A_1}(\phi_1(\tilde{\omega}; \omega)) E_{\tilde{\omega}} f_{A_2}(\phi_2(\tilde{\omega}; \omega))$$
$$= (\langle \widehat{A}_1 \otimes \widehat{A}_2\rangle_\Psi + \mathrm{Tr}\widehat{A}_1\mathrm{Tr}\widehat{A}_2)(\langle \widehat{A}_3\rangle_\Psi + \mathrm{Tr}\widehat{A}_3) + \langle \widehat{A}_1 \otimes \widehat{A}_2 \otimes \widehat{A}_3\rangle_\Psi.$$

We denote the average with respect to ω by brackets $\langle \ldots \rangle$ and the average with respect to $\tilde{\omega}$ by brackets $\langle \ldots \rangle$. The above equality can be written as

$$\langle f_{A_1} f_{A_2} f_{A_3}\rangle - \langle f_{A_3} < f_{A_1} >< f_{A_2} >\rangle$$
$$= (\langle \widehat{A}_1 \otimes \widehat{A}_2\rangle_\Psi + \mathrm{Tr}\widehat{A}_1\mathrm{Tr}\widehat{A}_2)(\langle \widehat{A}_3\rangle_\Psi + \mathrm{Tr}\widehat{A}_3) + \langle \widehat{A}_1 \otimes \widehat{A}_2 \otimes \widehat{A}_3\rangle_\Psi. \tag{7.40}$$

The left-hand side of this equality contains only classical averages (with respect to fluctuations of the prequantum field) and the right-hand side contains only quantum averages. Of course, this coupling between classical and quantum tri-correlations is quite complicated. Nevertheless, this equality shows that, in spite of the common opinion, quantum correlations for a triparticle system can be expressed as correlations of three classical random signals. We now find pairwise correlations for quadratic forms of these signals:

$$\langle f_{A_1} f_{A_2}\rangle \equiv E f_{A_1} f_{A_2} = \int dP_3(\omega)\int dP_{12}(\tilde{\omega}) f_{A_1}(\phi_1(\tilde{\omega};\omega) f_{A_2}(\phi_2(\tilde{\omega};\omega))$$
$$= \int dP_3(\omega) F_{A_1,A_2}(\omega),$$

where

$$F_{A_1,A_2}(\omega) = (\mathrm{Tr}\widehat{\phi}_{12}(\omega)\widehat{\phi}^*_{12}(\omega)\widehat{A}_1 + \mathrm{Tr}\widehat{A}_1)(\mathrm{Tr}(\widehat{\phi}^*_{12}(\omega)\widehat{\phi}_{12}(\omega)\widehat{A}_2$$
$$+\mathrm{Tr}\widehat{A}_2) + 2(\widehat{A}_1 \otimes \widehat{A}_2\phi_{12}(\omega), \phi_{12}(\omega)).$$

We find the last term (in fact, 1/2 of it):

$$\int dP_3(\omega) f_{A_1\otimes\widehat{A}_2}(\phi_{12}(\omega)) = \mathrm{Tr}\widehat{\Psi}_3\widehat{\Psi}^*_3\widehat{A}_1 \otimes \widehat{A}_2 + \mathrm{Tr}\widehat{A}_1\mathrm{Tr}\widehat{A}_2$$
$$= \langle\widehat{A}_1 \otimes \widehat{A}_2\rangle_\Psi + \mathrm{Tr}\widehat{A}_1\mathrm{Tr}\widehat{A}_2.$$

By using (7.38) and (7.39) we get

$$E f_{A_1} f_{A_2} - E_\omega(E_{\tilde{\omega} f_{A_1}}(\phi_1(\tilde{\omega};\omega))E_{\tilde{\omega}} f_{A_2}(\phi_2(\tilde{\omega};\omega))$$
$$= 2(\mathrm{Tr}\widehat{A}_1\mathrm{Tr}\widehat{A}_2 + \langle\widehat{A}_1 \otimes \widehat{A}_2\rangle_\Psi),$$

or in shorter notation,

$$\frac{1}{2}\left[\langle f_{A_1} f_{A_2}\rangle - \langle < f_{A_1} >< f_{A_2} >\rangle\right] = \mathrm{Tr}\widehat{\mathrm{A}}_1\mathrm{Tr}\widehat{\mathrm{A}}_2 + \langle\widehat{\mathrm{A}}_1 \otimes \widehat{\mathrm{A}}_2\rangle_\Psi. \tag{7.41}$$

We also find

$$\langle f_{A_1} f_{A_3}\rangle = \int dP_3(\omega) f_{A_3}(\phi_3(\omega))\int dP_{12}(\tilde{\omega}) f_{A_1}(\phi_1(\tilde{\omega};\omega))$$
$$= \int dP_3(\omega) f_{A_3}(\phi_3(\omega))(\mathrm{Tr}\widehat{\phi}_{12}(\omega)\widehat{\phi}^*_{12}(\omega) + I)\widehat{A}_1.$$

We have, for any two orthonormal bases, $\{e_k\}$ in H_1 and $\{g_j\}$ in H_2,

$$\mathrm{Tr}\widehat{\phi}_{12}\widehat{\phi}^*_{12}\widehat{A}_1 = \mathrm{Tr}[(\mathrm{Tr}_{H_2}\phi_{12} \otimes \phi_{12})\widehat{A}_1]$$
$$= \sum_k\sum_j(\phi_{12} \otimes \phi_{12}\widehat{A}_1 e_k \otimes g_j, e_k \otimes g_j)$$
$$= \sum_{k,j}(\phi_{12}, \widehat{A}_1 e_k \otimes g_j)(\phi_{12} e_k \otimes f_j)$$
$$= \sum_{kj}(\widehat{A}_1 \otimes I\phi_{12}, e_k \otimes g_j)(e_k \otimes f_j, \phi_{12})$$
$$= (A_1 \otimes I\phi_{12}, \phi_{12}) = f_{A_1\otimes I}(\phi_{12}).$$

Thus,

$$\langle f_{A_1} f_{A_3}\rangle = \mathrm{Tr}\widehat{A}_1\langle f_{A_3}\rangle + \langle f_{A_1\otimes I} f_{A_3}\rangle$$
$$= \mathrm{Tr}\widehat{A}_1\langle f_{A_3}\rangle + \langle f_{A_1\otimes I}\rangle\langle f_{A_3}\rangle + 2\langle A_1 \otimes A_3\rangle_\Psi$$

or

$$\frac{1}{2}(\langle f_{A_1} f_{A_3}\rangle - \langle f_{A_1\otimes I}\rangle\langle f_{A_3}\rangle - \mathrm{Tr}\widehat{A}_1\langle f_{A_3}\rangle) = \langle A_1 \otimes A_3\rangle_\Psi. \tag{7.42}$$

7.8.1 *Symmetrization*

All these formulas coupling classical and quantum correlations are asymmetric with respect to $\widehat{A}_1$, $\widehat{A}_2$, $\widehat{A}_3$ and f_{A_1}, f_{A_2}, f_{A_3}. The reason is clear. In the previous construction we unified subsystems S_1 and S_2 in one system $S_{12} = (S_1, S_2)$ and represented the correlations between degrees of freedom of three systems by averaging the correlations inside S_{12} with respect to fluctuations in S_3. To make the classical–quantum coupling symmetric, we should repeat this construction by using permutations of subsystems and then take the symmetric mixture of three probabilities. Hence, except for the probability space with $\Omega_{12,3}$, $P_{12,3}$, we also consider the probability spaces with $\Omega_{13,2}$, $P_{12,2}$ and $\Omega_{23,1}$, $P_{23,1}$. We set

$$\Omega_{123} = \Omega_{12,3} \cup \Omega_{13,2} \cup \Omega_{23,1}$$

and take the symmetric mixture of probabilities

$$P_{123} = \frac{1}{3}[P_{12,3} + P_{13,2} + P_{23,1}].$$

By using the equality for tri-correlations of classical variables and quantum observables, see (7.40), and its cyclic permutations, we obtain

$$\langle f_{A_1} f_{A_2} f_{A_3} \rangle_{123} - \frac{1}{3}[\langle f_{A_3} < f_{A_1} >< f_{A_2} > \rangle_{12,3} + \langle f_{A_2} < f_{A_1} >< f_{A_3} > \rangle_{13,2}$$

$$+ \langle f_{A_1} < f_{A_2} >< f_{A_3} > \rangle_{23,1}] = \langle \widehat{A}_1 \otimes \widehat{A}_2 \otimes \widehat{A}_3 \rangle_\Psi + \mathrm{Tr} \widehat{A}_1 \otimes \widehat{A}_2 \otimes \widehat{A}_3$$

$$+ \frac{1}{3}[\langle \widehat{A}_1 \otimes \widehat{A}_2 \rangle_\Psi \langle \widehat{A}_3 \rangle_\Psi + \langle \widehat{A}_1 \otimes \widehat{A}_3 \rangle_\Psi \langle \widehat{A}_2 \rangle_\Psi + \langle \widehat{A}_2 \otimes \widehat{A}_3 \rangle_\Psi \langle \widehat{A}_1 \rangle_\Psi]$$

$$+ \frac{1}{3}[\mathrm{Tr} \widehat{A}_3 \langle \widehat{A}_1 \otimes \widehat{A}_2 \rangle_\Psi + \mathrm{Tr} \widehat{A}_2 \langle \widehat{A}_1 \otimes \widehat{A}_3 \rangle_\Psi + \mathrm{Tr} \widehat{A}_1 \langle \widehat{A}_2 \otimes \widehat{A}_3 \rangle_\Psi]$$

$$+ \frac{1}{3}[\mathrm{Tr} \widehat{A}_1 \otimes \widehat{A}_2 \langle \widehat{A}_3 \rangle_\Psi + \mathrm{Tr} \widehat{A}_1 \otimes \widehat{A}_3 \langle \widehat{A}_2 \rangle_\Psi + \mathrm{Tr} \widehat{A}_2 \otimes \widehat{A}_3 \langle \widehat{A}_1 \rangle_\Psi].$$

We now find the pairwise correlations of prequantum variables (quadratic forms) corresponding to three quantum observables $\widehat{A}_j \in \mathcal{L}(H_j)$, $j = 1, 2, 3$. We shall use the special representation of probability spaces given in Remark 8.1. It matches better with the physical situation (integration with respect to classical prequantum fields).

We start with the bi-correlation $Ef_{A_1}f_{A_2}$. By using the representation (7.41) we obtain

$$\langle f_{A_1} f_{A_2}\rangle = \int_{H_1\times H_2\times H_3} d\mu_{12,3}(x_1,x_2,x_3) f_{A_1}(x_1) f_{A_2}(x_2)$$
$$= \langle < f_{A_1} >< f_{A_2} >\rangle + 2(\langle \widehat{A}_1\otimes\widehat{A}_2\rangle_\Psi + \mathrm{Tr}\widehat{A}_1\otimes\widehat{A}_2).$$

We remark that

$$\mathrm{Tr}\widehat{\phi}_{12}\widehat{\phi}_{12}^*\widehat{A}_1 = (\widehat{A}_1^*\otimes I\phi_{12},\phi_{12}) = f_{A_1\otimes I}(\phi_{12});$$
$$\mathrm{Tr}\widehat{\phi}_{12}^*\widehat{\phi}_{12}\widehat{A}_2 = f_{I\otimes A_2}(\phi_{12}).$$

Set

$$F(\phi_{12}) = f_{A_1\otimes I}(\phi_{12}) f_{I\otimes A_2}(\phi_{12}) + \mathrm{Tr}\widehat{A}_1 f_{I\otimes A_2}(\phi_{12})$$
$$+ \mathrm{Tr}\widehat{A}_2 f_{A_1\otimes I}(\phi_{12}) + \mathrm{Tr}\widehat{A}_1\mathrm{Tr}\widehat{A}_2.$$

Then

$$\langle < f_{A_1} >< f_{A_2} >\rangle = \int_{H_{12}\times H_3} dP_3(\phi_{12},\phi_3)\; F(\phi_{12})$$

$$= \mathrm{Tr}\widehat{A}_1(\mathrm{Tr}\widehat{\Psi}_3\widehat{\Psi}_3^* I\otimes\widehat{A}_2 + \mathrm{TrI}\otimes\widehat{\mathrm{A}}_2)$$

$$+\mathrm{Tr}\widehat{A}_2(\mathrm{Tr}\widehat{\Psi}_3\widehat{\Psi}_3^*\widehat{A}_1\otimes I + \mathrm{Tr}\widehat{A}_1\otimes I) + \mathrm{Tr}\widehat{A}_1\otimes\widehat{A}_2$$

$$+(\mathrm{Tr}\widehat{\Psi}_3\widehat{\Psi}_3^*\widehat{A}_1\otimes I + \mathrm{Tr}\widehat{A}_1\otimes I)(\mathrm{Tr}\widehat{\Psi}_3\widehat{\Psi}_3^* I\otimes\widehat{A}_2 + \mathrm{Tr}I\otimes\widehat{A}_2)$$

$$+2\mathrm{Tr}(\widehat{\Psi}_3\widehat{\Psi}_3^* + I)(I\otimes\widehat{A}_2)(\widehat{\Psi}_3\widehat{\Psi}_3^* + I)(\widehat{A}_1\otimes I)$$

$$= \mathrm{Tr}\widehat{A}_1(\langle I\otimes\widehat{A}_2\rangle_\Psi + \mathrm{Tr}(I\otimes\widehat{A}_2)) + \mathrm{Tr}\widehat{A}_2(\langle \mathrm{A}_1\otimes\mathrm{I}\rangle_\Psi + \mathrm{Tr}(\widehat{\mathrm{A}}_1\otimes\mathrm{I}))$$

$$\mathrm{Tr}\widehat{A}_1\otimes\widehat{A}_2 + (\langle\widehat{A}_1\otimes I\rangle_\Psi + \mathrm{Tr}\widehat{A}_1\otimes I)(\langle I\otimes\widehat{A}_2\rangle_\Psi + \mathrm{Tr}I\otimes\widehat{A}_2)$$

$$+2(2\langle\widehat{A}_1\otimes\widehat{A}_2\rangle_\Psi + \mathrm{Tr}\widehat{A}_1\otimes I\,\mathrm{Tr}I\otimes\widehat{A}_2 + \mathrm{Tr}\rho^{(12)}\widehat{A}_2\rho^{(12)}\widehat{A}_1$$

$$= \mathrm{Tr}\widehat{A}_1(\langle\widehat{A}_2\rangle_\Psi + \mathrm{Tr}I\otimes\widehat{A}_2) + \mathrm{Tr}\widehat{A}_2(\langle\widehat{A}_1\rangle_\Psi + \mathrm{Tr}\widehat{A}_1\otimes I) + \mathrm{Tr}\widehat{A}_1\otimes\widehat{A}_2$$

$$+(\langle\widehat{A}_1\rangle_\Psi + \mathrm{Tr}\widehat{A}_1\otimes I)(\langle\widehat{A}_2\rangle_\Psi + \mathrm{Tr}I\otimes\widehat{A}_2)$$

$$+4\langle\widehat{A}_1\otimes\widehat{A}_2\rangle_\Psi + 2\mathrm{Tr}(\widehat{A}_1\otimes I)\mathrm{Tr}(I\otimes\widehat{A}_2) + 2\mathrm{Tr}\rho^{(12)}\widehat{A}_2\rho^{(12)}\widehat{A}_1$$

$$= \langle\widehat{A}_1\rangle_\Psi\langle\widehat{A}_2\rangle_\Psi + c_1\langle\widehat{A}_1\rangle_\Psi + c_2\langle\widehat{A}_2\rangle_\Psi + c_0 \qquad (7.43)$$

$$+4\langle\widehat{A}_1\otimes\widehat{A}_2\rangle_\Psi + 2\mathrm{Tr}\rho^{(12)}\widehat{A}_2\rho^{(12)}\widehat{A}_1,$$

where

$$c_1 = \mathrm{Tr}\widehat{A}_2 + \mathrm{Tr}I \otimes \widehat{A}_2,\ c_2 = \mathrm{Tr}\widehat{A}_1 + \mathrm{Tr}A_1 \otimes I, \tag{7.44}$$

$$c_0 = \mathrm{Tr}\widehat{A}_1\mathrm{Tr}I \otimes \widehat{A}_2 + \mathrm{Tr}\widehat{A}_2\mathrm{Tr}\widehat{A}_1 \otimes I + \mathrm{Tr}\widehat{A}_1 \otimes \widehat{A}_2 + 3\mathrm{Tr}(\widehat{A}_1 \otimes I)\mathrm{Tr}(\widehat{A}_2 \otimes I). \tag{7.45}$$

We have used the following operator equalities

$$\mathrm{Tr}\rho_\Psi^{(12)}(I \otimes \widehat{A}_2)\,(\widehat{A}_1 \otimes I) = \langle \widehat{A}_1 \otimes \widehat{A}_2\rangle_\Psi; \tag{7.46}$$

$$\mathrm{Tr}(I \otimes \widehat{A}_2)\,\rho_\Psi^{(12)}\,(\widehat{A}_1 \otimes I) = \langle \widehat{A}_1 \otimes \widehat{A}_2\rangle_\Psi. \tag{7.47}$$

The first one is trivial; we prove the second. Let $\{e_m\}$ be an orthonormal basis in $H_1 \otimes H_2$ consisting of the eigenvectors of $\rho_\Psi^{(12)}$ with eigenvalues p_n. Then

$$\begin{aligned}\mathrm{Tr}(I \otimes \widehat{A}_2)\,\rho_\Psi^{(12)}\,(\widehat{A}_1 \otimes I) &= \sum_k \sum_m p_m(e_m \otimes e_m\widehat{A}_1 \otimes I e_k,\ I \otimes \widehat{A}_2 e_k)\\ &= \sum_{k,m} p_m(e_m,\ \widehat{A}_1 \otimes I e_k)(e_m,\ I \otimes \widehat{A}_2 e_k)\\ &= \sum_m p_m \sum_k (\widehat{A}_1 \otimes I e_m,\ e_k)(e_k,\ I \otimes \widehat{A}_2 e_m)\\ &= \sum_m p_m(\widehat{A}_1 \otimes \widehat{A}_2 e_m,\ e_m)\\ &= \mathrm{Tr}\rho_\Psi^{(12)}\widehat{A}_1 \otimes \widehat{A}_2.\end{aligned}$$

We continue with the bi-correlation $Ef_{A_1}f_{A_3}$. Formula (7.42) is merely a cryptogram of the coupling between classical and quantum bi-correlations. It does not help us so much. Therefore we perform direct calculations

$$\langle f_{A_1} f_{A_3}\rangle = \int_{H_1\times H_2\times H_3} d\mu_{12,3}(x_1, x_2, x_3) f_{A_1}(x_1) f_{A_3}(x_3)$$

$$\int_{H_{12}\times H_3} dP_3(\phi_{12}, \phi_3) f_{A_3}(\phi_3) \int_{H_1\times H_2} dP_{12}(\phi_1, \phi_2; \phi_{12}, \phi_3) f_{A_1}(\phi_1)$$

$$= \int_{H_{12}\times H_3} dP_3(\phi_{12}, \phi_3) f_{A_3}(\phi_3)(\mathrm{Tr}\widehat{\phi}_{12}\widehat{\phi}_{12}^*\widehat{A}_1 + \mathrm{Tr}\widehat{A}_1)$$

$$= \int_{H_{12}\times H_3} dP_3(\phi_{12}, \phi_3)(f_{\widehat{A}_1\otimes I}(\phi_{12}) f_{A_3}(\phi_3) + \mathrm{Tr}\widehat{A}_1 f_{A_3}(\phi_3))$$

$$= [(\mathrm{Tr}\widehat{\Psi}_3\widehat{\Psi}_3^*\widehat{A}_1\otimes I+\mathrm{Tr}\widehat{A}_1\otimes I)(\mathrm{Tr}\widehat{\Psi}_3^*\widehat{\Psi}_3\widehat{A}_3+\mathrm{Tr}\widehat{A}_3)+2(\widehat{A}_1\otimes I\widehat{\otimes}A_3\Psi,\Psi)]$$

$$+\mathrm{Tr}\widehat{A}_1(\mathrm{Tr}\widehat{\Psi}_3^*\widehat{\Psi}_3\widehat{A}_3+\mathrm{Tr}\widehat{A}_3) = (\langle\widehat{A}_1\rangle_\Psi+\mathrm{Tr}\widehat{A}_1\otimes I)(\langle\widehat{A}_3\rangle_\Psi+\mathrm{Tr}\widehat{A}_3)$$

$$+2\langle\widehat{A}_1\otimes\widehat{A}_3\rangle_\Psi+\mathrm{Tr}\widehat{A}_1(\langle\widehat{A}_3\rangle_\Psi+\mathrm{Tr}\widehat{A}_3)$$

$$= \langle\widehat{A}_1\rangle_\Psi\langle\widehat{A}_3\rangle_\Psi+k_1\langle\widehat{A}_1\rangle_\Psi+k_3\langle\widehat{A}_3\rangle_\Psi+k_0+2\langle\widehat{A}_1\otimes\widehat{A}_3\rangle_\Psi, \quad (7.48)$$

where

$$k_1=\mathrm{Tr}\widehat{A}_3,\ k_3=\mathrm{Tr}\widehat{A}_1\otimes I+\mathrm{Tr}\widehat{A}_1,\ k_0=\mathrm{Tr}\widehat{A}_3(\mathrm{Tr}\widehat{A}_1\otimes I+\mathrm{Tr}\widehat{A}_1). \quad (7.49)$$

We now perform symmetrization by using the permutation of indices and constructing the corresponding uniform mixture of three probabilities. To be more concrete, we select probability spaces as proposed in Remark 8.1. Set

$$\mathbf{H}=H_1\times H_2\times H_3,\ \phi=(\phi_1,\phi_2,\phi_3)\in\mathbf{H}.$$

We consider three probability spaces with sets of random parameters

$$\Omega_{12,3}=H_{12}\times\mathbf{H},\ \Omega_{13,2}=H_{13}\times\mathbf{H},\ \Omega_{23,1}=H_{23}\times\mathbf{H},$$

and probabilities

$$dP_{12,3}(\phi_{12},\phi_3)=dP_3(\phi_{12};\phi_3)dP_{12}(\phi_1,\phi_2,\phi_{12},\phi_3),$$

$$dP_{13,2}(\phi_{13},\phi_2)=dP_2(\phi_{13},\phi_2)dP_{13}(\phi_1,\phi_3;\phi_{13},\phi_2),$$

$$dP_{23,1}(\phi_{23},\phi_1)=dP_1(\phi_{23},\phi)dP_{23}(\phi_2,\phi_3;\phi_{23},\phi_1).$$

Set now $\Omega_{123}=\Omega_{12,3}\bigcup\Omega_{13,2}\bigcup\Omega_{23,1}$ (disjoint union of the sets). Let $A_1\in\mathcal{F}_{12,3}$, $A_2\in\mathcal{F}_{13,2}$, $A_3\in\mathcal{F}_{23,1}$, where $\mathcal{F}_{ik,j}$ are Borel σ-algebras of the corresponding Hilbert spaces. For the set $A=A_1\bigcup A_2\bigcup A_3(\subset\Omega_{123})$, we put $P(A)=[P_{12,3}(A_1)+P_{13,2}(A_2)+P_{23,1}(A_3)]/3$. Then P is a probability measure and, e.g.,

$$\langle f_{A_1}f_{A_2}\rangle=Ef_{A_1}f_{A_2}=\frac{1}{3}[\langle f_{A_1}f_{A_2}\rangle_{12,3}+\langle f_{A_1}f_{A_2}\rangle_{13,2}+\langle f_{A_1}f_{A_2}\rangle_{23,1}].$$

The last term is given by (7.43)–(7.46). To find the second term, we should change $3\to2$ in (7.48) and (7.49). We find the corresponding coefficients

$$K_1=\mathrm{Tr}\widehat{A}_2,\ K_2=\mathrm{Tr}\widehat{A}_1\otimes I+\mathrm{Tr}\widehat{A}_1,\ K_0=\mathrm{Tr}\widehat{A}_2(\mathrm{Tr}\widehat{A}_1\otimes I+\mathrm{Tr}\widehat{A}_1).$$

Finally, to get the first average, we change $3 \to 1$ and $1 \to 2$ in (7.48) and (7.49). We find the corresponding coefficients

$$\tilde{K}_1 = \mathrm{Tr} I \otimes \widehat{A}_2 + \mathrm{Tr}\widehat{A}_2,\ \tilde{K}_2 = \mathrm{Tr}\widehat{A}_1,\ \tilde{K}_0 = \mathrm{Tr}\widehat{A}_1(\mathrm{Tr} I \otimes \widehat{A}_2 + \mathrm{Tr}\widehat{A}_2).$$

Thus we obtain

$$\begin{aligned}\langle f_{A_1} f_{A_2}\rangle &= \langle \widehat{A}_1\rangle\langle \widehat{A}_2\rangle_\Psi + d_1\langle \widehat{A}_1\rangle_\Psi + d_2\langle \widehat{A}_2\rangle_\Psi + d_0 \\ &\quad + 8\langle A_1 \otimes \widehat{A}_2\rangle_\Psi + 2\mathrm{Tr}\rho^{(12)}\widehat{A}_2\rho^{(12)}\widehat{A}_1]/3,\end{aligned}$$

where

$$\begin{aligned} d_1 &= 3\mathrm{Tr}\widehat{A}_2 + 2\mathrm{Tr} I \otimes \widehat{A}_2,\ d_2 = 3\mathrm{Tr}\widehat{A}_1 + 2\mathrm{Tr}\widehat{A}_1 \otimes I, \\ d_0 &= 2(\mathrm{Tr}\widehat{A}_1\mathrm{Tr} I \otimes \widehat{A}_2 + \mathrm{Tr}\widehat{A}_2\mathrm{Tr}\widehat{A}_1 \otimes I) \\ &\quad + 3(\mathrm{Tr}\widehat{A}_1 \otimes \widehat{A}_2 + \mathrm{Tr}(\widehat{A}_1 \otimes I)\mathrm{Tr}(\widehat{A}_2 \otimes I)).\end{aligned}$$

We now find the symmetrized correlation

$$\langle f_{A_1} f_{A_3}\rangle = \frac{1}{3}\left[\langle f_{A_1} f_{A_3}\rangle_{23,1} + \langle f_{A_1} f_{A_3}\rangle_{13,2} + \langle f_{A_1} f_{A_3}\rangle_{12,3}\right].$$

We start with $\langle f_{A_1} f_{A_3}\rangle_{13,2}$. We should change $3 \to 2$, $2 \to 3$ in (7.43)–(7.46). We have

$$\begin{aligned}\langle f_{A_1} f_{A_3}\rangle_{13,2} &= \langle \widehat{A}_1\rangle_\Psi\langle \widehat{A}_3\rangle_\Psi + a_1\langle \widehat{A}_1\rangle_\Psi + a_3\langle \widehat{A}_3\rangle_\Psi + a_0 \\ &\quad + 4\langle \widehat{A}_1 \otimes \widehat{A}_3\rangle_\Psi + 2\mathrm{Tr}\rho^{(13)}\widehat{A}_3\rho^{(13)}\widehat{A}_1,\end{aligned}$$

where

$$a_1 = \mathrm{Tr}\widehat{A}_3 + \mathrm{Tr} I \otimes \widehat{A}_3,\quad a_3 = \mathrm{Tr}\widehat{A}_1 + \mathrm{Tr} A_1 \otimes I,$$

$$a_0 = \mathrm{Tr}\widehat{A}_1\mathrm{Tr} I \otimes \widehat{A}_3 + \mathrm{Tr}\widehat{A}_3\mathrm{Tr}\widehat{A}_1 \otimes I + \mathrm{Tr}\widehat{A}_1 \otimes \widehat{A}_3 + 3\mathrm{Tr}(\widehat{A}_1 \otimes I)\mathrm{Tr}(I \otimes \widehat{A}_3).$$

To find $\langle f_{A_1} f_{A_3}\rangle_{12,3}$, we use (7.48) and (7.49):

$$\begin{aligned}\langle f_{A_1} f_{A_3}\rangle_{12,3} &= \langle \widehat{A}_1\rangle_\Psi\langle \widehat{A}_3\rangle_\Psi + \mathrm{Tr}\widehat{A}_3\langle \widehat{A}_1\rangle_\Psi + (\mathrm{Tr}\widehat{A}_1 \otimes I + \mathrm{Tr}\widehat{A}_1)\langle \widehat{A}_3\rangle_\Psi \\ &\quad + \mathrm{Tr}\widehat{A}_3(\mathrm{Tr}\widehat{A}_1 \otimes I + \mathrm{Tr}\widehat{A}_1) + 2\langle \widehat{A}_1 \otimes \widehat{A}_3\rangle_\Psi.\end{aligned}$$

To find $\langle f_{A_1} f_{A_3}\rangle_{23,1}$, we change $3 \to 1$ and $1 \to 3$in (7.48) and (7.49):

$$\begin{aligned}\langle f_{A_1} f_{A_3}\rangle_{23,1} &= \langle \widehat{A}_1\rangle_\Psi\langle \widehat{A}_3\rangle_\Psi + (\mathrm{Tr} I \otimes \widehat{A}_3 + \mathrm{Tr}\widehat{A}_3)\langle \widehat{A}_1\rangle_\Psi + \mathrm{Tr}\widehat{A}_1\langle \widehat{A}_3\rangle_\Psi \\ &\quad + \mathrm{Tr}\widehat{A}_1(\mathrm{Tr} I \otimes \widehat{A}_3 + \mathrm{Tr}\widehat{A}_3) + 2\langle \widehat{A}_1 \otimes \widehat{A}_3\rangle_\Psi.\end{aligned}$$

In summary, we obtain

$$\begin{aligned}\langle f_{A_1} f_{A_3}\rangle_{123} &= \langle \widehat{A}_1\rangle_\Psi\langle \widehat{A}_3\rangle_\Psi + [g_1\langle \widehat{A}_1\rangle_\Psi + g_3\langle \widehat{A}_3\rangle_\Psi + g_0 \\ &\quad + 8\langle \widehat{A}_1 \otimes \widehat{A}_3\rangle_\Psi + 2\mathrm{Tr}\rho^{(13)}\widehat{A}_3\rho^{(13)}\widehat{A}_1]/3,\end{aligned}$$

where

$$g_1 = 3\text{Tr}\widehat{A}_3 + 2\text{Tr}I \otimes \widehat{A}_3,\ g_3 = 3\text{Tr}\widehat{A}_1 + 2\text{Tr}\widehat{A}_1 \otimes I,$$
$$g_0 = 2(\text{Tr}\widehat{A}_3\text{Tr}\widehat{A}_1 \otimes I + \text{Tr}\widehat{A}_1\text{Tr}I \otimes \widehat{A}_3)$$
$$+3(\text{Tr}\widehat{A}_1 \otimes \widehat{A}_3 + \text{Tr}(\widehat{A}_1 \otimes I)\text{Tr}(I \otimes \widehat{A}_3))$$

We see that the expressions for the coefficients g_i coincide with the expressions for the coefficients d_i after changing $2 \to 3$. The asymmetry of pairwise correlations for the probabilities $P_{ij,k}$ disappeared in the symmetric mixture of their probabilities P_{123}.

We can say that *the use of the probability P_{123} is the "prequantum trace" of indistinguishability of three quantum systems.* The use of, e.g., the probability $P_{12,3}$ means that the system S_3 was selected as the "reference frame." Here we find the correlation between observables on subsystems S_1 and S_2 under the condition that the prequantum state ϕ_3 of S_3 was fixed.

7.9 PCSFT Representation for a Mixed State

Consider now a composite system $S = (S_1, S_2)$ in a quantum state given by a density operator $\rho \in L_s(H_1 \otimes H_2, H_1 \otimes H_2)$, $\rho \geq 0$, $\text{Tr}\rho = 1$. Let p_k be eigenvalues of ρ corresponding to the eigenvectors $\Psi_k \in H_1 \otimes H_2$. Thus $\rho = \sum_k p_k \Psi_k \otimes \Psi_k$, where $p_k \in [0, 1]$, $\sum p_k = 1$. Set $\rho^{(1)} = \text{Tr}_{H_2}\rho$, $\rho^{(2)} = \text{Tr}_{H_1}\rho$; set

$$D_k = \begin{pmatrix} \rho^{(1)} + \epsilon I & \widehat{\Psi}_k \\ \widehat{\Psi}_k^* & \rho^{(2)} + \epsilon I \end{pmatrix},$$

where $\epsilon > 0$ is large enough (so that $D_k \geq 0$). We stress that the diagonal blocks do not depend on k.

Denote by μ_k the Gaussian measure with zero mean value and the covariance operator D_k. Consider the mixture of these Gaussian measures

$$\mu \equiv \mu_\rho = \sum_k p_k \mu_k.$$

We remark that μ is a probability measure $\mu(H_1 \times H_2) = \sum_k p_k \mu_k(H_1 \times H_2) = 1$. It is not the Gaussian measure. We can find its characteristic function

$$\tilde{\mu}(z) = \sum_k p_k \exp\{-\frac{1}{2}(D_k z, z)\}$$

and its covariance operator

$$(Dz_1, z_2) = \sum p_k(D_k z_1, z_2) = \left(\sum_k p_k D_k z_1, z_2\right).$$

Thus,

$$D = \begin{pmatrix} \rho_1 + \epsilon I & \sum_k p_k \widehat{\Psi}_k \\ \sum_k p_k \widehat{\Psi}_k^* & \rho_2 + \epsilon I \end{pmatrix}.$$

We now find

$$\begin{aligned} E f_{A_1}(\phi_1(\omega)) f_{A_2}(\phi_2(\omega)) &= \int_{H_1 \times H_2} f_{A_1}(\phi_1) f_{A_2}(\phi_2) d\mu(\phi) \\ &= \sum_k p_k \int_{H_1 \times H_2} f_{A_1}(\phi_1) f_{A_2}(\phi_2) d\mu_k(\phi) \\ &= \sum_k p_k \mathrm{Tr}(\rho_1 + I\epsilon)\widehat{A}_1 \mathrm{Tr}(\rho_2 + I\epsilon)\widehat{A}_2 \\ &\quad + 2\sum_k p_k \mathrm{Tr}\widehat{\Psi}_k \widehat{A}_2 \widehat{\Psi}_k^* \widehat{A}_1 \\ &= [\langle \widehat{A}_1 \rangle_\rho + \epsilon \mathrm{Tr}\widehat{A}_1][\langle \widehat{A}_2 \rangle_\rho + \epsilon \mathrm{Tr}\widehat{A}_2] \\ &\quad + 2\sum_k p_k(\widehat{A}_1 \otimes \widehat{A}_2 \Psi_k, \Psi_k). \end{aligned}$$

The latter term is given by

$$\begin{aligned} \sum_k p_k \mathrm{Tr}(\Psi_k \otimes \Psi_k)\widehat{A}_1 \otimes \widehat{A}_2 &= \mathrm{Tr}\sum_k p_k(\Psi_k \otimes \Psi_k)\widehat{A}_1 \otimes \widehat{A}_2 \\ &= \mathrm{Tr}\rho \widehat{A}_1 \otimes \widehat{A}_2 = \langle \widehat{A}_1 \otimes \widehat{A}_2 \rangle_\rho. \end{aligned}$$

Thus

$$E f_{A_1} f_{A_2} = [\langle \widehat{A}_1 \rangle_\rho + \epsilon \mathrm{Tr}\widehat{A}_1][\langle \widehat{A}_2 \rangle_\rho + \epsilon \mathrm{Tr}\widehat{A}_2] + 2\langle A_1 \otimes \widehat{A}_2 \rangle_\rho.$$

We also remark that

$$E f_{A_i} = \mathrm{Tr}(p\rho_i + \epsilon I)\widehat{A}_i, \, i = 1, 2.$$

Thus,

$$\frac{1}{2}\mathrm{cov}(f_{A_1}, f_{A_2}) = \langle A_1 \otimes A_2 \rangle_\rho.$$

Theorem. Let $\epsilon \geq 1/4$. Then each $D_k \geq 0$.

Proof. We remark that

$$\rho^{(1)} = \mathrm{Tr}_{H_2} \sum_k p_k \Psi_k \otimes \Psi_k = \sum_k p_k \mathrm{Tr}_{H_2} \Psi_k \otimes \Psi_k = \sum_k p_k \widehat{\Psi}_k \widehat{\Psi}_k^*.$$

In the same way,

$$\rho^{(2)} = \sum_k p_k \widehat{\Psi}_k^* \widehat{\Psi}_k.$$

Thus,

$$(\rho^{(1)} u, u) \geq (\widehat{\Psi}_k \widehat{\Psi}_k^* u, u), (\rho^{(2)} v, v) \geq (\widehat{\Psi}_k^* \widehat{\Psi}_k v, v).$$

For $w = (u, v)$, we have

$$(D_k w, w) \geq \|\widehat{\Psi}_k^* u\|^2 + \epsilon \|u\|^2 + 2(\Psi_k^* u, v) + \|\widehat{\Psi}_k v\|^2 + \epsilon \|v\|^2 \geq 0,$$

if $\epsilon \geq 1/4$.

Chapter 8

Phenomenological Detection Model

In this chapter a measurement model for PCSFT will be presented [201]. We explore the basic idea of PCSFT: a quantum state (e.g., a pure state) is simply a label for an ensemble of classical fields—*classical random field.* We model the process of interaction of a detector with a random field. By using another terminology, we can speak about classical random signals[a] and, in particular, about switching to time averages (see Section 3.1.5). The model developed in this chapter is of the phenomenological type. It is based on (heuristically justified) the coupling between the energy distribution of a random field and the probability distribution of clicks of a detector. This model will be concreted in Chapter 9 by considering the detectors of the threshold type. In the later model we shall be able to describe the process of the interaction of a random signal with a threshold-type detector.

First we consider the simplest model of measurement: a detector reacts only to the square-type nonlinearities of the prequantum field.[b] In this model the probability of the selection of a field (from

[a]This terminology (in particular, used in Sections 8.9–8.10) is motivated by the paper by Ohya and Watanabe [252] on the interplay of the classical and quantum signal theories.

[b]In PCSFT the average of any prequantum random field $\phi(x, \omega)$ is the "zero field": $\phi_0(x) = 0$ for any $x \in \mathbf{R}^3$. The dispersion of $\phi(x, \omega)$ is very small. Therefore for

Beyond Quantum
Andrei Khrennikov

ISBN 978-981-4411-73-8 (Hardcover), 978-981-4411-74-5 (eBook)
www.panstanford.com

an ensemble of fields representing realizations of a random field) is proportional to its energy, i.e., its L_2-norm. We call this approach the *field's energy detection theory.* The energy of a classical prequantum field plays the crucial role. It will be shown that in this detection model the *Born's rule* is a consequence of the well-known *Bayes formula* for conditional probability.[c]

Considering experiments (see Section 3.11) in which the contributions of higher powers of the prequantum field, e.g., $|\phi(x)|^4$, have impacts to the detectors, one can expect *deviations from the Born's rule.* In Section 8.5 we estimate the magnitude of these deviations.

We state again that the dimension of the space of classical fields is infinite. This is the source of mathematical complexity of the detection model. To escape the mathematical difficulties and at the same time to illustrate all distinguishing features of our model of detection of classical fields, we start with a toy model with finite-dimensional state space (Section 8.1).

In PCSFT there are two main reasons to proceed with the Gaussian prequantum random fields:

(a) the one-to-one correspondence between the basic objects of QM and PCSFT: density operators and prequantum random fields;
(b) we were able to obtain coupling between classical and quantum correlations, see equality (3.17), only for the Gaussian random fields.

However, we should not overestimate the role of the arguments presented in (a) and (b). In principle, there is no reason to expect that QM provides a one-to-one image of prequantum randomness. It may well happen that QM simply ignores differences inside a huge class of random fields having the same covariance operator. (In the QM formalism this operator is represented by the corresponding density operator.) Concerning (b), it seems that formulas similar

---

the "statistical majority" of realizations of $\phi(x, \omega)$, they amplitudes are very small. Thus, the term based on $|\phi(x)|^2$ makes the main contribution to the detector output. The contributions of higher order nonlinearities, $|\phi(x)|^n$, $n > 2$, are much smaller. The first approximation of detection is based on taking into account only quadratic nonlinearity $|\phi(x)|^2$.

[c]Here everything is classical: fields, detectors; nevertheless, output probabilities are the same as in QM.

to (3.17) could be derived even for non-Gaussian random fields. Therefore, in Sections 8.1–8.7 we proceed with arbitrary random fields. However, the basic formula (3.65), Section 8.5, was derived under the assumption that prequantum random fields are Gaussian. We recall that this formula is used in Section 3.11 for the experimental verification of violation of the Born's rule.

In Sections 8.1–8.7 we do not take into account the presence of the background random field. This is simplification of the real experimental situation: the noisy component of a prequantum random field is always nontrivial. In Section 8.9.1 we show that quantum mechanical probabilities can be derived as a result of cutoff of this noisy component. A mathematical description of this cutoff operation is nontrivial, especially for real physical random fields described by L_2-valued random variables. It will be presented in Section 8.9. This is a kind of a renormalization procedure which is similar to the procedures of QFT. The presence of the noisy component of the white noise type generates infinities which should be discarded to obtain QM predictions. In QFT the presence of divergences is considered as one of its main problems. (We remark that divergences do not appear in QM.) Some top experts in quantum theory including P. Dirac were extremely critical to the theory which manipulates with infinities to obtain proper answers. They considered this situation as unacceptable for a scientific theory. In some way PCSFT justifies the presence of infinities in quantum theory by clarification their origin, namely, the singularity of the background field. At the subquantum level nature is very singular and noisy. To obtain reasonable answers, experimenters have to discard the contribution of the background field. As a consequence of singularity, averages with respect to the background noise are divergent. Thus there is nothing mystical in the appearance of infinities. In Sections 8.1–8.7 we present a PCSFT measurement model for one special (but the most important) quantum observable, the position observable; the case of the momentum observable was studied in detail in Ref. [178]; quantum observables with purely discrete spectra are considered in Section 8.8. The case of noisy signals (Section 8.9), is studied for quantum observables given by orthogonal projectors.

In this chapter, the similarity between QM and the classical signal theory becomes especially evident. In fact, we apply the classical signal theory to reproduce probabilistic predictions of conventional QM. We remark that the coupling of classical Gaussian signals with QM has already been discussed in Ref. [252]: a proper notion of the entropy of a classical Gaussian random signal was introduced as the entropy of the density operator constructed on the basis of the covariance operator of this signal.[a] There was introduced the entropy functional of an input source and the mutual entropy functional; a fundamental inequality for communication processes was proved. In contrast to Ref. [252], in the usual communication theory, one takes *differential entropy* as the definition of the entropy (information) carried by an input source. However, for an input Gaussian process, the differential entropy for an input source is not compatible with the mutual information in Shannon's communication theory; so the differential entropy is not good for discussing the Gaussian communication processes. This possibility to use the quantum entropy in classical signal theory was considered by the authors of Ref. [252] simply as a mathematical coincidence.

8.1 Finite-Dimensional Model

Let us consider systems (fields) with the state space H_n, where H_n is the n-dimensional real space: $H_n = \mathbf{R} \times \cdots \times \mathbf{R}$. In this model the state of a field is given by the vector $v = (v_1, \ldots, v_n)$. We set $||v||^2 = \sum_{j=1}^n v_j^2$. The Euclidean norm $||v||^2$ of the state v is interpreted as *field's energy.*

Suppose a measurement device (detector) is given. For a field in the state v, its output is one of the values

$$j = j(v), \tag{8.1}$$

where $j = 1, \ldots, n$. These are indexes encoding the coordinates of the state vector. Such a measurement is the discrete analog of the position measurement.

[a]They proceeded the other way around: they used quantum information theory to improve the classical signal theory.

Suppose also that *the probability to obtain a fixed value j_0 is proportional to $v_{j_0}^2$ and the coefficient of proportionality does not depend on j_0 (but it depends on the state v)*. For a field in the state v, the probability that such a detector produces the result $j = j_0$ can be found as

$$P(j = j_0|v) = k_v v_{j_0}^2. \tag{8.2}$$

The coefficient k_v is determined by the normalization of the total probability by one:

$$\sum_{j_0=1}^{n} P(j = j_0|v) = k_v \sum_{j_0=1}^{n} v_{j_0}^2 = 1. \tag{8.3}$$

Thus $k_v = \frac{1}{||v||^2}$ and

$$P(j = j_0|v) = \frac{v_{j_0}^2}{||v||^2}. \tag{8.4}$$

Consider now *an ensemble of fields,* a random field. Mathematically a random field is described as the probability measure μ on H_n or as the random vector $\xi(\omega) = (\xi_1(\omega), \ldots, \xi_n(\omega))$. It is assumed that $m_\mu = \int_{H_n} v d\mu(v) = 0$, zero mean value. Its dispersion, $\sigma^2(\mu) = \int_{H_n} ||v||^2 d\mu(v) = \alpha$, will play an important role in future consideration. We consider the following procedure of measurement of the quantity j, the discrete analog of position:

(a) first the detector selects a field in the state v from the ensemble representing a random field;
(b) then, for this selected field, the detector performs measurement of $j(v)$ and statistics of outputs is based on the rule which was formulated above (mathematically this is the equality (8.4)).

Consider a *special procedure of selection* of a fixed field from the ensemble. Our basic assumption is that the detector works in the following way. The probability to select a field in the state v from the ensemble is proportional to the *square of the norm of the state,* field's energy:

$$P_\mu(v) = K_\mu \, ||v||^2 = K_\mu \sum_{j=1}^{n} v_j^2.$$

The coefficient of proportionality can again be found from the normalization of probability by one: $1 = \int_{H_n} dP_\mu(v) = K_\mu \int_{\mathcal{H}_n} ||v||^2 d\mu(v)$. Thus,

$$K_\mu = \frac{1}{\alpha} \tag{8.5}$$

and

$$dP_\mu(v) = \frac{||v||^2}{\alpha} d\mu(v). \tag{8.6}$$

Let μ be a discrete probability measure. It is concentrated in finite number of points, $v^{(1)}, \ldots, v^{(m)}$, of the state space. It is determined by the positive weights $\mu(v^k)$, $k = 1, \ldots, m$, where

$$\sum_{k=1}^{m} \mu(v^k) = 1, v^{(1)}, \ldots, v^{(m)} \in H_n.$$

Now the formula (8.6) takes the form

$$P_\mu(v^{(i)}) = \frac{||v^{(i)}||^2}{\alpha} \mu(v^{(i)}). \tag{8.7}$$

This is the probability that a measurement device selects the subsequent j-measurement, a field in the state $v^{(i)} = (v_1^{(i)}, \ldots, v_n^{(i)})$.

In the discrete case we now find the final probability to obtain the concrete result $j = j_0 : \mathbf{p}_\mu(j = j_0)$. Let us apply the Bayes formula. The probability $\mathbf{p}_\mu(j = j_0)$ is obtained as the result of the combination of two probabilities: the probability $P_\mu(v^{(i)})$ to select a field in the state $v = v^{(i)}$ from the ensemble of fields and the probability $P(j = j_0|v^{(i)})$ to obtain the result $j = j_0$ for measurement for the field in the state $v^{(i)}$:

$$\mathbf{p}_\mu(j = j_0) = \sum_i P_\mu(v^{(i)})\ P(j = j_0|v^{(i)}). \tag{8.8}$$

Putting the probabilities given by (8.7) and (8.4) in this formula, we obtain

$$\mathbf{p}_\mu(j = j_0) = \frac{1}{\alpha} \sum_i |v_{j_0}^{(i)}|^2 \mu(v^{(i)}). \tag{8.9}$$

We now generalize (8.9) to the case of the continuous distribution μ. We have $P(j = j_0|v) = \frac{v_{j_0}^2}{||v||^2}$ and $P(v) = \frac{||v||^2}{\alpha} \mu(v)$. Again by using the Bayes formula we find the probability to obtain $j = j_0$ for measurement of the random field with the probability distribution μ :

$$\mathbf{p}_\mu(j = j_0) = \int_{H_n} P(j = j_0|v) dP_\mu(v) = \frac{1}{\alpha} \int v_{j_0}^2 d\mu(v).$$

8.1.1 *Pure States*

Let us take the vector $u \in H_n$ such that $||u|| = \alpha$. Consider now the Gaussian measure $\mu \equiv \mu_u$ having zero mean value and the covariance operator

$$\rho_u = u \otimes u,$$

i.e., $(\rho_u y, y) = (u, y)^2$. This Gaussian measure is concentrated on the one-dimensional subspace $L_u = \{z = cu : c \in \mathbf{R}\}$. Thus,

$$d\mu_u(v) = e^{-p^2/2\alpha} dp/\sqrt{2\pi\alpha}, \ p = (v, w),$$

where $w = \frac{u}{\sqrt{\alpha}}$. Thus $||w|| = 1$. Then

$$\mathbf{p}_\mu(j = j_0) = \frac{1}{\alpha} \int_{\mathbf{R}^n} v_{j_0}^2 d\mu_u(v) = w_{j_0}^2.$$

This is nothing else than the *Born's rule* of QM. The normalized vector w can be interpreted as a "pure quantum state." In our approach it is nothing else than the symbolic representation of the Gaussian random vector with the probability distribution μ_u. We can proceed the other way around by starting directly with an arbitrary normalized vector $w \in H_n$, a "pure state." We remark that the Gaussian random vector was chosen only for simplicity. The same result is obtained for any random vector having the same covariance matrix. The detectors under consideration are not able to distinguish between two random vectors with the same covariance matrix.

8.1.2 *Born's Rule*

To write this rule similarly to the original Born's formula, we consider the state space H_n as a space of functions $v : X_n \to \mathbf{R}$, where $X_n = \{x_1, \ldots, x_n\}$ is some discrete set. Thus, instead of the set of labels $\{j = 1, 2, \ldots, n\}$, we now consider an arbitrary discrete set X_n, the space of results of measurement. For any vector $v \equiv v(x)$, we can say that the measurement process under consideration is the X-measurement: $X = X(v)$. We have

$$\mathbf{p}_\mu(X = x) = \frac{1}{\alpha} \int v^2(x) d\mu(v)$$

or

$$\mathbf{p}_\mu(x \in I) = \frac{1}{\alpha} \sum_{X \in I} \int v^2(x) d\mu(v)$$
$$= \frac{1}{\alpha} \int \sum_{x \in I} v^2(x) d\mu(v),\ x \in X_n,\ I \subset X_n.$$

In particular, for the Gaussian measure μ_u, we have

$$\mathbf{p}_{\mu_u}(X = x) = w^2(x), \quad w = \frac{u}{\sqrt{\alpha}}.$$

By choosing $n = 2k$ and considering the complex representation of the real state space H_{2k}, namely, $H_{2k} = \mathbf{C}^k$, we obtain the theory of measurement, where (as in the previous consideration)

$$P(X = x|v) = \frac{|v(x)|^2}{||v||^2},\ P(v)) = \frac{||v||^2}{\alpha}\mu(v),$$

and, finally,

$$\mathbf{p}_\mu(X = x) = \frac{1}{\alpha} \int_{\mathbf{C}^n} |v(x)|^2 d\mu(v),$$

in particular,

$$\mathbf{p}_{\mu_u}(X = x) = |w(x)|^2.$$

In principle, we can consider as the state space any space $\mathcal{H}_n = T \times \cdots \times T$, where T is a number field with the valuation $|\cdot|_T$. For example, T can be chosen as the field of p-adic numbers [132, 133]. In this case we consider a probability μ on Q_p^n and

$$P(X = x|v) = \frac{|v(x)|_p^2}{||v||^2},\ ||v||^2 = \sum_{k=1}^{n} |v(x_k)|_p^2.$$

We emphasize that the theory of measurement under consideration is purely classical. No noncommutative mathematics was involved in the description. The crucial point was the use of the Bayes formula (and nothing else!). From the physical viewpoint, all "nonclassical features" are induced by special functioning of measurement devices (detectors).

8.2 Position Measurement for the Prequantum Field

8.2.1 *Classical Random Fields*

As everywhere in this book we consider the configuration space of complex random fields $H = L_2(\mathbf{R}^3)$, the space of square-integrable complex fields, $\phi : \mathbf{R}^3 \to \mathbf{C}$. It is endowed with the norm $\|\phi\|^2 = \int_{\mathbf{R}^3} |\phi(x)|^2 dx$. A random field is the H-valued random variable $\omega \to \phi(\omega) \in H$. As always, we consider only random fields with zero mean value: $E\phi(\omega) = 0$. This equality simply means that, for any function $f \in H$,

$$E \int_{\mathbf{R}^3} f(x)\overline{\phi(x,\omega)}dx = 0.$$

Covariance of a random field is defined as

$$D_\mu(f, g) = E\Big(\int_{\mathbf{R}^3} f(x)\overline{\phi(x,\omega)}dx\Big)\Big(\int_{\mathbf{R}^3} \overline{g(x)}\phi(x,\omega)dx\Big).$$

The corresponding operator is denoted by D_μ. The dispersion is given by $\sigma^2(\mu) = E\,\|\phi(\omega)\|^2$. As always, we consider the dispersion as a parameter, say α, of our model.

8.2.2 *Position Measurement*

We denote by $\mathcal{E}_\mu$ the ensemble of fields represented by the probability measure μ on H. It is the probability distribution of the random field $\phi(\omega)$. The ensemble $\mathcal{E}_\mu$ consists of realizations of this random field. We recall once again that in PCSFT each quantum system is just a symbolic representation of the corresponding "prequantum random field."

Let X be the position observable. This observable is considered as an observable on fields. Thus $X(\omega) = X(\phi(\omega))$ is a random variable. It takes values in $\mathbf{R}$. Our aim is to derive the probability distribution of this random variable from the probability distribution of the random field.

We consider random fields interacting with detectors. Suppose that the position observable X is given by a measurement device M_X. For example, M_X can be chosen as a collection of detectors located at all points $x \in \mathbf{R}^3$. For any point x_0, we can consider the observable X_{x_0} given by a detector $M_X(x_0)$ located at x_0. For any sufficiently regular set $I \subset \mathbf{R}^3$, we can consider the observable X_I given by a collection of detectors $M_X(I)$ located in the domain $I \subset \mathbf{R}^3$.

8.3 Field's Energy Detection Model

In our model of detection the measurement process consists of two steps:

(a) selection of a field $\phi \in \mathcal{E}_\mu$;
(b) measurement for this field: $X(\phi)$.

We assume that the detectors are sensitive to the *energy* of a field. The energy density of the field ϕ at the point x_0 is defined as

$$\pi_2(x_0, \phi) = |\phi(x_0)|^2,$$

field's energy in the domain $I \subset \mathbf{R}^3$ as

$$\pi_2(I, \phi) = \int_I |\phi(x)|^2 dx,$$

and, finally, the total energy of the field ϕ as

$$\pi_2(\phi) = \|\phi\|^2 = \int_{\mathbf{R}^3} |\phi(x)|^2 dx.$$

We now formulate the fundamental feature of the class of detectors under consideration, namely, *sensitivity to the energy of a field,* in the form of two postulates:

Postulate 1. *The probability P_μ to select the fixed field ϕ from the random field $\phi(x, \omega)$ (the ensemble $\mathcal{E}_\mu$) is proportional to the total energy of ϕ :*

$$dP_\mu(\phi) = K_\mu \pi_2(\phi) d\mu(\phi). \tag{8.10}$$

The coefficient of proportionality K_μ can be found from the normalization of probability by one

$$K_\mu = \frac{1}{\int_H \pi_2(\phi) d\mu(\phi)} = \frac{1}{\sigma^2(\mu)} = \frac{1}{\alpha}.$$

(We state again that $\alpha = \mathrm{Tr} D_\mu$, where D_μ is the covariance operator of μ.) Thus, we get

$$dP_\mu(\phi) = \frac{\pi_2(\phi)}{\alpha} d\mu(\phi), \tag{8.11}$$

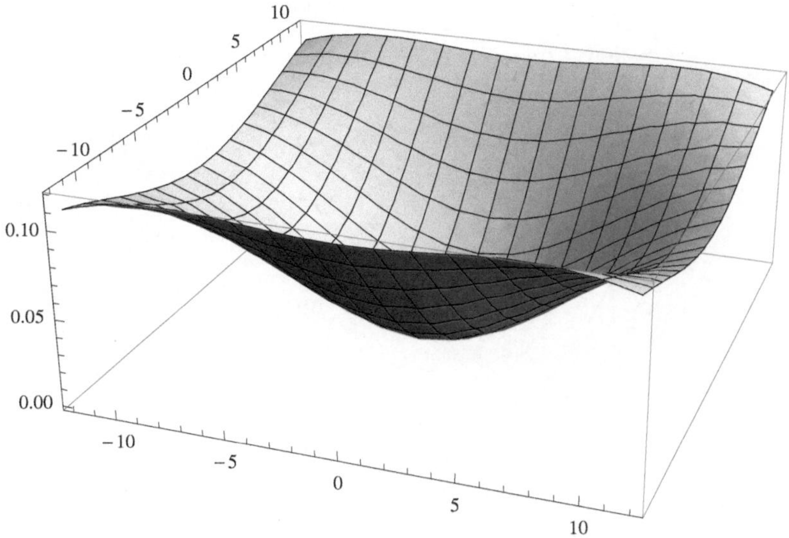

Figure 8.1 Bowl-type distribution, I.

and, for any Borel subset $U \subset H$, we have

$$P_\mu(\phi \in U) = \frac{1}{\alpha}\int_U ||\phi||^2 d\mu(\phi), \tag{8.12}$$

or in the random field notation:

$$P_\mu(\phi \in U) = \frac{1}{\alpha}E\left(\chi_U(\phi(\omega))||\phi(\omega)||^2\right), \tag{8.13}$$

where $\chi_U(\phi)$ is the characteristic function of the set U.

What does the probability P_μ look like? The probability P_μ is absolutely continuous with respect to the measure μ with the density $\phi \to \frac{\|\phi\|^2}{\alpha}$, $\alpha = \mathrm{Tr} D_\mu$. Thus the detector produces cutoff of a random field: its realizations with relatively low levels of the energy, $\pi_2(\phi) = \|\phi\|^2 \ll 1$, are discarded (cf. Chapter 9). High-energy realizations of a random field are statistically negligible. We apply the Chebyshov inequality. For any random field with finite dispersion,[a] the probability of realization of a field of high energy,

$$\pi_2(\phi) = \|\phi\|^2 \gg 1,$$

[a]Finiteness of dispersion of prequantum random fields is one of the basic assumptions of PCSFT

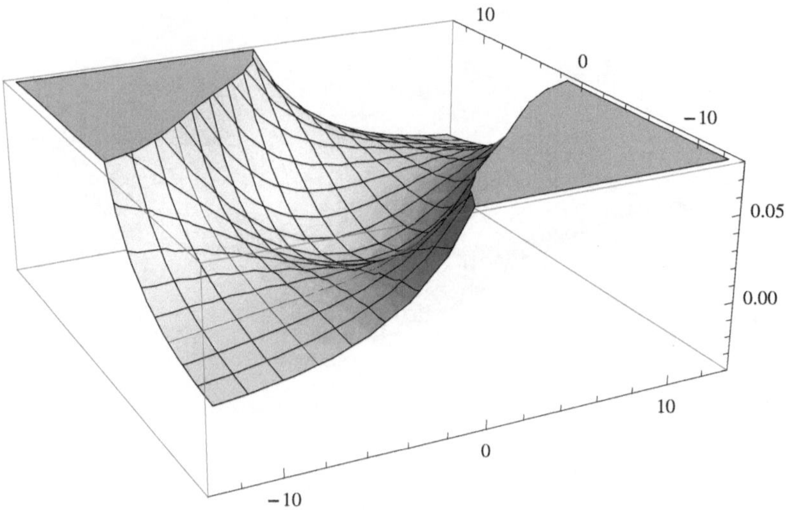

Figure 8.2 Bowl-type distribution, II.

is small:

$$P_\mu(\pi_2(\phi) > C) \leq \frac{\alpha}{C} \to 0, \quad C \to +\infty. \tag{8.14}$$

We proceed under the assumption that prequantum random fields are sharply concentrated around the vacuum. In our model the vacuum is an ideal state represented by the field ϕ, which is zero as an element of L_2, i.e., it is zero almost everywhere. To get a very small probability, in the Chebyshov inequality a constant $C > 0$ need not be too large. For example, take the Gaussian distribution μ.

The distribution P_μ (the energy of the random field) has the form of a *bowl* (see Figs. 8.1 and 8.2). (This is the two-dimensional illustration. The real physical distribution is infinite-dimensional. The zero-points of the graphs in Figs. 8.1 and 8.2. correspond to vacuum: the field $\phi = 0$ as an element of L_2.) The diameters of these bowls are very small. Motivated by these illustrations, we shall call the distribution P_μ a *bowl-type distribution* even in the case of an arbitrary (i.e., in general, non-Gaussian) measure μ.

The selection procedure of the concrete field from a random field was formalized by Postulate 1. This postulate is intuitively attractive: fields with relatively high energy are selected more often. We now formalize the b-step of the X-measurement in the following form.

Postulate 2. *The probability $P(X = x_0|\phi)$ to obtain the result $X = x_0$ for the fixed field ϕ (where $\phi \neq 0$ as an element of L_2) is proportional to the energy density $\pi_2(x_0, \phi)$ of this field at x_0. The coefficient of proportionality does not depend on x_0, so $k(x_0|\phi) \equiv k_\phi$.*

The coefficient of proportionality $k(x_0|\phi)$ can be obtained from the normalization of probability by one: $1 = \int_{\mathbf{R}^3} P(X = x|\phi)dx = k_\phi \int_{\mathbf{R}^3} |\phi(x)|^2 dx$. Thus,

$$k_\phi = \frac{1}{||\phi||^2} \tag{8.15}$$

and hence,

$$P(X = x_0|\phi) = \frac{|\phi(x_0)|^2}{\|\phi\|}, \phi \neq 0. \tag{8.16}$$

For a random field with the distribution μ, the probability to obtain the result $X = x_0$, where $x_0 \in \mathbf{R}^3$, can be obtained by using the classical Bayes' formula

$$\mathbf{p}_\mu(X = x_0) = \int_H P(x_0|\phi)dP_\mu(\phi) = \int_H \frac{|\phi(x_0)|^2}{||\phi||^2} dP_\mu(\phi). \tag{8.17}$$

Thus we finally have

$$\mathbf{p}_\mu(X = x_0) = \frac{1}{\alpha}\int_H |\phi(x_0)|^2 d\mu(\phi). \tag{8.18}$$

In fact, the quantity $\mathbf{p}_\mu(X = x)$ is the probability density:

$$\mathbf{p}_\mu(X \in I) = \int_I \mathbf{p}_\mu(X = x)dx = \frac{1}{\alpha}\int_H \left(\int_I |\phi(x)|^2 dx\right) d\mu(\phi), \tag{8.19}$$

where I is a Borel subset of $\mathbf{R}^3$, e.g., a cube.

8.4 Coupling between Probabilities of Detection of Classical Random Fields and Quantum Particles

In the previous section we presented a model of detection of classical random fields. For the position measurement, the probability of detection was derived, see (8.19). On the other hand, the conventional quantum model also provides the probability of detection (in particular, for the position measurement). Since formulas given by QM work well and have been confirmed by numerous experiments, to justify the PCSFT model of measurement, we have to couple the two mathematical models of the position measurement.

Let B be a Borel subset of the real line, e.g., an interval. We introduce the projectors $\widehat{B}\phi(x) = \chi_B(x)\phi(x)$, where χ_B is the characteristic function of this set.

Theorem 4.1. *The probability measure* $\mathbf{p}_\mu$ *can be represented in the following operator form:*

$$\mathbf{p}_\mu(I) = Tr\ \rho\ \widehat{B}, \tag{8.20}$$

where $\rho = D_\mu/\alpha$.

The probability distribution $\mathbf{p}_\mu$ on $\mathbf{R}^3$ given by PCSFT coincides with the probability distribution given by QM.

Let Ψ be the normalized vector—the pure state of QM. We consider the measure μ_Ψ on H with zero mean value and the covariance operator $D_\Psi = \Psi \otimes \Psi$. In particular, we can choose the Gaussian measure. Then we easily find that

$$\mathbf{p}_\Psi(x) \equiv \mathbf{p}_{\mu_\Psi}(x) = |\Psi(x)|^2. \tag{8.21}$$

Thus,

$$\mathbf{p}_\Psi(B) = \int_B |\Psi(x)|^2 dx. \tag{8.22}$$

This is nothing else than the *Born's rule.*

We find the mean value of the position x with respect to the probability measure $\mathbf{p}_\mu$. We restrict our consideration to the one-dimensional case. We have

$$\langle x \rangle_{\mathbf{p}_\mu} = \int_{-\infty}^{+\infty} x d\mathbf{p}_\mu(x) = \mathrm{Tr}\ \rho\ \widehat{x},$$

where $\widehat{x}$ is the position operator in Schrödinger's representation of QM.

Thus the energy detection model for PCSFT implies the same probability distributions and averages as the conventional QM model. As we shall see, there is even a possibility to go beyond QM.

8.5 Deviation from Predictions of Quantum Mechanics

Position measurements of higher precision arise very naturally by generalization of the energy detection model: through consideration of prequantum physical variables, $\phi \to f(\phi)$, given by nonquadratic functionals of the classical field. The appearance of additional terms in (5.51) (see Chapter 5; see also Section 3.1.4), induce deviations from the predictions of QM for averages. Now we would like to find the corresponding deviations for the probabilities of detection, namely, the deviations from the *Born's rule.*

From the PCSFT viewpoint, the conventional quantum observables correspond to measurements of quadratic forms of the prequantum field. The quadratic forms arise naturally in the oscillatory approximation of smooth functionals of the field. Therefore the quadratic forms give the main contribution to averages and probabilities. However, by increasing the precision of measurements, by approaching the fine subquantum time and space scales, one can expect that the contributions of nonlinearities of higher orders will be detectable. The aim of this section is to estimate the magnitudes of such contributions.

We restrict our modeling to the case of the fourth-order polynomials of classical fields. The main point is that, instead of the energy of the field ϕ, namely, $\pi_2(\phi) = \|\phi\|^2$, we shall consider its perturbation by integral of the fourth power of $\phi(x)$. And we repeat the PCSFT scheme for the position measurement in this framework. To simplify the presentation and to escape introduction of additional parameters to have functionals, $\phi \to f(\phi)$, having the physical dimension of energy, we proceed with dimensionless fields and the position variable x.

The measurement process over a random field again consists of the two steps: (a) selection of the field $\phi \in \mathcal{E}_\mu$ from the ensemble of fields $\mathcal{E}_\mu$; (b) measurement for this selected field: $X(\phi)$.

We assume that measurement devices (detectors) are sensitive to the magnitude of the following functional of the classical field. We define this functional of the field ϕ at the point x_0 as

$$\pi_{2,4}(x_0, \phi) = |\phi(x_0)|^2 + |\phi(x_0)|^4.$$

For the domain $I \subset \mathbf{R}^3$, we set

$$\pi_{2,4}(I, \phi) = \int_I (|\phi(x)|^2 + |\phi(x)|^4) dx,$$

in particular, we set

$$\pi_{2,4}(\phi) = \int_{\mathbf{R}^3} (|\phi(x)|^2 + |\phi(x)|^4) dx.$$

We remark once again that, since the dispersion α of the random field $\phi(x, \omega)$ is considered as a small parameter of the model (statistically the field is concentrated in a small neighborhood of the point $\phi \equiv 0$), the additional perturbation term $\int_{\mathbf{R}^3} |\phi(x, \omega)|^4 dx$ is small (from the point of view of random fluctuations).

We now formulate the fundamental feature of the class of detectors under consideration, namely, *sensitivity to the magnitude of the functional* $\pi_{2,4}$ in the following two postulates:

Postulate 1 ($\pi_{2,4}$). *The probability* P_μ *to select the fixed field* ϕ *from the random field* $\phi(x, \omega)$ *(the ensemble* $\mathcal{E}_\mu$*) is proportional to* $\pi_{2,4}(\phi)$*:*

$$dP_\mu(\phi) = K_\mu \pi_{2,4}(\phi) d\mu(\phi). \tag{8.23}$$

The coefficient of proportionality K_μ can be found from the normalization of the total probability by one: $K_\mu = \frac{1}{\int_H \pi_{2,4}(\phi) d\mu(\phi)}$. Thus we obtain

$$dP_\mu(\phi) = \frac{\pi_{2,4}(\phi)}{\int_H \pi_{2,4}(\phi) d\mu(\phi)} d\mu(\phi), \tag{8.24}$$

and, for any Borel subset $U \subset H$, we have

$$P_\mu(\phi \in U) = \frac{1}{\int_H \pi_{2,4}(\phi) d\mu(\phi)} \int_U \pi_{2,4}(\phi) d\mu(\phi), \tag{8.25}$$

or in the random field notation,

$$P_\mu(\phi \in U) = \frac{1}{E\pi_{2,4}(\phi(\omega))} E\left(\chi_U(\phi(\omega))\pi_{2,4}(\phi(\omega)\right), \tag{8.26}$$

where $\chi_U(\phi)$ is the characteristic function of the set U.

The selection procedure from a random field was formalized by Postulate 1. We now formalize step-b of the X-measurement in the following form.

Postulate 2 ($\pi_{2,4}$). *The probability* $P(X = x_0|\phi)$ *to obtain the result* $X = x_0$ *for the fixed field* ϕ *is proportional to the quantity*

$\pi_{2,4}(x_0, \phi)$. *The coefficient of proportionality does not depend on* x_0. *Thus* $k(x_0|\phi) \equiv k_\phi$.

The coefficient of proportionality $k(x_0|\phi)$ can be obtained from the normalization of the total probability by one: $1 = \int_{\mathbf{R}^3} P(X = x|\phi)dx = k_\phi \int_{\mathbf{R}^3} \pi_{2,4}(x, \phi)dx$. Thus,

$$k_\phi = \frac{1}{\pi_{2,4}(\phi)}. \tag{8.27}$$

The probability to obtain the result $X = x_0$ is obtained by using the classical Bayes' formula

$$\mathbf{p}_\mu(X = x_0) = \int_H P(x_0|\phi)dP_\mu(\phi) = \int_H \frac{|\phi(x_0)|^2 + |\phi(x_0)|^4}{\pi_{2,4}(\phi)} dP_\mu(\phi). \tag{8.28}$$

Finally, we derive the formula

$$\mathbf{p}_\mu(X = x_0) = \frac{1}{\int_H \pi_{2,4}(\phi)d\mu(\phi)} \int_H (|\phi(x_0)|^2 + |\phi(x_0)|^4)d\mu(\phi). \tag{8.29}$$

In fact, the quantity $\mathbf{p}_\mu(X = x)$ is the density of probability

$$\mathbf{p}_\mu(X \in I) = \int_I \mathbf{p}_\mu(X = x)dx$$

$$= \frac{1}{\int_H \pi_{2,4}(\phi)d\mu(\phi)} \int_H \left(\int_I (|\phi(x)|^2 + |\phi(x)|^4)dx \right) d\mu(\phi), \tag{8.30}$$

where I is a Borel subset of $\mathbf{R}^3$, e.g., a cube. It is convenient to make the field scaling and to move from the probability μ having the dispersion α to the corresponding normalized probability ν.

By this scaling we find the direct dependence of probabilities of detection on the small parameter α.

Then we can represent the coefficient of proportionality as

$$K_\mu = \frac{1}{\alpha + \alpha^2 c_4},$$

where

$$c_4 = \int_H \int_{-\infty}^{\infty} |\phi(x)|^4 dx d\nu(\phi).$$

We find the following dependence of the probability of the position detection on the small parameter α (the dispersion of random fluctuations):

$$\mathbf{p}_\mu(X = x_0) = \frac{1}{\alpha + \alpha^2 c_4} \int_H (\alpha|\phi(x_0)|^2 + \alpha^2|\phi(x_0)|^4) d\nu(\phi)$$
$$= \frac{1}{1 + \alpha c_4} \int_H (|\phi(x_0)|^2 + \alpha|\phi(x_0)|^4) d\nu(\phi). \quad (8.31)$$

If we make scaling in formula (8.18), we obtain

$$\mathbf{p}_\mu(X = x_0) = \int_H |\phi(x_0)|^2 d\nu(\phi). \quad (8.32)$$

The same result can be obtained by considering the limit $\alpha \to 0$ in (8.31). Thus the model presented in this section is really the $O(\alpha)$ perturbation of the model considered in Section 8.2 (and hence of QM).

We turn to our model of detection which takes into account the fourth power of the field. We have

$$\mathbf{p}_\mu(X \in I) = \frac{\int_H \int_I |\phi(x)|^2 dx d\nu(\phi) + \alpha \int_H \int_I |\phi(x)|^4 dx d\nu(\phi)}{1 + \alpha c_4}.$$

Hence,

$$\mathbf{p}_\mu(X \in I) \approx (1 - \alpha c_4)\Big(\int_H \int_I |\phi(x))|^2 dx d\nu(\phi$$
$$+ \alpha \int_H \int_I |\phi(x)|^4 dx) d\nu(\phi)\Big)$$
$$= \int_H \int_I |\phi(x)|^2 dx d\nu(\phi)) + \alpha\Big[\int_H \int_I |\phi(x)|^4 dx d\nu(\phi)$$
$$- \int_H \int_I |\phi(x)|^2 dx d\nu(\phi) \times \int_H \int_{\mathbf{R}^3} |\phi(x)|^4 dx d\nu(\phi)\Big].$$

The first summand gives the well-known Born's rule (the conventional QM prediction). The second summand (we denote it by $\Delta(I, \mu, \alpha)$) gives the deviation from the Born's rule.

We consider now this deviation in the case of a pure state. We take Ψ, $||\Psi|| = 1$. The normalized measure ν can be selected as the *Gaussian measure* $\nu \equiv \nu_\Psi$ with the covariance operator $\rho_\Psi = \Psi \otimes \Psi$. We have

$$\Delta(I, \Psi, \alpha) = \alpha\Big[\int_I |\Psi(x)|^4 dx - \int_I |\Psi(x)|^2 dx \int_{\mathbf{R}^3} |\Psi(x)|^4 dx\Big] \quad (8.33)$$

This formula was used in the examples considered in Section 3.11 to calculate the deviations from the Born's rule for the concrete wave functions.

8.6 Averages

We consider the nonquadratic functional of the prequantum field

$$f_x(\phi) = \int_{-\infty}^{+\infty} x(|\phi(x)|^2 + |\phi(x)|^4)dx.$$

It is mapped onto the position operator by the map T : PCSFT→ QM. PCSFT-average has the form

$$\begin{aligned}\langle f_x\rangle_\mu &= \int_{-\infty}^{+\infty} x \int_H (\phi(x)|^2 + |\phi(x)|^4)d\mu(\phi)dx \\ &= \alpha \int_{-\infty}^{+\infty} x \int_H |\phi(x)|^2 d\nu(\phi)dx + \alpha^2 \int_{-\infty}^{+\infty} x \int_H |\phi(x)|^4 d\nu(\phi)dx.\end{aligned}$$

On the other hand, by using the probability distribution which was derived on the basis of the energy's detection model for PCSFT we obtain

$$\begin{aligned}\langle x\rangle_{\mathbf{p}_\mu} &= \int_{-\infty}^{+\infty} x d\mathbf{p}_\mu(x) \\ &= \frac{\alpha \int_{-\infty}^{+\infty} x \int_H |\phi(x)|^2 d\nu(\phi)dx + \alpha^2 \int_{-\infty}^{+\infty} x \int_H |\phi(x)|^4 d\nu(\phi)dx}{\alpha + \alpha^2 \int_{-\infty}^{+\infty} \int_H |\phi(x)|^4 d\nu(\phi)dx}.\end{aligned}$$

We use the normalization based on average of the functional $\pi_{2,4}(\phi)$. Then

$$\frac{\langle f\rangle_\mu}{\langle \pi_{2,4}\rangle_\mu} = \int_{-\infty}^{+\infty} x d\mathbf{p}_\mu(x). \tag{8.34}$$

We now consider the functional $\pi_2(\phi) = ||\phi||^2$. The basic asymptotic equality of PCSFT can be written in the form

$$\frac{\langle f\rangle_\mu}{\langle \pi_2\rangle_\mu} = \langle T(f)\rangle_{T(\mu)} + O(\alpha), \alpha \to 0.$$

The model of measurement in which measurements of quantum systems are represented through interactions of classical random fields with detectors is presented. One of predictions of this model is a possibility to design measurements of functionals of the prequantum fields such that the detection probability deviates from the probability given by the Born's rule. The latter is only an approximate rule.

8.7 Local Measurements

To obtain a local measurement scheme, we should take into account that any detector is located in a concrete domain, say O, of space. More generally, if a detector measures some variable, e.g., energy, it operates only in the special range of variation of this variable. This fundamental fact of measurement theory should be taken into account, cf. Haag [101], and especially Ref. [102], and our PCSFT should be modified. We consider again position measurement. As before, the measurement process for a random field consists of the following two steps: (a) selection of the fixed field ϕ from the random prequantum field; (b) measurement for this field.

We proceed under the assumption that the detector operates in the domain O.

Postulate 1. (Local) *The probability to select the fixed field ϕ from the random field $\phi(x, \omega)$ is proportional to the energy of the field ϕ in the domain O :*

$$dP_\mu(\phi|O) = K_{\mu,O}\, \pi_2(O;\phi)\, d\mu(\phi).$$

The coefficient of proportionality K_μ can be found from the normalization of the total probability by one:

$$K_{\mu,O} = \frac{1}{\alpha_O},$$

where

$$\alpha_O = \int_H \pi_2(O;\phi) d\mu(\phi).$$

Thus we obtain

$$dP_\mu(\phi|O) = \frac{\pi_2(O;\phi)}{\alpha_O} d\mu(\phi),$$

Postulate 1 is intuitively attractive: fields which contain larger portions of the energy in the domain of detection O are selected more often. We now again formalize step-b.

Postulate 2. (Local) *The probability $P(X = x_0|\phi, O)$ to obtain the result $X = x_0$, where $x_0 \in O$, for the fixed field ϕ is proportional to the density of the energy of this field at x_0, i.e., to $\pi_2(x_0, \phi)$. The coefficient of proportionality does not depend on x_0, so $k(x_0|\phi) \equiv k_{\phi,O}$.*

This coefficient can be obtained from the normalization of the total probability by one:

$$1 = \int_O P(X = x|\phi, O)dx = k_{\phi,O} \int_O |\phi(x)|^2 dx.$$

Thus,

$$k_{\phi,O} = \frac{1}{\pi_2(O;\phi)}.$$

The probability to obtain the result $X = x_0, x_0 \in O$, for the random field with the probability distribution μ can be again derived by using the classical Bayes' formula

$$\mathbf{p}_\mu(X = x_0|O) = \int_H P(x_0|\phi, O)dP_\mu(\phi|O) = \int_H \frac{|\phi(x_0)|^2}{\pi_2(O;\phi)} dP_\mu(\phi|O).$$

Finally, the following formula is derived:

$$\mathbf{p}_\mu(X = x_0|O) = \frac{1}{\alpha_O} \int_H |\phi(x_0)|^2 d\mu(\phi). \tag{8.35}$$

The quantity $\mathbf{p}_\mu(X = x|O)$ has to be interpreted as the density of probability. For the domain $I \subset O$, we have

$$\mathbf{p}_\mu(X \in I|O) = \int_I \mathbf{p}_\mu(X = x|O)dx = \frac{1}{\alpha_O} \int_H \left(\int_I |\phi(x)|^2 dx \right) d\mu(\phi). \tag{8.36}$$

8.8 Measurement of Observables with Discrete Spectra

Let $f(\phi)$ be a physical variable of PCSFT (functional of the prequantum fields). Consider the corresponding QM observable, the operator $\widehat{A}$. For simplicity, suppose that $\widehat{A}$ has the purely discrete nondegenerate spectrum. In our model the quantum observable represented by the operator $\widehat{A}$ performs approximate measurement of the classical variable f_A. (We expect that in future it would be possible to measure prequantum random fields and, hence, arbitrary functionals of these fields.) By our prequantum model detectors operationally described by the formalism of QM work in the following way.

Let $a \in \mathbf{R}$ belong to the spectrum of $\widehat{A}$ and let e_a be the corresponding eigenvector: $\widehat{A}e_a = ae_a$. We consider the field functional corresponding to this eigenvector

$$\pi_2^{(\widehat{A})}(a, \phi) = |\langle\phi, e_a\rangle|^2.$$

Take the interval $I \subset \mathbf{R}$ and set

$$\pi_2^{(\widehat{A})}(I, \phi) = \sum_{a \in I} \pi_2^{(\widehat{A})}(a, \phi)$$

and, finally,

$$\pi_2^{(\widehat{A})}(\phi) = \sum_{a \in \mathbf{R}} \pi_2^{(\widehat{A})}(a, \phi) \equiv ||\phi||^2.$$

We repeat briefly our scheme of measurement by applying it to the interaction of a prequantum field with the $\widehat{A}$-detector:

(a) selection of the concrete field signal ϕ with the probability proportional to the square of norm;
(b) production of some value $\widehat{A} = a$ with the probability proportional to the quantity $\pi_2^{(\widehat{A})}(a, \phi)$.

These postulates and the Bayes' formula imply:

$$\mathbf{p}_\mu^{(\widehat{A})}(a) = \frac{1}{\alpha} \int_H \pi_2^{(\widehat{A})}(a, \phi) d\mu(\phi) = \frac{1}{\alpha} \int_H |\langle\phi, e_j\rangle|^2 d\mu(\phi).$$

Let now $\mu = \mu_\Psi$, $||\Psi|| = 1$. It is a measure (e.g., Gaussian) with zero mean value and the covariance operator $\rho_\Psi = \Psi \otimes \Psi$. Then

$$\int_H |\langle\phi, e_a\rangle|^2 d\mu_\Psi(\phi) = |\langle\Psi, e_a\rangle|^2.$$

Hence,

$$\mathbf{p}_{\mu_\Psi}^{(\widehat{A})}(a) = |\langle\Psi, e_a\rangle|^2.$$

This is the Born's rule.

8.9 Classical Field Treatment of Discarding of Noise Contribution in Quantum Detectors

Consider a random field $\phi(\omega)$ taking values in H. Suppose that this field is distributed in the space H with the probability distribution μ_D (having zero mean value and the covariation operator D). By Postulate 1, a quantum detector selects a field $\phi \in L_2$ from a random field with probability P_μ given by (8.10): for a Borel set $O \subset H$,

$$P_{\mu_D}(\phi \in O) = \frac{\int_O ||\phi||^2 d\mu_D(\phi)}{\int_H ||\phi||^2 d\mu_D(\phi)} = \frac{1}{\mathrm{Tr} D} \int_O ||\phi||^2 d\mu_D(\phi).$$

We call such random fields *bowl random fields.* Since later we shall work with a few different covariance operators, we index carefully the probability distribution of the input random field and the induced probability distribution of a detector by their covariance operators.

In our theory, measurement apparata are mathematically described by subspaces of the space $H = L_2(\mathbf{R}^3)$ (the space of all possible classical signals). Take a (closed) linear subspace $L \subset H$. We consider a measurement device called *L-antenna* such that for a fixed (i.e., nonrandom) signal $\phi = \phi(x)$, the probability to detect this signal by the L-antenna is proportional to

$$q_L(\phi) = ||\widehat{P}_L \phi||^2.$$

Thus by our model of the signal detection each L-antenna reacts randomly to the input signal. It can be either detected or not and the probability of detection is proportional to $q_L(\phi)$.

Thus such an antenna integrates the energy of the projection of the signal on the L subspace:

$$q_L(\phi) = \int_{\mathbf{R}^3} |\widehat{P}_L \phi(x)|^2 dx,$$

where $\widehat{P}_L : H \to L$ is the orthogonal projector. A higher magnitude of $q_L(\phi)$ induces a higher probability of detection. Thus a signal with high energy concentration in the L-subspace will be detected with a high probability. But if the L-component of this signal has a low amplitude, then the probability of detection is low. In particular, if $\widehat{P}_L \phi = 0$, then the signal will not be detected at all.

The $q_L(\phi)$ has to be normalized and we obtain the probability of detection:

$$\mathbf{P}_L(\text{detection}|\phi) = \frac{q_L(\phi)}{||\phi||^2}, \ \phi \neq 0. \qquad (8.37)$$

For example, consider a one-dimensional antenna. It is given by a vector $u \in H$, $||u|| = 1$. Then the probability to detect a signal is given by

$$\mathbf{P}_u(\text{detection}|\phi) = \frac{|\langle \phi, u \rangle|^2}{||\phi||^2}, \ \phi \neq 0.$$

Take a point $x_0 \in \mathbf{R}^3$ and a ball $B_r(x_0)$ of a small radius $r > 0$. Set

$$u(x) = \chi_{B_r(x_0)}(x_0)/\sqrt{V},$$

where V is the volume of the ball. Consider a device which can measure the average of the prequantum field $\phi(x)$ on the ball $B_r(x_0)$:

$$\langle \phi \rangle_{\text{space}}(x_0) = \frac{1}{V} \int_{B_r(x_0)} \phi(x) dx = \frac{1}{\sqrt{V}} \langle \phi, u \rangle, \qquad (8.38)$$

i.e.,

$$\langle \phi, u \rangle = \sqrt{V} \ \langle \phi \rangle_{\text{space}}(x_0).$$

Thus,

$$\mathbf{P}_u(\text{detection}|\phi) = \frac{|\langle \phi, u \rangle|^2}{||\phi||^2} = V \frac{|\langle \phi \rangle_{\text{space}}(x_0)|^2}{||\phi||^2}.$$

If ϕ is a continuous field, then $\langle \phi \rangle_{\text{space}}(x_0) \approx \phi(x_0)$. Thus,

$$\mathbf{P}_u(\text{detection}|\phi) = V \frac{|\phi(x_0|^2)}{||\phi||^2} \to 0, \ V \to 0.$$

Thus the prequantum field cannot be smooth, it has to be singular to produce a nonnegligible probability. We shall see that it is really the case, due to the contribution of the background field of the white noise type. In other words, the prequantum mean field has to be strong enough in domains of integration with $V \to 0$.

We remark that the presented model of detection can be applied to the macroscopic electromagnetic field. In this case the mean field devices described by (8.38) are available. If the classical field is strong enough in $B_r(x_0)$, $r > 0$ and r is very small, we mimic the PCSFT scheme of quantum measurements.

Consider now a random signal with the probability distribution μ_D and a quantum detector which detects a field ϕ (in the ensemble of fields $\mathcal{E}_{\mu_D}$ corresponding to this random signal) with the probability distribution P_{μ_D} of the bowl-type. Then by the classical Bayes formula we have

$$\mathbf{P}_L(\text{detection}|\mu_D) = \int_H \mathbf{P}_L(\text{detection}|\phi) d P_{\mu_D}(\phi)$$
$$= \frac{1}{\text{Tr}D} \int_H ||\widehat{P}_L\phi||^2 d\mu_D(\phi) = \frac{1}{\text{Tr}D} \text{Tr}D\ \widehat{P}_L.$$

Thus the classical prequantum probability (of detection of a classical prequantum signal) is coupled with the quantum probability in the following way:

$$\mathbf{P}_L^{\text{CL}} = \text{Tr}\rho \hat{P}_L = \mathbf{P}_L^{\text{QM}}, \tag{8.39}$$

where

$$\rho = \frac{D}{\text{Tr}D}. \tag{8.40}$$

We have obtained the quantum formula for detection from rather natural model of detection of classical signals. The main postulate is that the probability of detection of the classical signal in a subspace $L \subset H$ is proportional to the intensity $q_L(\phi)$ of the L-component of this signal.

8.9.1 *Prequantum Signal Processing through Noisy Information Channel*

Starting with this section, we consider *Gaussian random signals.*

As was emphasized, a consistent formalism for PCSFT can be constructed only under the assumption that physical space is filled with the background field, the white-noise type field (vacuum fluctuations). In this section we present a model of detection of prequantum fields on such a noisy background. The prequantum field for, e.g., electron contains two components:

(a) pure (or filtered) signal;
(b) background noise.

Our model of detection of such noisy signals is closely coupled to the presence of thresholds in quantum detectors (cf. Chapter 9). QM can be interpreted as a theory of filtration of signals (called in QM quantum systems) from vacuum fluctuations.

To escape mathematical difficulties, we begin our study in the finite-dimensional case. Consider now a prequantum random field with the probability distribution μ_D containing the contribution of the background field of the white noise type. Here

$$D = D_0 + D_{\text{noise}}, \tag{8.41}$$

where $D_{\text{noise}} = \epsilon I$, $\epsilon > 0$, corresponds to white noise (scaled by $\sqrt{\epsilon}$) and D_0 corresponds to the prequantum field in the absence of noise. We remark that the Gaussian measure μ_D can be represented as the convolution

$$\mu_D = \mu_{D_{\text{noise}}} * \mu_{D_0}. \tag{8.42}$$

Thus μ_D can be considered as the output of the classical Gaussian channel with the input μ_{D_0} and the noise which is distributed by the law $N(0, D_{\text{noise}})$ and, finally, the linear transportation operator ϵI, see Ref. [252], p. 199. However, we are interested in the channel transforming μ_{D_0} into P_{μ_D} : in the process of propagation through this channel a Gaussian input is transformed into an output of the bowl type. Set

$$\lambda(\phi, O) = \frac{1}{\text{Tr } D} \int_H \chi_O(u)||u||^2 d\mu^{\phi}_{D_{\text{noise}}}(u),$$

where $\chi_O(u)$ is the characteristic function of the Borel set $O \subset H$ and $\mu^{\phi}_{D_{\text{noise}}}$ is the Gaussian measure distributed $\sim N(\phi, D_{\text{noise}})$, i.e., having the mean value ϕ and the covariance operator D_{noise}. Then the bowl-type measure P_{μ_D} can be written as

$$P_{\mu_D}(O) = \int_H \lambda(\phi, O) d\mu_{D_0}(\phi). \tag{8.43}$$

We have

$$\begin{aligned}
\int_H \lambda(\phi, O) d\mu_{D_0}(\phi) &= \frac{1}{\text{Tr} D} \int_H \left(\int_H \chi_O(u)||u||^2 d\mu^{\phi}_{D_{\text{noise}}}(u) \right) d\mu_{D_0}(\phi) \\
&= \frac{1}{\text{Tr} D} \int_{H \times H} \chi_O(\phi_1 + \phi_2)||\phi_1 \\
&\quad + \phi_2||^2 d\mu_{D_0} \otimes \mu_{D_{\text{noise}}}(\phi_1, \phi_2) \\
&= \frac{1}{\text{Tr} D} \int_H \chi_O(v)||v||^2 d\mu_{D_0} \star \mu_{D_{\text{noise}}}(v) \\
&= \frac{1}{\text{Tr} D} \int_O ||v||^2 d\mu_{D_0 + D_{\text{noise}}}(v) = P_{\mu_D}(O).
\end{aligned}$$

This transformation can be considered as a classical noisy channel amplifying the signal by the factor $||\phi||^2$. This is PCSFT description of a quantum detector.

8.9.2 *Detection of the Prequantum Signal in the Presence of the Background Noise*

In our interpretation QM can be considered as a formalism of extraction of the input signal from the noisy output. QM describes cutoff of the contribution of noise. Consider a random signal $\phi(\omega)$ with the Gaussian distribution μ_D, where $D = D_0 + D_{\text{noise}}$. We apply our measurement model. Then the probability of detection of this (noisy) signal by the L-antenna is given by

$$\begin{aligned}
\mathbf{P}(\text{detection}|\mu_D) &= \frac{1}{\text{Tr} D}\int_H ||\widehat{P}_L\phi||^2 d\mu_D(\phi) \\
&= \frac{\text{Tr} D\widehat{P}_L}{\text{Tr} D} = \frac{\text{Tr} D_0 \widehat{P}_L}{\text{Tr} D} + \frac{\text{Tr} D_{\text{noise}} \widehat{P}_L}{\text{Tr} D} \\
&= \frac{\text{Tr} D_0 \widehat{P}_L}{\text{Tr} D_0}\frac{\text{Tr} D_0}{\text{Tr} D} + \frac{\text{Tr} D_{\text{noise}} \widehat{P}_L}{\text{Tr} D_{\text{noise}}}\frac{\text{Tr} D_{\text{noise}}}{\text{Tr} D}. \qquad (8.44)
\end{aligned}$$

The proportions $\frac{\text{Tr} D_0}{\text{Tr} D}$ and $\frac{\text{Tr} D_{\text{noise}}}{\text{Tr} D}$ are probabilities of the pure signal (i.e., filtered from of noise) and noise contributions into the random signal ϕ :

$$\mathbf{P}(\text{pure signal}|\mu_D) = \frac{\int_H ||\phi||^2 d\mu_{D_0}(\phi)}{\int_H ||\phi||^2 d\mu_D(\phi)},$$

$$\mathbf{P}(\text{noise}|\mu_D) = \frac{\int_H ||\phi||^2 d\mu_{D_{\text{noise}}}(\phi)}{\int_H ||\phi||^2 d\mu_D(\phi)}.$$

The proportions

$$\frac{\text{Tr} D_0 \hat{P}_L}{\text{Tr} D_0}, \quad \frac{\text{Tr} D_{\text{noise}} \hat{P}_L}{\text{Tr} D_{\text{noise}}}$$

are probabilities of the L-detection of two components of the classical prequantum signal: (a) the first one is obtained by filtered of the background noise; (b) the second is the noise component;

$$\mathbf{P}(\text{detection}|\mu_{D_0}) = \frac{\int_H ||\hat{P}_L\phi||^2 d\mu_{D_0}(\phi)}{\int_H ||\phi||^2 d\mu_{D_0}(\phi)},$$

$$\mathbf{P}(\text{detection}|\mu_{D_{\text{noise}}}) = \frac{\int_H ||\hat{P}_L\phi||^2 d\mu_{D_{\text{noise}}}(\phi)}{\int_H ||\phi||^2 d\mu_{D_{\text{noise}}}(\phi)}.$$

We remark that

$$\mathbf{P}(\text{detection}|\mu_{D_0}) = \text{Tr}\rho\ \widehat{P}_L\rho = \frac{D_0}{\text{Tr}D_0}, \tag{8.45}$$

coincides with the QM probability. We also notice that equality (8.44) can be written as the formula of the total probability:

$$\mathbf{P}(\text{detection}|\mu_D) = \mathbf{P}(\text{detection}|\mu_{D_0})\mathbf{P}(\text{pure signal}|\mu_D) + \mathbf{P}(\text{detection}|\mu_{D_{\text{noise}}})\mathbf{P}(\text{noise}|\mu_D). \tag{8.46}$$

Thus prequantum measurement theory induces the following interpretation of quantum probability.

This is the probability of detection based on the procedure of filtration of noise. QM predicts not the probability of detection of the total signal, $\mathbf{P}(detection|\mu_D)$, but the probability of detection of its cleaned (from the contribution of noise) component. Since a sufficiently strong noise component is present in all measurements for microsystems, the QM predictions match so well with the collected statistical data.

8.9.3 *Infinite-Dimensional Noise*

We recall that the previous consideration concerned the finite-dimensional Hilbert space. Let us proceed to the real physical situation, i.e., $H = L_2(\mathbf{R}^3)$. In this case the Gaussian measure μ_D is not well defined on H. To define it properly, one should consider a larger state space, e.g., $H_- = \mathcal{S}'(\mathbf{R}^3)$, the space of Schwartz generalized functions. However,

$$\int_{\mathcal{S}'(\mathbf{R}^3)} ||\phi||^2 d\mu_D(\phi) = \infty,$$

and the probability distribution P_{μ_D} is not well defined.

Nevertheless, we can proceed by considering P_{μ_D} as a "singular probability distribution" on $\mathcal{S}'(\mathbf{R}^3)$, a generalized function of infinitely many variables [109].

To escape mathematical difficulties, we shall use the following simple way of introduction of infinite-dimensional generalized functions, namely, through approximations by finite-dimensional distributions.

Consider an increasing chain of finite-dimensional subspaces $H_m \subset H : H_m \subset H_{m+1}$ such that $\cup_m H_m$ is dense in H. Denote the corresponding orthogonal projectors by $\widehat{P}_m \equiv \widehat{P}_{H_m}$. Set $D_m = \widehat{P}_m D \widehat{P}_m = \widehat{P}_m D_0 \widehat{P}_m + \widehat{P}_m D_{\text{noise}} \widehat{P}_m = D_{0,m} + D_{\text{noise},m}$.

Consider a random signal $\phi_m(\omega) \in H_m$ having the probability distribution

$$P_{\mu_{Dm}} = \frac{||\phi||^2}{\text{Tr} D_m} \mu_{D_m}.$$

We can repeat everything for this process and write the formula of the total probability (8.46):

$$\mathbf{P}(\text{detection}|\mu_{D_m}) = \mathbf{P}(\text{detection}|\mu_{D_{0,m}})\mathbf{P}(\text{pure signal}|\mu_{D_m}) + \mathbf{P}(\text{detection}|\mu_{D_{\text{noise},m}})\mathbf{P}(\text{noise}|\mu_{D_m}).$$

We remark that

$$\lim_{m\to\infty} \mathbf{P}(\text{detection}|\mu_{D_{0,m}}) = \mathbf{P}_L^{\text{QM}}(\text{detection}) = \text{Tr } p\widehat{P}_L, \ \rho = \frac{D_0}{\text{Tr} D_0}.$$

But

$$\lim_{m\to\infty} \mathbf{P}(\text{pure signal}|\mu_{D_m}) = \frac{\text{Tr} D_{0,m}}{\text{Tr} D_m} \to 0, \ m \to \infty.$$

Thus the statistical contribution of the background field is dominating. (The noise of vacuum fluctuations is dominating.) By creation of detectors which are sensitive to signals of very small energy we shall detect essentially only noise.

Thus the quantum probability is not the probability of the L-detection of the total prequantum signal. The quantum probability is the conditional probability:

$$\mathbf{P}(\text{detection}|\mu_{D_0}) = \mathbf{P}(\text{detection after discarding of background}).$$

We considered white noise, i.e., $D_{\text{noise}} = \epsilon I$. However, all the previous considerations are valid for any sort of Gaussian noise, with any operator $D_{\text{noise}} \geq 0$, $D^*_{\text{noise}} = D_{\text{noise}}$ and $D_{\text{noise}} : H \to H$ is bounded.

8.9.4 *Characteristic Function of Bowl-Type Distribution*

Let D be a trace-class operator, $D^* = D$, $D \geq 0$, on H. Consider the measure P_{μ_D} and find its characteristic function (Fourier transform). To make computations shorter, consider a real Hilbert space. We have

$$\tilde{P}_{\mu_D}(\alpha) = \int_H e^{i(\alpha,\phi)}||\phi||^2 d\mu_D(\phi) = \sum_{j=1}^{\infty} \int_H (\phi, e_j)^2 e^{i(\alpha,\phi)} d\mu_D(\phi)$$

$$= -\sum_{j=1}^{\infty} \frac{\partial^2}{\partial \alpha_j^2} e^{-\frac{1}{2}(D\alpha,\alpha)} = (\mathrm{Tr} D + ||D\alpha||^2) e^{-\frac{1}{2}(D\alpha,\alpha)}.$$

8.10 Quantum Channels as Linear Filters of Classical Signals

In this section we show that any quantum channel can be represented as a linear filter for classical signals. This induces the following PCSFT interpretation of quantum measurements. Quantum systems are classical random signals; quantum measurements are linear filters of this signals. Eigenvalues of a quantum observable (elements of its spectrum) are labels for outputs of a system of linear filters spitting a classical signal.

We again proceed with *Gaussian* prequantum fields.

Let H_i, $i = 1, 2$, be a complex Hilbert spaces. Consider a linear bounded operator $V : H_1 \to H_2$ (we remark that $V^* : H_2 \to H_1$) and the corresponding linear filter

$$\phi_{\text{out}}(\omega) = V\phi_{\text{in}}(\omega), \tag{8.47}$$

where ϕ_{in} is the H_1-valued random field which is distributed $N(0, D_{\text{in}})$. Then the H_2-valued random field $\phi_{\text{out}} \sim N(0, D_{\text{out}})$, where

$$D_{\text{out}} = V D_{\text{in}} V^*. \tag{8.48}$$

Example 10.1. (Unitary evolution) The solution of the Schrödinger equation (1.13), see Section 1.2.4, with the initial condition ϕ_{in}, a Gaussian random field, can be represented in the form of the linear filter (8.47), where $V = U_t = \exp\{-it\widehat{\mathcal{H}}/h\}$. This filter preserves the norm of the prequantum field

$$\|\phi_{\text{out}}(t, \omega)\|^2 = \|\phi_{\text{in}}(\omega)\|^2. \tag{8.49}$$

Hence, this filter preserves even the dispersion of a classical signal

$$\sigma^2(\phi_{\rm out})(t) = {\rm Tr}\, D_{\rm out}(t) = \sigma^2(\phi_{\rm in}) = {\rm Tr}\, D_{\rm in}. \qquad (8.50)$$

Example 10.2. (von Neumann–Lüders projection). Let $L = P$ be an orthogonal projector $P : H \to E$, where E is a linear subspace of H. Here the E-valued (Gaussian) random field $\phi_{\rm out}(\omega) = P\phi_{\rm in}(\omega)$ describes the output after the von Neumann–Lüders projection measurement (with filtration with respect to the value corresponding to the projector P). The von Neumann–Lüders filter does not preserve even the dispersion

$${\rm Tr}\, D_{\rm out} = {\rm Tr}\, P\, D_{\rm in}\, P \neq {\rm Tr}\, D_{\rm in}.$$

Moreover,

$${\rm Tr}\, D_{\rm out} \leq {\rm Tr}\, D_{\rm in},$$

i.e., the dispersion of a signal always decreases.

Example 10.3. (the von Neumann–Lüders measurement). Consider a dichotomous quantum observable $\widehat{A}$ with eigenvalues α_1 and α_2 and eigenspaces E_1 and E_2. Denote the corresponding projectors by P_1 and P_2. They are orthogonal. Take two *independent random fields* $\phi_k \sim N(0, D)$, $k = 1, 2$, where the corresponding quantum state $\rho = D/{\rm Tr} D$. We form the $H \times H$ valued random variable $\phi_{\rm in}(\omega) = (\phi_1(\omega), \phi_2(\omega)) \sim N(0, D \times D)$. We define a linear operator $V : H \times H \to H$, $V(x, y) = P_1 x + P_2 y$ and the corresponding linear filter

$$\phi_{\rm out}(\omega) = V\phi_{\rm in}(\omega) = P_1\phi_1(\omega) + P_2\phi_2(\omega). \qquad (8.51)$$

Since ϕ_1 and ϕ_2 are independent and Gaussian, their linear transforms $P_1\phi_1$ and $P_2\phi_2$ are also independent. Hence the covariance operator of the random field $\phi_{\rm out}$ is equal to the sum of covariance operators of the latter two random fields, i.e.,

$$D_{\rm out} = P_1 D_{\rm in} P_1 + P_2 D_{\rm in} P_2. \qquad (8.52)$$

We remark that the dispersion is preserved:

$${\rm Tr}\, D_{\rm out} = {\rm Tr}\, D_{\rm in}. \qquad (8.53)$$

This example can be easily generalized to an arbitrary quantum observable with a purely discrete spectrum, $\widehat{A} = \sum_k \alpha_k P_k$; moreover, to any POVM. For simplicity, we present the case of measurement with finitely many results.

Example 10.4. (POVM measurement as classical linear filter) Consider a POVM $\{Q_i = V_i V_i^*\}_{i=1}^n$, where

$$\sum_{i=1}^{n} Q_i = I. \tag{8.54}$$

Define the corresponding linear map $V : H_1 \times \cdots \times H_n \to H$, $V(x_1, \ldots, x_n) = V_1 x_1 + \cdots + V_n x_n$. Consider also a quantum state, density operator ρ. Take a vector $\phi_{\text{in}}(\omega) = (\phi_1(\omega), \ldots, \phi_n(\omega))$ consisting of equally distributed independent Gaussian random fields of $N(0, D)$-type, where the quantum state under consideration $\rho = D/\text{Tr}D$. We define a linear filter corresponding to the map V:

$$\phi_{\text{out}}(\omega) = V\phi_{\text{in}}(\omega) = \sum_i V_i \phi_i(\omega). \tag{8.55}$$

We have

$$D_{\text{out}} = \sum_i V_i D_{\text{in}} V_i^*. \tag{8.56}$$

This is nothing else than a completely positive map, see, e.g., Ref. [120]. Hence it preserves the trace, i.e., the equality (8.53) holds. Therefore equality (8.56) for covariance operators is transformed into the equality for quantum states, density operators

$$\rho_{\text{out}} = \frac{D_{\text{out}}}{\text{Tr}D_{\text{out}}} = \frac{D_{\text{out}}}{\text{Tr}D_{\text{in}}} = \sum_i \frac{\text{Tr}V_i D_{\text{in}} V_i^*}{\text{Tr}D_{\text{in}}} \rho_{\text{out},i}, \tag{8.57}$$

where

$$\rho_{\text{out},i} = \frac{V_i D_{\text{in}} V_i^*}{\text{Tr}V_i D_{\text{in}} V_i^*}. \tag{8.58}$$

Set $\rho_{\text{in}} = D_{\text{in}}/\text{Tr}D_{\text{in}}$. Then

$$\rho_{\text{out},i} = \frac{V_i \rho_{\text{in}} V_i^*}{\text{Tr}V_i \rho_{\text{in}} V_i^*} \tag{8.59}$$

and

$$\rho_{\text{out}} = \sum_i (\text{Tr}V_i \rho_{\text{in}} V_i^*)\, \rho_{\text{out},i}. \tag{8.60}$$

Since any *quantum channel* can represented (the Kraus representation [120]) in the form (8.56), in general, with the infinite number of terms, we demonstrated (generalization to the infinite number of terms in the Kraus decomposition is straightforward) that any quantum channel can be represented as the linear filter of classical signals of the form (8.55). This is a step in the same direction as Ref. [252]: exploring the analogy between quantum information theory and classical signal theory.

Chapter 9

Quantum Individual Events

All previous chapters of this book were devoted to the development of the theory of classical (prequantum) random fields reproducing quantum averages (including correlations for entangled systems) as averages of quadratic forms of fields. Since the basic model of prequantum fields is the Gaussian one, these fields are continuous. Hence, in the same way as in the classical signal theory, averages are calculated for variables with a *continuous* range of values. This model, PCSFT, is really a *prequantum* model. In accordance with the Bohr's viewpoint, we consider QM as an operational formalism describing (predicting) results of measurements on microsystems. QM cannot describe intrinsic physical processes in the microworld, but only measurements performed by macroscopic classical devices. In contrast, PCSFT describes intrinsic processes in the microworld. However, it does not describe results of measurements on the level of *individual events,* clicks of detectors. As we pointed in Section 1.1.8, in principle one might be satisfied with creation of a prequantum model which reproduces only quantum probabilistic predictions, i.e., without establishing a direct connection with theory of measurement on the level of individual events. This approach would match with views of Schrödinger and the Bild concept in general. However, historical development of QM demonstrated that

Beyond Quantum
Andrei Khrennikov

ISBN 978-981-4411-73-8 (Hardcover), 978-981-4411-74-5 (eBook)
www.panstanford.com

the Bild concept was not attractive for the majority of physicists. Since the experimental verification is considered as the basic counterpart of any physical theory, Bild-like prequantum theories are considered as metaphysical. Measurement theory connecting PCSFT with experiment on the level of individual events was developed in my paper [211] (see also Refs. [209, 210, 212, 213]). In this paper I elaborated a scheme of discrete measurements of classical random signals which reproduces the basic rule of QM, *the Born's rule.* This is the scheme of *threshold-type detection*: such a detector clicks after it has "eaten" a special portion of energy ϵ of a prequantum random signal. We call such measurement theory threshold detection (TSD) model. TSD is a classical measurement model which delivers the same predictions as QM. In contrast to QM, which is an operational formalism[a], TSD provides a detailed description of the energy balance in the process of interaction of a classical random field (fluctuating at a fine time scale) with a detector, as well as the conditions inducing an individual quantum event, a click. As was pointed in Section 1.3, Devil is in detectors (cf. [11, 13, 14, 222, 223, 267–270].) Thus QM can be interpreted as theory of measurements of classical random signals with the aid of threshold-type detectors. They produce discretization of continuous classical fields. This discretization is the essence of quantum phenomena. Hence, "quantumness is created in detectors," there is no quantumness without measurement. In particular, this viewpoint contradicts the views of early Einstein who claimed that the electromagnetic field is quantized not only in the process of the energy exchange with material bodies, but even in vacuum (see Section 1.1.7). In the PCSFT/TSD model photons appear only on the level of detection, so they are nothing else than the clicks of detectors. Surprisingly this viewpoint on the notion of photon coincides with the views of some top level experimenters working in quantum optics, e.g., A. Zeilinger, A. Migdal, and S. Polyakov also identify photon with a click of a detector (their talks at Växjö conferences). Of course, the majority of experts in quantum optics

[a]In QM, measurement devices are simply black boxes which are symbolically represented by self-adjoint operators or more generally (and even more formally) by POVM.

(including aforementioned) proceed under the Bohrian assumption on completeness of QM.

Our TSD model is fundamentally based on the *assumption of ergodicity* of prequantum random signals, i.e., the time and ensemble averages coincide (see also Section 3.1.5). We stress that this ergodicity is related to the subquantum processes, i.e., to the intrinsic physical processes in the microworld. Another important assumption, cf. Sections 3.1.1, 3.1.5, and 5.15, is that the prequantum random signals fluctuate on a very fine time scale (comparing with the time scale of lab measurements). The duration of measurement, i.e., the interaction of a random field with a detector, is huge on the prequantum time scale. However, it is sufficiently small with respect to the time scale of lab measurements.

The rigorous mathematical justification of the derivation of the Born's rule in the TSD framework is rather complicated. First we present its scheme at the heuristic level and then we proceed to the rigorous mathematical derivation (Section 9.7) based on the theory of ergodic stochastic processes.

We remark that ergodic processes are *stationary.* In the process of interaction with a detector, a prequantum field (emitted by a source and evolving in the free space in accordance with the Schrödinger equation with a random initial condition, see Section 1.2.4) approaches a steady state, the field becomes stationary. This first phase is very short. Then the stationary stochastic process transmits its energy to the detector and this stage is very long (infinitely long in the pure mathematical framework). We remark that this discussion is about the prequantum time scale.

Max Born postulated the probabilistic rule coupling the wave function with quantum detection probabilities [42]: the probability equals to the squared wave function. Whether it is possible to derive this rule from some natural physical principles is still the subject of intensive debates,[a] see, e.g., Deutsch [73] and 't Hooft [289] for attempts to derive the Born's rule.

In this chapter we shall derive the Born's rule in the PCSFT/TSD framework: threshold-type detectors interacting with random

[a]"The conclusion seems to be that no generally accepted derivation of the Born rule has been given to date, but this does not imply that such a derivation is impossible in principle" [225].

signals of a special class produce a statistics of clicks which is (approximately) described by the Born's rule. Of course, this framework is based on PCSFT coupling between the covariance operator of a random field and the density operator of the quantum system corresponding to this random field.[a]

PCSFT/TSD also proposed very natural solution of quantum *measurement problem* [296]. Continuous evolution of the prequantum random field induces discrete jumps corresponding to approaching the detection threshold. Such jumps are formally described by the von Neumann projection postulate. However, this solution of the measurement problem is also far from the expectations of people excited by quantum exotics. Moreover, since, to solve the measurement problem, we go beyond QM, our solution would not be considered as the solution of the problem formulated by von Neumann [296]. The latter is expected to be solved in the conventional quantum framework.

We start our presentation of TSD without any coupling to PCSFT, i.e., at the beginning we shall not consider random signals as corresponding to quantum systems. We present a very general detection scheme based on threshold detectors. We shall see that the Born's rule arises the under natural restrictions on random signals and detectors, see conditions (a)–(d) in Section 9.2. Then, in Section 9.3, we shall connect TSD with PCSFT by coupling covariance operators of classical signals with quantum density operators (see Section 1.2.2).

9.1 Classical Random Signals: Ensemble and Time Representations of Averages

We start with brief repetition of the presentation of Section 3.1.5 on interrelation between ensemble and time averages and the role played by the assumption of ergodicity in this framework.

The state space of classical signal theory is the L_2-space $H = L_2(\mathbf{R}^3)$. Elements of H are classical fields $\phi : \mathbf{R}^3 \to \mathbf{C}^n$. We consider

[a] This derivation of the fundamental rule of QM which establishes the basic relation of the theoretical predictions of QM with experiment definitely would not match with the expectation of those who consider QM as theory which is full of mysteries.

complex-valued fields; for example, for the classical electromagnetic field we use Riemann–Silberstein representation, $\phi(x) = E(x) + iB(x)$. A random field (signal) is a field (signal) depending on a random parameter ω, $\phi(x, \omega)$. In the measure-theoretic framework (Section 2.5) it is represented as H-valued random variable, $\omega \to \phi(\omega) \in H$. Its probability distribution is denoted by the symbol μ on H. Consider functionals of fields, $f : H \to \mathbf{C}, \phi \to f(\phi)$. These are physical observables for classical signals. For example, the energy of the classical electromagnetic field is given by the quadratic functional

$$f(\phi) \equiv f(E, B) = \int_{\mathbf{R}^3} |\phi(x)|^2 dx = \int_{\mathbf{R}^3} (E^2(x) + B^2(x))dx.$$

The average of an observable can be written as the integral over the space of fields

$$\langle f \rangle = \int_H f(\phi) d\mu(\phi).$$

To find $\langle f \rangle$, we consider an ensemble (in theory infinite) of realizations of the random field and calculate the average of $f(\phi)$ with respect to this ensemble. This measure-theoretic (ensemble) representation is very convenient in theoretical considerations [246, 265]. However, in practice in *classical signal theory* one never produces an ensemble of different realizations of a random signal (e.g., a random radio signal). Instead of this, a single time dependent realization, $\phi(s, x)$, is under investigation. It is measured at different instances of time. Finally, the time average is calculated. The latter is given by

$$\bar{f} = \lim_{\Delta \to \infty} \frac{1}{\Delta} \int_0^\Delta f(\phi(s))ds. \tag{9.1}$$

In classical signal theory [246, 265] the ensemble and time averages are coupled by the *ergodicity assumption*. Under this assumption we obtain that

$$\bar{f} = \langle f \rangle, \tag{9.2}$$

i.e.,

$$\int_H f(\phi) d\mu(\phi) = \lim_{\Delta \to \infty} \frac{1}{\Delta} \int_0^\Delta f(\phi(s))ds \approx \frac{1}{\Delta} \int_0^\Delta f(\phi(s))ds, \tag{9.3}$$

for sufficiently large Δ.

In coming consideration we shall operate only with observables given by quadratic functionals of classical signals:

$$\phi \to f_A(\phi) = \langle \widehat{A}\phi, \phi \rangle, \tag{9.4}$$

where $\widehat{A}$ is a self-adjoint operator. Moreover, to describe a procedure of the position detection we need only functionals of the form

$$\phi \to |\phi(x_0)|^2, \tag{9.5}$$

where $x_0 \in \mathbf{R}^3$ is a fixed point which determines the quadratic functional (later x_0 will be considered as the position of a detector).

In what follows we consider only random signals with covariance operators of the type

$$D_\psi = |\psi\rangle\langle\psi|, \tag{9.6}$$

where $\psi \in H$ is arbitrary vector (i.e., it need not be normalized by 1). Although in (9.6) we used Dirac notation, for a moment there is no direct connection with QM. This is simply a random signal which covariance operator is given by an orthogonal projector to *one-dimensional subspace.* Of course, this is a very special class of signals. However, these are purely classical entities. Moreover, we shall start our derivation of the detection probabilities for the threshold-type detectors for random signals of this type only to obtain the most well-known version of Born's rule, for the wave function (see Section 9.3). (The probability is the squared absolute value of the wave function.) In principle, this rule can be derived in its most general form (i.e., for quantum density operators) by starting with random signals with arbitrary covariance operators (see Section 9.4).

For the corresponding probability distribution[a] $\mu \equiv \mu_\psi$, we have

$$\langle f_{x_0} \rangle = \int_H |\phi(x_0)|^2 d\mu_\psi(\phi) = |\psi(x_0)|^2. \tag{9.7}$$

And under the assumption of ergodicity, we obtain

$$|\psi(x_0)|^2 = \lim_{\Delta\to\infty} \frac{1}{\Delta} \int_0^\Delta |\phi(s, x_0)|^2 ds \approx \frac{1}{\Delta} \int_0^\Delta |\phi(s, x_0)|^2 ds, \tag{9.8}$$

for sufficiently large Δ. Consider the functional

$$\pi(\phi) = \|\phi\|^2 = \int_{\mathbf{R}^3} |\phi(x)|^2 dx. \tag{9.9}$$

[a]We remark that, since the covariance operator is the one-dimensional projector, this probability distribution is concentrated on the corresponding one-dimensional subspace of the complex Hilbert space.

In PCSFT it represents the *total energy* of a signal. From previous chapters, we know its average:

$$\langle \pi \rangle = \int_H \pi(\phi) d\mu(\phi) = \mathrm{Tr} D_\mu. \tag{9.10}$$

In particular, for $\mu = \mu_\psi$,

$$\langle \pi \rangle = \int_H \pi(\phi) d\mu_\psi(\phi) = \|\psi\|^2. \tag{9.11}$$

By ergodicity

$$\langle \pi \rangle = \|\psi\|^2 = \lim_{\Delta\to\infty} \frac{1}{\Delta} \int_0^\Delta \|\phi(s)\|^2 ds \approx \frac{1}{\Delta} \int_0^\Delta ds \int_{\mathbf{R}^3} dx |\phi(s, x)|^2, \tag{9.12}$$

for sufficiently large Δ.

If, as usual in signal theory, the quantity $|\phi(s, x)|^2$ has the physical dimension of the energy density, i.e., energy/volume, then by selecting some unit of time denoted γ we can interpret the quantity

$$\frac{1}{\gamma} \int_0^\Delta |\phi(s, x_0)|^2 ds dV, \tag{9.13}$$

as the energy which can be collected in the volume dV during the time interval Δ (from the random signal $\phi(s) \in H$). In the same way

$$\frac{1}{\gamma} \int_0^\Delta ds \int_{\mathbf{R}^3} dx |\phi(s, x)|^2, \tag{9.14}$$

is the total energy which can be collected during the time interval Δ. Its time average can be represented in the form (9.12).

9.2 Discrete-Counts Model for Detection of Classical Random Signals

We consider the following model of detector's functioning. Its basic parameter is detection threshold energy $\epsilon \equiv \epsilon_{\mathrm{click}}$. The detector under consideration clicks after it has collected[a] the energy

$$E_{\mathrm{collected}} \approx \epsilon. \tag{9.15}$$

[a]For a large class of detectors used in quantum optics (based on the photoelectric effect, e.g., photomultipliers), the ϵ is given by the total energy of the photon. By our PCSFT terminology this is the total energy of the input signal. Since PCSFT is intrinsically statistical theory, it operates with average energy. Thus in our measurement scheme this situation corresponds to the choice $\epsilon = \langle \pi \rangle = \|\psi\|$, However, in general we proceed with a threshold of an arbitrary (nonzero) magnitude.

Select γ as, e.g., 1 second. This is the lab unit of time. It is very large in units of the subquantum time scale. Consider such a detector located in small volume dV around a point $x_0 \in \mathbf{R}^3$. In average it clicks each Δ seconds, where Δ is determined form the approximative equality

$$\frac{1}{\gamma}\int_0^{\Delta} |\phi(s, x_0)|^2 ds dV \approx \epsilon, \tag{9.16}$$

or

$$\frac{\Delta}{\gamma}\left(\frac{1}{\Delta}\int_0^{\Delta} |\phi(s, x_0)|^2 ds\right) dV \approx \epsilon, \tag{9.17}$$

or

$$\frac{\Delta}{\gamma}|\psi(x_0)|^2 dV \approx \epsilon. \tag{9.18}$$

Thus at the point x_0 such a detector clicks (in average) with the frequency

$$\lambda(x_0) = \frac{\gamma}{\Delta} \approx \frac{|\psi(x_0)|^2 dV}{\epsilon}. \tag{9.19}$$

This frequency of clicks coincides with the probability of detection at the point x_0.

Consider a large interval of time, say T. The number of clicks at x_0 during this interval is given by

$$n_T(x_0) = \frac{T|\psi(x_0)|^2 dV}{\epsilon\gamma}. \tag{9.20}$$

The same formula is valid for any point $x \in \mathbf{R}^3$. Hence, the probability of detection at x_0 is

$$P(x_0) = \frac{n_T(x_0)}{\int n_T(x)dx} \approx \frac{|\psi(x_0)|^2 dV}{\int |\psi(x)|^2 dx} = |\Psi(x_0)|^2 dV, \tag{9.21}$$

where the normalized function

$$\Psi(x) = \psi(x)/\|\psi\|, \tag{9.22}$$

i.e., $\|\Psi\|^2 = 1$.

Thus, starting with classical signal theory, we obtain a kind of the wave function, $\Psi(x)$, a normalized vector of the L_2-space. This vector encodes information about correlations in the corresponding random signal. Ergodicity implies (as we proved)[a] that these

[a] It seems that this is a new result in theory of ergodic stochastic processes.

correlations also encode the detection probabilities for the detectors of the threshold type.

> *We stress that the resulting probability does not depend on the threshold ϵ.*

In Section 9.7 we represent this measurement scheme in the rigorous mathematical framework; in particular, we estimate the deviation of the probability of detection from Born's rule. We also define the probability space of the classical prequantum random-field process, $\phi(s, x; \omega)$.

Hence, *the Born's rule is valid for probabilities of "discretized detection" of classical random signals under the following assumptions:*

(a) *ergodicity;*
(b) *a detector clicks after it "has eaten" approximately a portion of energy ϵ;*
(c) *the energy is collected by this detector through time integration of signal's energy;*
(d) *the interval of integration Δ is long enough from the viewpoint of the internal time scale of a signal.*

The assumption (d) is necessary to match (a). We remark that the internal time scale of a signal, i.e., the scale of its random fluctuations, has to be distinguished from the time scale of macroscopic measurement (observer's time scale). The former is essentially finer than the latter.

We presented a natural scheme of discrete detections which is based on time integration of signal's energy by a detector. This scheme applied to classical random signals reproduces the Born's rule for *discrete clicks, individual events of QM*

How can this detection scheme be applied to QM?

In the presented scheme of threshold detection we treat prequantum random fields of PCSFT representing quantum systems as classical random signals.

9.3 Quantum Probabilities from Threshold-Type Detectors

In PCSFT quantum systems are represented by classical random fields. Hence, quantum measurements have to be interpreted as measurements of classical random signals. We now explore the measurement scheme of Section 9.2.

Take a prequantum random field ϕ with zero average and the covariance operator given by (9.6): $D_\psi = |\psi\rangle\langle\psi|$. Then we can introduce the wave function Ψ by normalization of ψ, see (9.22). We now consider quantum measurements for systems in the pure state Ψ as measurements of the corresponding classical signal ϕ and we derive the Born's rule for QM.

Thus we presented a model of discrete detection of prequantum random fields (corresponding to quantum systems) which reproduces the basic rule of QM, the Born's rule.

It is important to remark that the temporal structure of a prequantum random signal is not uniquely determined by its covariance operator. The latter is time independent, because we consider only stationary signals. Therefore the quantum wave function does not determine uniquely the corresponding prequantum process. Different ergodic prequantum processes can produce the same statistics of clicks of detectors. However, as we have already discussed a few times, the class of the prequantum processes might be restricted by physical reasons, e.g., to Gaussian processes.

9.4 The Case of an Arbitrary Density Operator

We now generalize the above scheme of classical signal measurement to signals with arbitrary covariance operators. (We recall that we had considered only a very special class of signals with covariance operators given by one-dimensional projectors, see (9.6).) By applying this scheme to classical signals corresponding to quantum systems we obtain the Born's rule for quantum states given by density operators.

Let $\phi(x, \omega)$ be a random field with the probability distribution μ and the covariance operator D. Since this is a trace-class operator, it can be represented as the integral operator:

$$D\phi(x) = \int_{\mathbf{R}^3} D(x, y)\phi(y)dy. \tag{9.23}$$

Let us consider the integral operator with the kernel

$$A_{x_0}(x, y) = \delta(x - x_0)\delta(y - x_0), \tag{9.24}$$

where $x_0 \in \mathbf{R}^3$ is a fixed point. Hence, $\widehat{A}_{x_0}\phi(x) = \phi(x_0)\delta(x - x_0)$ and

$$\langle \widehat{A}_{x_0}\phi, \phi\rangle = \phi(x_0)\int \delta(x - x_0)\overline{\phi(x)}dx = |\phi(x_0)|^2. \tag{9.25}$$

Set $\widehat{B} = D\widehat{A}_{x_0}$. Then it has the kernel

$$B(u, v) = \int D(u, z)A_{x_0}(z, v)dz = D(u, x_0)\delta(v - x_0).$$

Hence

$$\mathrm{Tr}\widehat{B} = \int B(u, u)du = D(x_0, x_0). \tag{9.26}$$

Thus

$$\int_H |\phi(x_0)|^2 d\mu(\phi) = \int_H \langle \widehat{A}_{x_0}\phi, \phi\rangle d\mu(\phi) = \mathrm{Tr}D\widehat{A}_{x_0} = D(x_0, x_0). \tag{9.27}$$

We also have

$$\langle \pi \rangle = \mathrm{Tr}D = \int_{\mathbf{R}^3} D(x, x)dx. \tag{9.28}$$

We now generalize the detection scheme of Section 9.2. Let $\phi(s, \omega)$ be an ergodic random signal (stochastic process) valued in complex Hilbert space $H = L_2$, i.e., $\phi = \phi(s, x, \omega)$. Its mean value and its covariance operator do not depend on time. We consider signals with zero mean value. Denote the covariance operator by D. (The signal is stationary and its probability distribution at the fixed instance of prequantum time does not depend on s.)

We can directly jump to the generalization of the relation (9.18):

$$\frac{\Delta}{\gamma} D(x_0, x_0)dV \approx \epsilon, \tag{9.29}$$

or

$$\frac{\gamma}{\Delta} \approx \frac{D(x_0, x_0)dV}{\epsilon}. \tag{9.30}$$

Thus the probability of signal detection at the point x_0 is given by

$$P(x_0) \approx \frac{D(x_0, x_0)dV}{\int_{\mathbf{R}^3} D(x, x)dx}. \tag{9.31}$$

To find coupling with QM, we set $\rho = D/\mathrm{Tr}D$. Then $\mathrm{Tr}\rho = 1$ and the above equality can be written in terms of the density operator ρ :

$$P(x_0) \approx \rho(x_0, x_0)dV. \tag{9.32}$$

This is nothing else than the Born's rule for the quantum state given by the density operator ρ. Thus we demonstrated that even for a mixed quantum state it is possible to find such a prequantum random field that the "discrete-click" measurement of this signal produces the basic probabilistic rule of QM.

9.5 The General Scheme of Threshold Detection of Classical Random Signals

Let $e(x)$ be a fixed L_2-function. We present a version of the discrete detection scheme of Section 9.2 by taking $e(x)$, instead of $\delta_{x_0}(x) = \delta(x - x_0)$. (In principle, we may proceed without the restriction that $e \in L_2$ and select e as a distribution; but we proceed with L_2-functions; on one hand this simplifies essentially mathematics of presentation; on the other hand, in this section we are mainly interested in a signal detection scheme underlying quantum measurements of observers with discrete spectra.) Consider a detector, "e-antenna," corresponding to the function $e(x)$, cf. Section 8.9. It works in the following way. As in the scheme of Section 9.2, it integrates the energy of a signal during some interval of time, until the threshold ϵ is approached. The only difference with Section 9.2 is that such a detector integrates the "energy along the direction $e \in L_2$," namely,

$$\frac{1}{\gamma}\int_0^{\Delta} |\langle\phi(s, x), e\rangle|^2 ds.$$

The rest of the detection scheme coincides with the one presented Section 9.2 (see also Section 9.4):

$$\frac{1}{\gamma}\int_0^{\Delta} |\langle\phi(s, x), e\rangle|^2 ds \approx \epsilon. \tag{9.33}$$

Then

$$\frac{\Delta}{\gamma}\Big(\frac{1}{\Delta}\int_0^\Delta |\langle\phi(s,x), e\rangle|^2 ds\Big) \approx \epsilon, \tag{9.34}$$

$$\frac{\Delta}{\gamma}\int_H |\langle\phi, e\rangle|^2 d\mu(\phi) \approx \epsilon, \tag{9.35}$$

$$\frac{\Delta}{\gamma}\langle De, e\rangle \approx \epsilon, \tag{9.36}$$

finally, for the probability of detection of this signal by the e-antenna, we get

$$P(e) \approx \frac{\langle De, e\rangle}{\mathrm{Tr}D}. \tag{9.37}$$

The normalization by the trace, the total average energy of the prequantum random signal, is a consequence of the normalization procedure which is used in QM: normalization is based on the total number of clicks collected in all detectors and not on the total number of systems (in our classical signal model, the number of pulses) emitted by a source, cf. with the normalization procedure in classical probability theory 2.22. The classical and quantum normalizations are different! In classical probability theory the normalization is by the total number of *possible elementary events.* In quantum probability theory the normalization is by the total number of *detected elementary events.*

In our detection scheme this normalization procedure is described in the following way. Consider the expansion of the signal $\phi(s, \omega)$ with respect to orthonormal basis $\{e_j\}$:

$$\phi(s,\omega) = \sum_j \langle\phi(s,\omega), e_j\rangle e_j.$$

Physically this mathematical procedure corresponds to splitting of the random signal into disjoint channels. Each channel is equipped by a detector of the threshold type. The only restriction is that all detectors have the same detection threshold $\epsilon > 0$. By (9.36), for each detector, we obtain

$$\frac{\Delta}{\gamma}\langle De_j, e_j\rangle \approx \epsilon. \tag{9.38}$$

Consider a large interval of time, say T. During this interval the number of clicks of the detector in the jth channel is given by

$$n_T(j) \approx \frac{T\langle De_j, e_j\rangle}{\epsilon\gamma}. \tag{9.39}$$

We now find the total number of clicks at all detectors:

$$n_T = \sum_i n_T(i) \approx \frac{T\sum_i \langle De_i, e_i\rangle}{\epsilon\gamma} = \frac{T\,\mathrm{Tr}D}{\epsilon\gamma}.$$

Therefore, for $e = e_j$, we obtain the formula (9.37). To proceed to QM, we first normalize the covariance operator of the random signal: $\rho = D/\mathrm{Tr}D$. Hence,

$$P(e) \approx \langle \rho e, e\rangle. \tag{9.40}$$

(We stress once again that the special normalization procedure, by the total number of *detected* systems (pulses), plays the crucial role to obtain Born's rule. Thus this rule is related to experimental counting of detection events. Of course, intrinsic properties of prequantum random fields are also important. These properties are encoded in the covariance operators.)

Consider a quantum observable $\widehat{A}$ with purely discrete nondegenerate spectrum and its eigenfunctions: $\widehat{A}e_k = \lambda_k e_k$. Consider the ensemble of detectors corresponding to the system of normalized eigenfunctions $\{e_k(x)\}$; ensemble of e_k-antennas. Each click induced by the e_k-antenna is associated with the value λ_k. Denote this classical observable by the symbol M_A. The average of M_A is given by

$$\langle M_A\rangle = \sum_k \lambda_k P(e_k) = \sum_k \lambda_k \langle \rho e_k, e_k\rangle. \tag{9.41}$$

Thus we obtained the quantum formula for averages:

$$\langle M_A\rangle = \mathrm{Tr}\rho\widehat{A}. \tag{9.42}$$

We now can extend our approach (representation of quantum measurements as "click-measurements" of classical random signals) to quantum observables with discrete degenerate spectra. Consider a subspace L of H and the corresponding L-antenna (cf. Section 8.9). Let $\widehat{P}_L$ be the corresponding orthogonal projector. The L-antenna produces a click if

$$\frac{1}{\gamma}\int_0^\Delta \|\widehat{P}_L\phi(s)\|^2 ds \approx \epsilon, \tag{9.43}$$

$$\frac{\gamma}{\Delta} \approx \mathrm{Tr} D \widehat{P}_L/\epsilon. \tag{9.44}$$

To complete the derivation of Born's rule, we start with decomposition of the Hilbert space H into the sum of mutually orthogonal subspaces:

$$H = L_1 \oplus L_2 \oplus \ldots \oplus L_n \oplus \ldots, \tag{9.45}$$

i.e., $\sum_j \widehat{P}_{L_j} = I$, where I is the unit operator. We can represent random signal $\phi(s, \omega)$ in the form:

$$\phi(s, \omega) = \sum_j \widehat{P}_{L_j} \phi(s, \omega).$$

Physically this mathematical expansion corresponds to splitting of the signal into disjoint channels. The only difference from previous considerations is that now in each channel the signal is not one dimensional (in general subspaces in (9.45) can have infinite dimension).

Consider a large interval of time, say T. During this interval the number of clicks at the detector in the jth channel is given by

$$n_T(L_j) \approx \frac{T \, \mathrm{Tr} D \widehat{P}_{L_j}}{\epsilon \gamma}. \tag{9.46}$$

We now find the total number of clicks at all detectors:

$$n_T = \sum_i n_T(L_i) \approx \frac{T \sum_i \mathrm{Tr} D \widehat{P}_{L_i}}{\epsilon \gamma} = \frac{T \, \mathrm{Tr} D}{\epsilon \gamma}.$$

Therefore

$$P(L_j) \approx \mathrm{Tr} D \widehat{P}_{L_j} / \mathrm{Tr} D = \mathrm{Tr} \rho \widehat{P}_{L_j}, \tag{9.47}$$

where ρ is the normalization of the covariance operator of the signal.

In this approach there is no reason to consider only orthogonal projectors and self-adjoint operators; a POVM, $\{\widehat{Q}_k\}$, $\sum_k \widehat{Q}_k = I$, $\widehat{Q}_k \geq 0$, arises naturally as the symbolic representation of generalization of aforementioned scheme (of measurement of classical random signals) to the family of Q_k-antennas: each integrates the Q_k-transform of a classical signal. The starting point of generalization is the relation:

$$\frac{1}{\gamma} \int_0^{\Delta} \|\widehat{Q}_k \phi(s)\|^2 ds \approx \epsilon. \tag{9.48}$$

Then we can proceed in the same way as with the orthogonal decomposition of Hilbert space. The crucial equality giving the right normalization is

$$\sum_k \mathrm{Tr} D\widehat{Q}_k = \mathrm{Tr} D.$$

Here the orthogonality of operators is not needed.

By starting with Dirac's approach to quantum observables (based on generalized eigenvectors) we can easily generalize our scheme to operators with continuous spectra and, in particular, obtain the position measurement as a special case.

9.6 Probability of Coincidence

We recall that Bohr elaborated his complementarity principle[a] from analysis of the two-slit experiment. On the one hand, quantum systems exhibit interference properties which are similar to properties of classical waves. On the other hand, these systems also exhibit particle properties. Wave properties (interference) are exhibited if both slits are open and experimenter does not try to control "which slit passing." At this experimental context one can be totally fine with a classical wave type model. However, if experimental context is changed and detectors are placed behind slits, then "wave features of quantum systems disappear and particle features are exhibited." What does the latter mean? Why is the usage of the wave picture impossible? Typically, it is claimed that, since classical wave is spatially extended, two detectors (behind both slits) can click simultaneously and produce double clicks. However, as it is commonly claimed, there are no double clicks at all; hence, the wave model has to be rejected (in the context of the presence of detectors). Bohr had not found any reasonable explanation of context dependent features of quantum systems and he elaborated the complementarity principle.

Of course, the claim that there are no double clicks at all is meaningless at the experimental level. There are always double clicks. The question is whether the number of double clicks

[a]This principle is often called "wave–particle" duality. However, we stress that Bohr had never used the latter terminology by himself.

is very small (comparing with the numbers of single clicks). Corresponding experiments have been done [95, 96] and it was shown that the number of double clicks is relatively small, i.e., the coefficient of second-order coherence is less than one, see also Section 1.3. Such experiments are considered as confirmation of Bohr's complementarity principle.[a]

We show that the absence of double clicks is a consequence of usage of detectors with sufficiently high thresholds.

Consider again a random signal ϕ with zero average and the covariance operator given by the orthogonal projector onto the vector ψ, see (9.6). In general, this vector can have nonunit norm; the only restriction is that it is nonzero. Consider two threshold-type detectors located in neighborhoods V_{x_j} of points x_j, $j = 1, 2$. We assume that they have the same detection threshold ϵ. It is convenient to represent the threshold as $\epsilon = C\|\psi\|$, where $C > 0$ is a constant.

For moments of clicks, we have two approximate equalities:

$$\frac{1}{\gamma}\int_0^{\Delta_C(x_j)}\int_{V_{x_j}} |\phi(s, x)|^2 dxds \approx \epsilon, \ j = 1, 2; \tag{9.49}$$

a double click corresponds to the (approximate) coincidence of moments of clicks

$$\Delta_C(x_1, x_2) = \Delta_C(x_1) = \Delta_C(x_2). \tag{9.50}$$

We shall use shorter notations: $\Delta_j \equiv \Delta_C(x_j)$, $\Delta \equiv \Delta_C(x_1, x_2)$. By taking into account the representation (9.18) of the detection conditions (9.49) and matching (9.50) we obtain the coincidence detection condition in the form

$$W_j \equiv \frac{\Delta_C}{\gamma}\int_{V_{x_j}} |\psi(x)|^2 dx \approx \epsilon, \ j = 1, 2. \tag{9.51}$$

To be more formal, let us choose a small constant $0 < \delta \ll 1$ such that

$$|W_j - \epsilon| \leq \delta, \ j = 1, 2. \tag{9.52}$$

[a] Born's complementarity principle matches well with our contextual viewpoint to the two-slit experiment: the classical FTP is violated, because the statistical data is collected under three different experimental contexts. However, opposite to adherents of the orthodox Copenhagen interpretation, we do not consider contextuality of the operational model, QM, as a barrier on the way beyond quantum. Noncontextual prequantum model can peacefully coexist with QM.

To estimate the probability of coincidence P_{12}, we have to proceed in more rigorous mathematical framework than in considerations leading to single-detection probabilities, P_1 and P_2 (for the points x_1 and x_2). As typical for rigorous probabilistic studies, we introduce the random parameter ω. To be completely rigorous, we have to introduce the Kolmogorov probability space $(\Omega, \mathcal{F}, P)$, where Ω is the set of random parameters, $\mathcal{F}$ is the σ-algebra of its subsets, and P is the probability measure on this σ-algebra (Section 9.7). All random quantities are written as dependent on these parameters, e.g., $\Delta \equiv \Delta(\omega)$. We consider the set of random parameters corresponding to coincidence detection, (9.52):

$$A_{12} = \{\omega \in \Omega : |W_j(\omega) - \epsilon| \leq \delta, \; j = 1, 2\}.$$

(A_{12} is the event of coincidence detection). We have to estimate its probability, $P_{12} = P(A_{12})$). We shall get a rather rough estimate which, nevertheless, will be sufficient for our purpose. However, we shall see that better estimates of this probability will clarify essentially interrelation between our "prequantum classical field theory," QM, and experiment. In principle, one may hope to derive an approximative expression for P_{12} as we did for probabilities P_j, $j = 1, 2$. However, this is a complicated probabilistic problem. We remark that A_{12} is a subset of the set

$$A_{12,\geq} = \{\omega \in \Omega : W_j(\omega) \geq \epsilon - \delta, \; j = 1, 2\}.$$

and in turn $A_{12,\geq}$ is a subset of the set

$$A_{1\times 2,\geq} = \{\omega : W_1(\omega) W_2(\omega) \geq (\epsilon - \delta)^2\}.$$

Hence, $P(A_{12}) \leq P(A_{1\times 2,\geq})$. We can estimate the latter probability (roughly) by using *Chebyshov inequality* (which is standard for such estimates). In the simplest form, for a random variable $u = u(\omega)$ and a constant $k > 0$, this inequality has the form: $P(\{\omega : u(\omega) \geq k\}) \leq \frac{E|u(\omega)|}{k}$. In our case $u = W_1 W_2$, $k = (\epsilon - \delta)^2$. We have

$$P(A_{12}) \leq \frac{E\, W_1 W_2}{(\epsilon - \delta)^2}, \tag{9.53}$$

where E denotes average with respect to ω :

$$\begin{aligned} E\, W_1 W_2 &= \int_{\Omega} W_1(\omega) W_2(\omega) dP(\omega) \\ &= \frac{E\, \Delta_C^2}{\gamma^2} \int_{V_{x_1}} |\psi(x)|^2 dx \int_{V_{x_2}} |\psi(x)|^2 dx. \end{aligned}$$

Set $E\Delta_{\tilde{C}}^2 = \overline{\Delta_{\tilde{C}}^2}$.

Now we obtain the estimate of the coincidence probability based on the Chebyshov inequality:

$$P(A_{12}) \leq \frac{\overline{\Delta^2}}{\gamma^2(\epsilon - \delta)^2} \int_{V_{x_1}} |\psi(x)|^2 dx \int_{V_{x_2}} |\psi(x)|^2 dx. \qquad (9.54)$$

We state again that $\|\psi\|^2$ represents the average energy of the field, and therefore it is natural to use it as the energy barrier scale, i.e., to select $\epsilon = C\|\psi\|$, $C > 0$. Set $\tilde{C} = C(1 - \frac{\delta}{C\|\psi\|})$. We obtain

$$P(A_{12}) \leq \frac{\overline{\Delta^2}}{\gamma^2 \tilde{C}^2} \int_{V_{x_1}} |\Psi(x)|^2 dx \int_{V_{x_2}} |\Psi(x)|^2 dx = \frac{\overline{\Delta^2} P_1 P_2}{\gamma^2 \tilde{C}^2}. \qquad (9.55)$$

Here $\Psi = \psi/\|\psi\|$. The final inequality has the form:

$$g^{(2)}(0) = \frac{P_{12}}{P_1 P_2} \leq \frac{\overline{\Delta^2}}{\gamma^2 \tilde{C}^2}. \qquad (9.56)$$

By increasing the parameter C (and hence $\tilde{C}$) the coefficient $g^{(2)}(0)$ can be done less than 1.

Our model predicts that quantum predictions match better experiment for higher value of threshold. It is interesting to compare this prediction with the real experiment. Unfortunately, it seems that Grangier-type experiment with detailed monitoring of dependence of the coincidence probability on the value of the threshold have never been done. More specifically, in Grangier's experiment, did the calibration of detectors play a crucial role to eliminate coincidences?

This is clearly a crucial question. It is hard to give a definite answer as to how exactly it was influencing the result at a fundamental level because Grangier [95, 96] didn't study the influence of threshold on the $g(2)$ parameter he was measuring. It is nevertheless clear that with a lower threshold, he would have gotten a less good result.

Let me translate the part of Grangier's thesis [96] where he explains for the first time the role of the threshold and how its level was chosen:

> [...] In this configuration, the threshold has a double role of acquisition of timing information and of selection of the pulses (the too weak pulses are not taken into account). The problems

> connected to the choice of the discriminator threshold and of the high voltage of the photomultipliers are discussed in detail in reference [22]. We have in the present experiment chosen a **rather high threshold**, which amounts to give the priority of the stability of the counting rates and the reproducibility of the results, rather than to the global detection efficiencies. *(I stressed with bold the important fact that Grangier proceeded with rather high threshold.)*

The measurement model, TSD, presented in this chapter provides a strong motivation to perform Grangier-type experiment with monitoring of dependence of the coincidence probability on the threshold.

Summary on coincidence probability: Classical random field model, PCSFT, combined with the threshold detection model, TSD, violates the prediction of the standard classical and semiclassical models on the magnitude of the coefficient of second-order interference. The prediction of PCSFT/TSD is compatible with the known experimental data (for sufficiently large threshold). More detailed experiments of the Grangier's type are needed.

9.7 Stochastic Process Description of Detection

We remark that nowadays theory of stochastic processes valued in Hilbert and more general topological vector spaces is the well-developed domain of probability theory, see Ref. [58] for introduction. In PCSFT we work with stochastic processes valued in $H = L_2(\mathbf{R}^3) : s \to \phi(s; \omega) \in H$, where s and ω are the time (prequantum) and chance parameters, respectively. Since, for any pair of these parameters, $\phi(s, \omega)$ is an L_2-function, we can write this process as the time dependent random field $\phi(s, x, \omega)$.

In accordance with the axiomatics of probability theory (Section 2.5) the random parameter ω belongs to some set Ω endowed with a σ-algebra of subsets, say $\mathcal{F}$, and a probability measure P defined on $\mathcal{F}$. It is important to have a way to construct probability spaces for stochastic processes. This problem was solved by Kolmogorov (1933) for finite-dimensional state spaces [58] for the infinite-dimensional case (and the L_2-space is infinite dimensional). In our case Ω is selected as the space of all functions from $[0, +\infty)$ to H. Thus ω is in fact a trajectory, $s \to \omega(s)$. Then by starting with finite-dimensional distributions (which are given) we define a probability measure on cylindric subsets of Ω, i.e., sets that are determined by selection of finite vectors of the time instances $(s_1, \ldots, s_n)$. By the Hilbert space version of Kolmogorov theorem [58] this cylindric probability can be extended to σ-additive probability measure.

In our case the concrete form of finite-dimensional probability distributions of a prequantum process does not play any role. This is the essence of PCSFT that different prequantum stochastic processes can reproduce the same quantum state Ψ.

To match our scheme of detection making, we need only the following:

(a) it has zero average and its covariance operator (which does not depend on time) is given by (9.6),
(b) the process is ergodic,
(c) the process has continuous trajectories.

Any process satisfying (a)–(c) can be chosen to perform considerations of Section 9.2 in the rigorous mathematical framework (see Ref. [58] for examples of such processes).

In the mathematical framework the approximate equality (9.16) can be treated in the following way. Take a neighborhood V_{x_0} of the point x_0. Consider a detector integrating field's energy in this domain. Let us fix a "threshold deviation" parameter σ (a small positive constant such that $\sigma \ll \epsilon$). We define $\Delta_{V_{x_0}} \equiv \Delta_{V_{x_0}}(\sigma, \omega)$ from the equality

$$\left|\frac{1}{\gamma}\int_0^\Delta ds \int_{V_{x_0}} |\phi(s, x; \omega)|^2 dx - \epsilon\right| < \sigma, \tag{9.57}$$

i.e., as one of Δs satisfying this constraint. We remark that if $\int_{V_{x_0}} |\psi(x)|^2 dx > 0$, then ergodicity implies that $\Delta_{V_{x_0}}(\sigma, \omega) < \infty$ for almost all ω.

We now apply the pointwise ergodic theorem to the function

$$\phi \to \int_{V_{x_0}} |\phi(x)|^2.$$

For $\sigma > 0$, there exists $\Delta'_{V_{x0}} \equiv \Delta'_{V_{x0}}(\sigma, \omega)$ such that $\Delta \geq \Delta'_{V_{x0}}$ implies that

$$\left|\frac{1}{\Delta}\int_0^{\Delta} ds \int_{V_{x0}} |\phi(s, x; \omega)|^2 dx - \int_{V_{x0}} |\psi(x)|^2 dx\right| < \sigma. \tag{9.58}$$

In particular, take $\Delta = \Delta'_{V_{x0}}$. We have

$$\begin{aligned}
&\frac{1}{\gamma}\int_0^{\Delta'_{V_{x0}}} ds \int_{V_{x0}} |\phi(s, x; \omega)|^2 dx \\
&= \frac{\Delta'_{V_{x0}}}{\gamma}\left[\frac{1}{\Delta'_{V_{x0}}}\int_0^{\Delta'_{V_{x0}}} ds \int_{V_{x0}} |\phi(s, x; \omega)|^2 dx\right] \\
&\leq \frac{\Delta'_{V_{x0}}}{\gamma}\left[\int_{V_{x0}} |\psi(x)|^2 dx + \sigma\right] \leq \frac{\Delta'_{V_{x0}}}{\gamma}(\|\psi\|^2 + \sigma).
\end{aligned}$$

The inequality (9.57) implies that

$$\frac{1}{\gamma}\int_0^{\Delta_{V_{x0}}} ds \int_{V_{x0}} |\phi(s, x; \omega)|^2 dx \geq \epsilon - \sigma. \tag{9.59}$$

Suppose that the threshold ϵ is large enough, so that

$$\frac{\Delta'_{V_{x0}}}{\gamma}(\|\psi\|^2 + \sigma) + \sigma \leq \epsilon \tag{9.60}$$

for any x_0 (almost everywhere). Then we obtain that

$$\int_0^{\Delta'_{V_{x0}}} ds \int_{V_{x0}} |\phi(s, x; \omega)|^2 dx \leq \int_0^{\Delta_{V_{x0}}} ds \int_{V_{x0}} |\phi(s, x; \omega)|^2 dx. \tag{9.61}$$

Hence,

$$\Delta'_{V_{x0}} \leq \Delta_{V_{x0}} \text{ (a.e.).} \tag{9.62}$$

This inequality in combination with ergodicity condition (9.58) imply

$$\left|\frac{1}{\Delta_{V_{x0}}}\int_0^{\Delta_{V_{x0}}} ds \int_{V_{x0}} |\phi(s, x; \omega)|^2 dx - \int_{V_{x0}} |\psi(x)|^2 dx\right| < \sigma. \tag{9.63}$$

Physically the condition (9.60) guarantees that the threshold energy ϵ is so large that the corresponding detection time at any x_0 (the time interval for energy integration) is large enough to guarantee the ergodic stabilization of the time average to the ensemble average.

Finally, we have

$$\left|\frac{\Delta_{V_{x_0}}}{\gamma}\int_{V_{x_0}}|\psi(x)|^2dx-\epsilon\right|$$

$$\leq\left|\frac{\Delta_{V_{x_0}}}{\gamma}\left[\int_{V_{x_0}}|\psi(x)|^2dx-\frac{1}{\Delta_{V_{x_0}}}\int_0^{\Delta_{V_{x_0}}}ds\int_{V_{x_0}}|\phi(s,x;\omega)|^2dx\right]\right|$$

$$+\left|\frac{1}{\gamma}\int_0^{\Delta_{V_{x_0}}}ds\int_{V_{x_0}}|\phi(s,x;\omega)|^2dx-\epsilon\right|\leq\sigma(1+\frac{\Delta_{V_{x_0}}}{\gamma}).$$

Hence,

$$\left|\frac{\Delta_{V_{x_0}}}{\gamma}\frac{\int_{V_{x_0}}|\psi(x)|^2dx}{\epsilon}-1\right|<\frac{\sigma(1+\frac{\Delta_{V_{x_0}}}{\gamma})}{\epsilon}.$$

Thus

$$\frac{\Delta_{V_{x_0}}}{\gamma\epsilon}\int_{V_{x_0}}|\psi(x)|^2dx=1+g_{V_{x_0}}(\omega),$$

where $g_{V_{x_0}}(\omega)=O(\frac{\sigma}{\epsilon})$. (We state again that $\Delta_{V_{x_0}}\equiv\Delta_{V_{x_0}}(\omega)$.) Therefore

$$\frac{\gamma}{\Delta_{V_{x_0}}}=\frac{1}{\epsilon(1+g_{V_{x_0}})}\int_{V_{x_0}}|\psi(x)|^2dx.$$

We proved, cf. (9.19),

$$\lambda(V_{x_0})=\frac{\gamma}{\Delta_{V_{x_0}}}\approx\frac{1}{\epsilon}\int_{V_{x_0}}|\psi(x)|^2dx, \tag{9.64}$$

where $\lambda(x_0)$ is the frequency of clicks for a threshold-type detector integrating the energy of a signal in the neighborhood V_{x_0} of the point x_0.

Take a large interval of time $[0, T]$. The number of clicks of the aforementioned detector during this interval is given by the quantity

$$n_T(V_{x_0})=\frac{T}{\Delta_{V_{x_0}}}=\frac{1}{\epsilon\gamma(1+g_{V_{x_0}})}\int_{V_{x_0}}|\psi(x)|^2dx. \tag{9.65}$$

In particular,

$$n_T(\mathbf{R}^3)=\frac{T}{\Delta_{V_{x_0}}}=\frac{T}{\epsilon\gamma(1+g_{\mathbf{R}^3})}\int_{\mathbf{R}^3}|\psi(x)|^2dx. \tag{9.66}$$

Hence, the probability of detection is given by

$$P(V_{x_0})=\frac{n_T(V_{x_0})}{n_T(\mathbf{R}^3)}=\frac{(1+g_{\mathbf{R}^3})}{(1+g_{V_{x_0}})}\frac{\int_{V_{x_0}}|\psi(x)|^2dx}{\int_{\mathbf{R}^3}|\psi(x)|^2dx}. \tag{9.67}$$

By introducing the normalized function Ψ we write

$$P(V_{x_0}) = \int_{V_{x_0}} |\Psi(x)|^2 dx(1 + O(\sigma/\epsilon)). \tag{9.68}$$

It is interesting that matching with Born's rule is better for larger threshold ϵ.

References

1. Aaronson, S. (2003). Is quantum mechanics an island in theoryspace? In Khrennikov, A. (ed.), *Quantum Theory: Reconsideration of Foundations-2*, Ser. Math. Modeling **10** (Växjö University Press, Växjö), pp. 15–28.
2. Accardi, L. (1970). The probabilistic roots of the quantum mechanical paradoxes. In Diner, S., Fargue, D., Lochak, G., Selleri, F. (eds.), *The Wave–Particle Dualism. A Tribute to Louis de Broglie on his 90th Birthday* (D. Reidel Publishing Company, Dordrecht), pp. 47–55.
3. Accardi, L. (1997). *Urne e Camaleoni: Dialogo sulla Realta, le Leggi del Caso e la Teoria Quantistica* (Il Saggiatore, Rome).
4. Accardi, L. (2005). Some loopholes to save quantum nonlocality. In Adenier, G., Khrennikov, A. (eds.), *Foundations of Probability and Physics-3*, Ser. Conference Proceedings **750** (American Institute of Physics, Melville, NY), pp. 1–20.
5. Accardi, L. (2006). Could one now convince Einstein? In Adenier, G., Khrennikov, A., Nieuwenhuizen, Th. M. (eds.), *Quantum Theory: Reconsideration of Foundations-3*, Ser. Conference Proceedings **810** (American Institute of Physics, Melville, NY), pp. 3–18.
6. Accardi, L., Khrennikov, A. (2007). Chameleon effect, the range of values hypothesis, and reproducing the EPR–Bohm correlations. In Adenier, G., Fuchs, C., Khrennikov, A. (eds.), *Foundations of Probability and Physics-3*, Ser. Conference Proceedings **889** (American Institute of Physics, Melville, NY), pp. 21–29.
7. Accardi, L., Adenier, G., Fuchs, C. A., Jaeger, G., Khrennikov, A., Larsson, J.-A., Stenholm, S. (eds.) (2009). *Foundations of Probability and Physics-5*, Ser. Conference Proceedings **1101** (American Institute of Physics, Melville, NY).
8. Adenier, G., Khrennikov, A., Nieuwenhuizen, Th. M. (eds.) (2006). *Quantum Theory: Reconsideration of Foundations-3*, Ser. Conference Proceedings **810** (American Institute of Physics, Melville, NY).

9. Adenier, G., Khrennikov, A. (2006). Anomalies in EPR-Bell experiments. In Adenier, G., Khrennikov, A., Nieuwenhuizen, Th.M. (eds.), *Quantum Theory: Reconsideration of Foundations-3*, Ser. Conference Proceedings **810** (American Institute of Physics, Melville, NY), pp. 283–293.
10. Adenier, G., Fuchs, C., Khrennikov, A. (eds.) (2007). *Foundations of Probability and Physics-3*, Ser. Conference Proceedings **889** (American Institute of Physics, Melville, NY).
11. Adenier, G., Khrennikov, A. Yu. (2007). Is the fair sampling assumption supported by EPR experiments? *J. Phys. B: At. Mol. Opt. Phys.* pp. 131–141.
12. Adenier, G., Khrennikov, A., Lahti, P., Manko, V. I., Nieuwenhuizen, Th. M. (eds.) (2008). *Quantum Theory: Reconsideration of Foundations-4*, Ser. Conference Proceedings **962** (American Institute of Physics, Melville, NY).
13. Adenier, G. (2008). *Local Realist Approach and Numerical Simulation of Nonclassical Experiments in Quantum Mechanics* (Växjö University Press, Växjö).
14. Adenier, G. (2008). A fair sampling test for EPR–Bell experiments. *J. Russ. Laser Res.* **29**, pp. 409–417.
15. Adler, S. L. (2004). *Quantum Theory as an Emergent Phenomenon* (Cambridge University Press, Cambridge).
16. D'Agostino, S. (1992). Continuity and completeness in physical theory: Schrödinger's return to the wave interpretation of quantum mechanics in the 1950s. In Bitbol, M., O. Darrigol (eds.), *E. Schrödinger: Philosophy and the Birth of Quantum Mechanics* (Editions Frontieres, Gif-sur-Yvette), pp. 339–360.
17. Aichele, T., Herzog, U., Scholz, M, Benson, O. (2005). Single photon generation and simultaneous observation of wave and particle properties. In Adenier, G., Khrennikov, A. (eds.), *Foundations of Probability and Physics-3*, Ser. Conference Proceedings **750** (American Institute of Physics, Melville, NY), pp. 35–41.
18. Allahverdyan, A. E., Balian, R., Nieuwenhuizen, Th. M. (2005). The quantum measurement process in an exactly solvable model. In Khrennikov, A. (ed.), *Foundations of Probability and Physics-3*, Ser. Conference Proceedings **750** (American Institute of Physics, Melville, NY), pp. 16–24.
19. Allahverdyan, A., Khrennikov, A. Yu., Nieuwenhuizen, Th.M. (2005). Brownian entanglement. *Phys. Rev.* A **71**, pp. 032102-1–032102-14.

20. Appleby, D. M. (2007). Concerning dice and divinity. In Adenier, G., Fuchs, C., Khrennikov, A. (eds.), *Foundations of Probability and Physics-3*, Ser. Conference Proceedings **889** (American Institute of Physics, Melville, NY), pp. 30–39.
21. Ashwanden, M., Philipp, W., Hess, K., Adenier, G. (2006). Local time dependent instruction-set model for the experiment of Pan *et al.* In Adenier, G., Khrennikov, A., Nieuwenhuizen, Th.M. (eds.), *Quantum theory: reconsideration of foundations-3*, Ser. Conference Proceedings **810** (American Institute of Physics, Melville, NY), pp. 437–446.
22. Aspect, A., Dalibard, J., Roger, G. (1982). Experimental test of Bell's inequalities using time-varying analyzers. *Phys. Rev. Lett.* **49**, pp. 1804–1806.
23. Atmanspacher, H., Primas, H. (2005). Epistemic and ontic quantum realities. In Adenier, G., Khrennikov, A. (eds.), *Foundations of Probability and Physics-3*, Ser. Conference Proceedings **750** (American Institute of Physics, Melville, NY), pp. 49–62.
24. Bacciagaluppi, G. (2003). Classical extensions, classical representations and Bayesian updating in quantum mechanics. In Khrennikov, A. (ed.), *Quantum Theory: Reconsideration of Foundations-2*, Ser. Mathematical Modeling **10** (Växjö University Press, Växjö), pp. 55–70.
25. Ballentine, L. E. (1989). The statistical interpretation of quantum mechanics. *Rev. Mod. Phys.* **42**, pp. 358–381,
26. Ballentine, L. E. (1989). *Quantum Mechanics* (Prentice Hall, Englewood Cliffs, NJ).
27. Ballentine, L. E. (1998). *Quantum Mechanics: A Modern Development* (WSP, Singapore).
28. Ballentine, L. E. (2001). Interpretations of probability and quantum theory. In Khrennikov, A. (ed.), *Foundations of Probability and Physics*, Quantum Probability and White Noise Analysis **13** (WSP, Singapore), pp. 71–84.
29. Ballentine, L. E. (2003). The classical limit of quantum mechanics and its implications for the foundations of quantum mechanics, Khrennikov, A. (ed.), *Quantum Theory: Reconsideration of Foundations-2*, Ser. Math. Modeling **10** (Växjö University Press, Växjö), pp. 71–82.
30. Belavkin, V. P. (2003). Visible information dynamics of hidden quantum mechanics. In Khrennikov, A. (ed.), *Foundations of Probability and Physics-2*, Ser. Math. Modeling **5** (Växjö University Press, Växjö), pp. 69–98.

31. Bell, J. (1987). *Speakable and Unspeakable in Quantum Mechanics* (Cambridge University Press, Cambridge).
32. Bialynicki-Birula, I., J. Mycielski, J. (1976). Nonlinear wave mechanics. *Ann. Phys.* **100**, pp. 62–93.
33. Bialynicki-Birula, I. (1994). On the wave function of the photon. *Acta Physica Polonica* A **86**, pp. 97–110.
34. Bialynicki-Birula, I. (2007). Rényi entropy and the uncertainty relations. In Adenier, G., Fuchs, C., Khrennikov, A. (eds.), *Foundations of Probability and Physics-3*, Ser. Conference Proceedings **889** (American Institute of Physics, Melville, NY), pp. 52–61.
35. Bitbol, M., Darrigol, O. (1992). *E. Schrödinger: Philosophy and the Birth of Quantum Mechanics* (Editions Frontieres, Gif-sur-Yvette).
36. Bohm, D., Hiley, B. (1993). *The Undivided Universe: an Ontological Interpretation of Quantum Mechanics* (Routledge and Kegan Paul, London).
37. Bohr, N. (1935). Can quantum-mechanical description of physical reality be considered complete? *Phys. Rev.* **48**, pp. 696–702.
38. Bohr, N. (1987). *The Philosophical Writings of Niels Bohr* (Woodbridge, Conn., Ox Bow Press).
39. Bohr, N., Kramers, H. A., Slater, J. C. (1920). Über die Quantentheorie der Strahlung. *Zeitschrift für Physik* **24**, pp. 69–87.
40. Boole, G. (1958). *An Investigation of the Laws of Thought* (Dover Edition, New York).
41. Boole, G. (1862). On the theory of probabilities. *Phil. Trans. R. Soc. Lond.* **152**, pp. 225–242.
42. Born, M. (1926). Zur Quantenmechanik der Stossvorgänge. *Zeitschrift für Physik* **37**, pp. 863–867.
43. Boyer, T. H. (1980). A brief survey of stochastic electrodynamics. In Barut, A. O. (ed.), *Foundations of Radiation Theory and Quantum Electrodynamics* (Plenum Press, New York), pp. 141–162.
44. Brida, G., Genovese, M., Gramegna, M., Novero, C., Predazzi, E. (2002). A first test of Wigner function local realistic model. *Phys. Lett. A* **121**, pp. 299–312.
45. Buonomano, V. (1987). The double slit experiment and non-ergodic interpretation of quantum mechanics. In Honig W., Paranella E., Kraft D. F. (eds.), *Proceedings of the International Conference on Quantum Uncertainties* (Plenum Press, New York), pp. 427–434.

46. Busch, P., Grabowski, M., Lahti, P. (1995). *Operational Quantum Physics* (Springer Verlag, Berlin).
47. Busch, P. (2003). Less (precision) is more (information): Quantum information in fuzzy probability theory, Khrennikov, A. (ed.), *Quantum Theory: Reconsideration of Foundations-2*, Ser. Math. Modeling **10** (Växjö University Press, Växjö), pp. 113–148.
48. Busemeyer, J. B., Wang, Z., Townsend, J. T. (2006). Quantum dynamics of human decision making. *J. Math. Psychol.* **50**, pp. 220–241.
49. Busemeyer, J. B., Wang, Z. (2007). Quantum information processing explanation for interactions between inferences and decisions. In Bruza, P. D., Lawless, W., van Rijsbergen, K., Sofge, D. A. (eds.), *Quantum Interaction, AAAI Spring Symposium*, Technical Report SS-07-08 (AAAI Press, Menlo Park, CA), pp. 91–97.
50. Busemeyer, J. R., Matthews, M., Wang, Z. (2006). A quantum information processing explanation of disjunction effects. In Sun, R., Myake, N. (eds.), *The 29th Annual Conference of the Cognitive Science Society and the 5th International Conference of Cognitive Science* (Erlbaum, Mahwah, NJ), pp. 131–135.
51. Busemeyer, J. R., Santuy, E., Lambert-Mogiliansky, A. (2008). Comparison of Markov and quantum models of decision making. In P. Bruza, W. Lawless, K. van Rijsbergen, D. A. Sofge, B. Coeke, S. Clark (eds.), *Quantum interaction: Proceedings of the Second Quantum Interaction Symposium* (College Publications, London), pp. 68–74.
52. Caticha, A. (2007). From objective amplitudes to Bayesian probability. In Adenier, G., Fuchs, C., Khrennikov, A. (eds.), *Foundations of Probability and Physics-3*, Ser. Conference Proceedings **889** (American Institute of Physics, Melville, NY), pp. 62–70.
53. Cole, D. (2006). Simulation results related to stochastic electrodynamics. In Adenier, G., Khrennikov, A. Yu., Nieuwenhuizen, Th.M. (eds.), *Quantum Theory: Reconsideration of Foundations-3*, Ser. Conference Proceedings **810** (American Institute of Physics, Melville, NY), pp. 99–113.
54. Conte, E., Todarello, O., Federici, A., Vitiello, F., Lopane, M., Khrennikov, A., Zbilut, J. P. (2006). Some remarks on an experiment suggesting quantum-like behavior of cognitive entities and formulation of an abstract quantum mechanical formalism to describe cognitive entity and its dynamics. *Chaos Solitons Fractals* **31**, pp. 1076–1088.
55. Conte, E., Khrennikov, A., Todarello, O., Federici, A., Mendolicchio, L, Zbilut, J. P. (2008). A preliminary experimental verification on the pos-

sibility of Bell inequality violation in mental states. *Neuroquantology* **6**, pp. 214–221.

56. Conte, E., Khrennikov, A., Todarello, O., Federici, A., Mendolicchio, L, Zbilut, J. P. (2009). Mental state follow quantum mechanics during perception and cognition of ambiguous figures. *Open Sys. Inf. Dyn.* **16**, pp. 1–17.
57. D'Agostino, S. (1992). Continuity and completeness in physical theories: Schrödinger's return to the wave representation of quantum mechanics in 1950s. In Bitbol, M., Darrigol, O. (eds.), *E. Schrödinger: Philosophy and the Birth of Quantum Mechanics* (Editions Frontieres, Gif-sur-Yvette), pp. 348–360.
58. Daletski, A. L., Fomin, S. V. (1991). *Measures and Differential Equations in Infinite-dimensional Spaces* (Kluwer, Dordrecht).
59. D'Ariano, G. M. (2007). Operational axioms for quantum mechanics. In Adenier, G., Fuchs, C., Khrennikov, A.(eds.), *Foundations of Probability and Physics-3*, Ser. Conference Proceedings **889** (American Institute of Physics, Melville, NY), pp. 79–105.
60. Davidson, M. P. (2007). Stochastic models of quantum mechanics - a perspective. In Adenier, G., Fuchs, C., Khrennikov, A. (eds.), *Foundations of Probability and Physics-3*, Ser. Conference Proceedings **889** (American Institute of Physics, Melville, NY), pp. 106–119.
61. Davies, E. B., Lewis, J. T. (1970). An operational approach to quantum probability. *Comm. Math. Phys.* **17**, pp. 239–260.
62. De Baere, W. (2002). Subquantum nonreproducibility and the complete local description of physical reality. In Khrennikov, A. (ed.), *Quantum Theory: Reconsideration of Foundations,* Ser. Math. Modeling **2** (Växjö University Press, Växjö), pp. 59–74.
63. De la Pena, L., Cetto, A. M. (1996). *The Quantum Dice: An Introduction to Stochastic Electrodynamics* (Kluwer, Dordrecht).
64. De la Pena, L. (1982). Does quantum mechanics accept a stochastic support? *Found. Phys.* **12**, pp. 1017–1037.
65. De la Pena, L. (1969). New formulation of stochastic theory and quantum mechanics. *J. Math. Phys.* **10**, pp. 1620–1630.
66. De la Pena, L., Cetto, A. M. (1971). *Phys. Rev.* D **3**, pp. 795–803.
67. De la Pena, L., Cetto, A. M. (2006). Recent development in linear stochastic electrodynamics. In Adenier, G., Khrennikov, A., Nieuwenhuizen, Th. M. (eds.), *Quantum Theory: Reconsideration of Foundations-3*, Ser. Conference Proceedings **810** (American Institute of Physics, Melville, NY), pp. 131–140.

68. De Martini, F., Fortunato, M., Tombesi, P., Vitali, D. (1999). Generating entangled superpositions of macroscopically distinguishable states within a parametric oscillator. *Phys. Rev.* A **60**, pp. 1636–1651.

69. De Muynck, W. M. (2001). Interpretations of quantum mechanics, and interpretations of violations of Bell's inequality. In Khrennikov, A. (ed.), *Foundations of Probability and Physics*, Series PQ-QP: Quantum Probability and White Noise Analysis **13** (WSP, Singapore), pp. 95–104.

70. De Muynck, W. M. (2002). *Foundations of Quantum Mechanics, an Empiricists Approach* (Kluwer, Dordrecht).

71. De Raedt, K., Keimpema, K., De Raedt, H., Michielsen, K., Miyashita, S. (2006).: A local realist model for correlations of the singlet state. *Eur. Phys. J.* B **53**, pp. 139–142.

72. De Raedt, H., De Raedt, K., Michielsen, K., Keimpema, K., Miyashita, S. (2007). Event-based computer simulation model of Aspect-type experiments strictly satisfying Einstein's locality conditions. *J. Phys. Soc. Jpn.* **76**, pp. 104005–104015.

73. Deutsch, D. (1999). Quantum theory of probability and decision. *Proc. R. Soc. Lond. A* **455**, pp. 3129–3137.

74. Dirac, P. A. M. (1995). *The Principles of Quantum Mechanics* (Clarendon Press, Oxford).

75. Doebner, H.-D., Goldin G. A. (1996). Introducing nonlinear gauge transformations in a family of nonlinear Schrödinger equations. *Phys. Rev. A* **54**, pp. 3764–3771.

76. Doebner, H.-D., Goldin G. A. (1992). On a general nonlinear Schrödinger equation admitting diffusion currents. *Phys. Lett. A* **162**, pp. 397–401.

77. Dubischar, D., Gundlach, V. M., Steinkamp, O., Khrennikov, A. (1999). The interference phenomenon, memory effects in the equipment and random dynamical systems over the fields of *p*-adic numbers. *Il Nuovo Cimento B* **114**, pp. 373–382.

78. Einstein, A. (1910). On the theory of light quanta and the question of localization of electromagnetic energy. *Archives des sceiences physiques et naturelles* **29**, pp. 525–528; Howard D. (ed.) (1993). *The Collected Papers of Albert Einstein. The Swiss Years: writings, 1909–1911* (Princeton University Press, Princeton, NJ), pp. 207–208.

79. Einstein, A., Infeld, L. (1961). *Evolution of Physics: The Growth of Ideas from Early Concepts to Relativity and Quanta* (Simon and Schuster, New York).

80. Einstein, A., Podolsky, B. Rosen, N. (1935). Can quantum-mechanical description of physical reality be considered complete? *Phys. Rev.* **47**, pp. 777–780.

81. Elze, T. (2009). Does quantum mechanics tell an atomistic spacetime? *J. Phys. Conf. Ser.* **174**, art. no. 012009.

82. Elze, T. (2007). Is there a relativistic nonlinear generalization of quantum mechanics? *J. Phys. Conf. Ser.* **67**, art. no. 012016.

83. Elze, T. (2008). The attractor and the quantum states. arXiv 0806.3408.

84. Fine, A. (1996). *The Shaky Game. Einstein Realism and the Quantum Theory* (The University of Chicago Press, Chicago and London).

85. Feynman, R., Hibbs, A. (1965). *Quantum Mechanics and Path Integrals* (McGraw-Hill, New York).

86. Folse, H. J. (2002). Bohr's conception of the quantum mechanical state of a system and its role in the framework of complementarity, Khrennikov, A. (ed.), *Quantum Theory: Reconsideration of Foundations,* Ser. Math. Modeling **2** (Växjö University Press, Växjö), pp. 83–98.

87. Fuchs, C. A. (2002). The anti-Växjö interpretation of quantum mechanics. In Khrennikov, A. (ed.), *Quantum Theory: Reconsideration of Foundations,* Ser. Math. Modeling **2** (Växjö University Press, Växjö), pp. 99–116.

88. Fuchs, C. A. (2002). Quantum mechanics as quantum information (and only a little more). In Khrennikov, A. (ed.), *Quantum Theory: Reconsideration of Foundations,* Ser. Math. Modeling **2** (Växjö University Press, Växjö), pp. 463–543.

89. Fuchs, C. (2007). Delirium quantum (or, where I will take quantum mechanics if it will let me). In Adenier, G., Fuchs, C., Khrennikov, A. (eds.), *Foundations of Probability and Physics-3,* Ser. Conference Proceedings *889* (American Institute of Physics, Melville, NY), pp. 438–462.

90. Garola, C., Khrennikov, A. (2010). Dialog on beyond quantum. In Khrennikov, A. (ed.), *Quantum Theory: Reconsideration of Foundations-5,* Ser. Conference Proceedings **1232** (American Institute of Physics, Melville, NY), pp. 355–359.

91. Garola, C., Pykacz, J., Sozzo, S. (2006). Quantum machine and semantic realism approach: A unified model. *Found. Phys.* **36**, pp. 862–882.

92. Gill, R. (2003). Time, finite statistics and Bell's fifth position. In Khrennikov, A. (ed.), *Foundations of Probability and Physics-2,* Ser. Math. Modeling **5** (Växjö University Press, Växjö), pp. 179–206.

93. Gisin, N. (1990). Weinberg's non-linear quantum mechanics and supraluminal communications. *Phys. Lett. A*, **143**, pp. 1–2.

94. Gisin, N., Rigo, M. (1995). Relevant and irrelevant nonlinear Schrodinger equations. *J. Phys. A: Math. Gen.* **28**, pp. 7375–7386.

95. Grangier, P., Roger, G., Aspect, A. (1986). Experimental evidence for a photon anticorrelation effect on a beam splitter: a new light on single-photon interferences. *Europhys. Lett.* **1**, pp. 173–179.

96. Grangier, P. (1986). *Etude expérimentale de propriétés non-classiques de la lumière: interférence à un seul photon* (Université de Paris-Sud, Centre D'Orsay).

97. Groessing, G. (2008). The vacuum fluctuation theorem: Exact Schrödinger equation via nonequilibrium thermodynamics, *Phys. Lett. A* **372**, pp. 4556–4563.

98. Groessing, G. (ed.) (2012). Emergent Quantum Mechanics (Heinz von Foerster Congress). *J. Phys. Conf. Ser.* **361**.

99. Gudder, S. P. (2001). An approach to quantum probability, Khrennikov, A. (ed.), *Foundations of Probability and Physics*, Quantum Probability and White Noise Analysis **13** (WSP, Singapore).

100. Dubischar, D., Gundlach, V. M., Steinkamp, O., Khrennikov, A. (1999). The interference phenomenon, memory effects in the equipment and random dynamical systems over the fields of p-adic numbers. *Il Nuovo Cimento B* **114**, pp. 373–382.

101. Haag, R. (1996). *Local Quantum Physics* (Springer, Heidelberg).

102. Haag, R. (2000). Questions in quantum physics: a personal view, ArXiv: hep-th/0001006.

103. Hardy, L. (2002). Quantum theory from intuitively reasonable axioms. In Khrennikov, A. (ed.), *Quantum Theory: Reconsideration of Foundations,* Ser. Math. Modeling **2** (Växjö University Press, Växjö), pp. 117–130.

104. Heisenberg, W. (1989). *The Physical Principles of the Quantum Theory* (Dover Publ., London).

105. Helland, I. S. (2002). Quantum theory as a statistical theory. In Khrennikov, A. (ed.), *Quantum Theory: Reconsideration of Foundations,* Ser. Math. Modeling **2** (Växjö University Press, Växjö), pp. 219–242.

106. Hess, K., Philipp, W. (2003). Exclusion of time in Mermin's proof of Bell-type inequalities. In Khrennikov, A. (ed.), *Quantum Theory: Reconsideration of Foundations-2,* Ser. Math. Modeling **10** (Växjö University Press, Växjö), pp. 243–254.

107. Hess, K., Philipp, W. (2005). Bell's theorem: critique of proofs with and without inequalities. In Adenier, G., Khrennikov, A. (eds.), *Foundations of Probability and Physics-3*, Ser. Conference Proceedings **750** (American Institute of Physics, Melville, NY), pp. 150–155.

108. Hess, K. (2007). In memoriam Walter Philipp. In Adenier, G., Fuchs, C., Khrennikov, A. (eds.), *Foundations of Probability and Physics-3*, Ser. Conference Proceedings **889** (American Institute of Physics, Melville, NY), pp. 3–6.

109. Hida, T. (2008). *Lectures On White Noise Functionals* (WSP, Singapore).

110. Hiley, B. (2003). Phase space distributions of quantum phenomena. In Khrennikov, A.(ed.), *Quantum Theory: Reconsideration of Foundations - 2*, Ser. Math. Modeling **10** (Växjö University Press, Växjö), pp. 267–286.

111. Hiesmayer, B. C. (2003). Bell inequalities for neutral kaon system. In Khrennikov, A. (ed.), *Quantum Theory: Reconsideration of Foundations -2*, Ser. Math. Modeling **10** (Växjö University Press, Växjö), pp. 255–266.

112. Hilbert, D., Nordheim, L., von Neumann, J. (1927). Über die Grundlagen der Quantenmekanik. *Mat. Annalen* **98**, pp. 1–30

113. Hofer, W. A. (2011). Unconventional approach to orbital-free density functional theory derived from a model of extended electrons. *Found. Phys.* **41**, pp. 754–791.

114. Hofmann, H. F. (1999). *Quantum Noise and Spontaneous Emission in Semiconductor Laser Devices* (Institut für Technische Physik, Deutsches Zentrum für Luft und Raumfahrt, Stuttgard).

115. Holevo, A. S. (1982). *Probabilistic and Statistical Aspects of Quantum Theory* (North-Holland, Amsterdam).

116. Holevo, A. S. (2001). *Statistical Structure of Quantum Theory* (Springer, Berlin-Heidelberg).

117. Holland, P. (1993). *The Quantum Theory of Motion* (Cambridge University Press, Cambridge).

118. Howard, D., Stachel, J. (1994). *Einstein: The Formative Years, 1879–1909* (Birkhöuser, Boston).

119. Hudson, R. (2004). Translation invariant phase space mechanics. In Khrennikov, A. (ed.), *Quantum Theory: Reconsideration of Foundations-2*, Ser. Math. Modeling **10** (Växjö University Press, Växjö), pp. 301–314.

120. Ingarden, R. S., Kossakowski, A., Ohya, M. (1997). *Information Dynamics and Open Systems: Classical and Quantum Approach* (Kluwer, Dordrecht).

121. Jaeger, G., Khrennikov, A., Schlosshauer, M., Weihs, G. (eds.) (2011). *Advances in Quantum Theory* (American Institute of Physics, Ser. Conference Proceedings **1327**, Melville, NY).

122. Jin, F., De Raedt, H., Michielsen, K. (2010). Particle-based simulation approach for single-particle interference experiments: Application to double-slit experiments. In Khrennikov, A. (ed.), *Quantum Theory: Reconsiderations of Foundations-5*, Ser. Conference Proceedings **1232** (American Institute of Physics, Melville, NY), pp. 93–104.

123. Kant, I. (2008). *The Critique of Pure Reason* (Penguin Classics, London).

124. Khrennikov, A. (1982). A theorem of the existence of the solution of a stochastic differential equation in nuclear Frechet space. *Theor. Probability Appl.* **1**, pp. 85–89.

125. Khrennikov, A. (1984). Ito's formula in nuclear Frechet space. *Vestnik Moscow State University, Ser. Math.* 1, pp. 9–13.

126. Khrennikov, A. (1985). Stochastic integrals in locally convex spaces. *Uspehi Matematicheskih Nauk* **37**, pp. 161–162.

127. Khrennikov, A. Yu. (1983). *Infinite-dimensional Pseudo-differential Operators* (Moscow State University Publ., Moscow) (in Russian).

128. Khrennikov, A. Yu. (1985). Second quantization and pseudo-differential operators. *Theor. Math. Phys.* **66**, pp. 339–349.

129. Khrennikov, A. Yu. (1985). The Feynman measure on the phase space and symbols of infinite-dimensional pseudo-differential operators, *Math. Notes* **37**, pp. 734–742.

130. Khrennikov, A. Yu. (1988). Infinite-dimensional pseudo-differential operators. *Math. USSR-Izvestiya* **31**, pp. 575–601.

131. Khrennikov, A. Yu. (1989). The principle of correspondence in quantum theories of field and relativistic bosonic string. *Matematicheskii Sbornik* **180**, pp. 763–786.

132. Khrennikov, A. Yu. (1994). *P-adic Valued Distributions and their Applications to the Mathematical Physics* (Kluwer, Dordreht).

133. Khrennikov A. (1997). *Non-Archimedean Analysis: Quantum Paradoxes, Dynamical Systems and Biological Models* (Kluwer, Dordreht).

134. Khrennikov, A. (1998). p-adic probability predictions of correlations between particles in the two slit and neuron interferometry experiments. *Il Nuovo Cimento B* **113**, pp. 751–760.

135. Khrennikov, A. (1999). *Interpretations of Probability* (VSP International Science Publishers, Utrecht); the second edition (corrected and completed): (2009). (De Gruyter, Berlin).

136. Khrennikov, A. (2000). A perturbation of CHSH inequality induced by fluctuations of ensemble distributions. *J. Math. Phys.* **41**, pp. 5934–5944.

137. Khrennikov, A. (2000). Non-Kolmogorov probability models and modified Bell's inequality. *J. Math. Phys.* **41**, pp. 1768–1777.

138. Khrennikov, A. Yu., Petersson, H. (2001). Theorem of Paley-Wiener for generalized analytic functions on infinite-dimensional spaces. *Izvestiya: Mathematics* **65**, pp. 403–424.

139. Khrennikov A. (2001). Contextualist viewpoint to Greenberger-Horne-Zeilinger paradox. *Phys. Lett. A* **278**, pp. 307–314.

140. Khrennikov A. (2001). Linear representations of probabilistic transformations induced by context transitions. *J. Phys. A: Math. Gen.* **34**, pp. 9965–9981.

141. Khrennikov, A. (ed.) (2001). *Foundations of Probability and Physics,* Series PQ-QP: Quantum Probability and White Noise Analysis **13** (WSP, Singapore).

142. Khrennikov A. (2002). Quantum statistics via perturbation effects of preparation procedures. *Il Nuovo Cimento B* **117**, pp. 267–281.

143. Khrennikov A. (2002). Frequency analysis of the EPR–Bell argumentation. *Found. Phys.* **32**, pp. 1159–1174.

144. Khrennikov, A. (ed.) (2002). *Quantum Theory: Reconsideration of Foundations,* Ser. Math. Modelling **2** (Växjö University Press, Växjö).

145. Khrennikov, A. (2002). On foundations of quantum theory. In Khrennikov, A. (ed.), *Quantum Theory: Reconsideration of Foundations,* Ser. Math. Modeling **2** (Växjö University Press, Växjö), pp. 163–196.

146. Khrennikov, A., Volovich, Ja. (2002). Discrete time leads to quantum-like interference of deterministic particles. In Khrennikov, A. (ed.), *Quantum Theory: Reconsideration of Foundations,* Ser. Math. Modeling **2** (Växjö University Press, Växjö), pp. 441–454.

147. Khrennikov, A., Volovich, I. (2002). Local realism, contextualism and loopholes in Bell experiments. In Khrennikov, A. (ed.), *Foundations of Probability and Physics-2,* Ser. Math. Modeling **5** (Växjö University Press, Växjö), pp. 325–344.

148. Khrennikov, A. (ed.) (2003). *Foundations of Probability and Physics-2,* Ser. Math. Modelling **5** (Växjö University Press, Växjö).

149. Khrennikov A. (2003). Contextual viewpoint to quantum stochastics. *J. Math. Phys.* **44**, pp. 2471–2478.

150. Khrennikov A. (2003). Representation of the Kolmogorov model having all distinguishing features of quantum probabilistic model. *Phys. Lett. A* **316**, pp. 279–296.

151. Khrennikov A. (2003). Hyperbolic quantum mechanics. *Adv. Appl. Clifford Algebras* **13**, pp. 1–9.

152. Khrennikov A. (2004). Växjö interpretation-2003: Realism of contexts. In Khrennikov, A. (ed.), *Quantum Theory: Reconsideration of Foundations-2,* Ser. Math. Modeling **10** (Växjö University Press, Växjö), pp. 323–338.

153. Khrennikov, A., Volovich, Ja. (2004). Discrete time dynamical models and their quantum-like context-dependent properties. *J. Mod. Opt.* **51**, pp. 113–114.

154. Khrennikov, A., Loubenets, E. (2004). On relation between probabilities in quantum and classical experiments. *Found. Phys.* **34**, pp. 689–704.

155. Khrennikov, A. (2004). Contextual approach to quantum mechanics and the theory of the fundamental prespace. *J. Math. Phys.* **45**, pp. 902–921.

156. Khrennikov, A. (ed.) (2004). *Quantum Theory: Reconsideration of Foundations-2,* Ser. Math. Modelling **10** (Växjö University Press, Växjö).

157. Khrennikov, A. (2004). On quantum-like probabilistic structure of mental information. *Open Systems and Information Dynamics* **11**, pp. 267–275.

158. Khrennikov, A. (2004). *Information Dynamics in Cognitive, Psychological, Social, and Anomalous Phenomena* (Kluwer, Dordreht).

159. Khrennikov, A. (ed.) (2005). *Foundations of Probability and Physics-3* (American Institute of Physics, Ser. Conference Proceedings 750, Melville, NY).

160. Khrennikov, A. (2005). The principle of supplementarity: A contextual probabilistic viewpoint to complementarity, the interference of probabilities, and the incompatibility of variables in quantum mechanics. *Found. Phys.* **35**, pp. 1655–1693.

161. Khrennikov, A. (2005). Hyperbolic quantum mechanics. *Dokl. Acad. Nauk* **402**, pp. 170–172. English Translation: (2005). *Doklady Mathematics* **71**, pp. 363–365.

162. Khrennikov, A. (2005). A pre-quantum classical statistical model with infinite-dimensional phase space. *J. Phys. A: Math. Gen.* **38**, pp. 9051–9073.

163. Khrennikov, A. (2005). Schrödinger dynamics as the Hilbert space projection of a realistic contextual probabilistic dynamics. *Europhys. Lett.* **69**, pp. 678–684.

164. Khrennikov, A. (2005). Linear and nonlinear analogues of the Schrödinger equation in the contextual approach to quantum mechanics. *Dokl. Akad. Nauk* **404**, pp. 33–36; *Doklady Mathematics* **72**, pp. 791–794.

165. Khrennikov, A. (2006). Generalizations of quantum mechanics induced by classical statistical field theory. *Found. Phys. Lett.* **18**, pp. 637–650.

166. Khrennikov, A. (2006). Nonlinear Schrödinger equations from prequantum classical statistical field theory. *Phys. Lett. A* **357**, pp. 171–176.

167. Khrennikov, A. (2006). Prequantum classical statistical field theory: Complex representation, Hamilton-Schrödinger equation, and interpretation of stationary states. *Found. Phys. Lett.* **19**, pp. 299–319.

168. Khrennikov, A. (2006). Classical representation of quantum means and quantum dynamics. *Dokl. Math.* **73**, pp. 445–448.

169. Khrennikov, A. (2006). On the problem of hidden variables for quantum field theory. *Nuovo Cimento B* **121**, pp. 505–521.

170. Khrennikov, A. (2006). Quantum mechanics from time scaling and random fluctuations at the "quick time scale," *Nuovo Cimento B* **121**, pp. 1005–1021.

171. Khrennikov, A. (2006). Representation of quantum field theory as classical statistical mechanics for field functionals. *Dokl. Math.* **74**, pp. 758–761.

172. Khrennikov, A. (2007). Prequantum classical statistical field theory: Time scale of fluctuations. *Dokl. Math.* **75**, pp. 456–459.

173. Khrennikov, A. (2007). Quantum mechanics as an approximation of statistical mechanics for classical fields. *Rep. Math. Phys.* **60**, pp. 453–484.

174. Khrennikov, A. (2007). Quantum mechanics as the quadratic Taylor approximation of classical mechanics: The finite-dimensional case. *Theor. Math. Phys.* **152**, pp. 1111–1121.

175. Khrennikov, A. (2007). To quantum averages through asymptotic expansion of classical averages on infinite-dimensional space. *J. Math. Phys.* **48**, art. no. 013512.

176. Khrennikov, A. (2007). Quantum mechanics for military officers. In Adenier, G., Fuchs, C., Khrennikov, A. (eds.), *Foundations of Probability*

and Physics-3, Ser. Conference Proceedings **889** (American Institute of Physics, Melville, NY), pp. 137–151.

177. Khrennikov, A. (2008). Algorithm for Quantum-like Representation: Transformation of Probabilistic Data into Vectors on Bloch's Sphere. *Open Systems and Information Dynamics* **15**, pp. 223–230.
178. Khrennikov, A. (2008). Quantum probabilities from detection theory for classical random fields. *Fluctuations Noise Lett.* **8**, pp. L393–L400.
179. Khrennikov, A. (2008). Born's rule from classical random fields. *Phys. Lett. A* **372**, pp. 6588–6592.
180. Khrennikov, A. (2008). Analysis of the role of von Neumann's projection postulate in the canonical scheme of quantum teleportation. *J. Russ. Laser Res.* **29**, pp. 296–301.
181. Khrennikov, A. (2008). Quantum-like model for classical random electromagnetic field. *J. Mod. Opt.* **55**, pp. 2257–2267.
182. Khrennikov, A. (2008). EPR–Bohm experiment and Bell's inequality: Quantum physics meets probability theory. *Theor. Math. Phys.* **157**, pp. 1448–1460.
183. Khrennikov, A. (2008). Quantum Randomness as a Result of Random Fluctuations at the Planck Time Scale? *Int. J. Theor. Phys.* **47**, pp. 114–124.
184. Khrennikov, A. (2008). The role of von Neumann and Luders postulates in the Einstein, Podolsky, and Rosen considerations: Comparing measurements with degenerate and nondegenerate spectra. *J. Math. Phys.* **49**, art. no. 052102.
185. Khrennikov, A. (2009). EPR "paradox," projection postulate, time synchronization "nonlocality," *Int. J. Quantum Information* **7**, pp. 71–81.
186. Khrennikov, A. (2009). Nonlocality as well as rejection of realism are only sufficient (but non-necessary!) conditions for violation of Bell's inequality. *Inf. Sci.* **179**, pp. 492–504.
187. Khrennikov, A. (2009). Detection model based on representation of quantum particles by classical random fields: Born's rule and beyond. *Found. Phys.* **39**, pp. 997–1022.
188. Khrennikov, A. (2009). Von Neumann and Luders postulates and quantum information theory. *Int. J. Quantum Inf.* **7**, pp. 1303–1311.
189. Khrennikov, A. (2009). Entanglement's dynamics from classical stochastic process. *Europhys. Lett.* **88**, pp. 1–6, art. no. 40005.

190. Khrennikov, A. (2009). Quantum correlations from classical Gaussian correlations. *J. Russ. Laser Res.* **30**, pp. 472–479.

191. Khrennikov, A. (2009). Bell's inequality, Einstein, Podolsky, Rosen arguments and von Neumann's projection postulate. *Laser Phys.* **19**, pp. 346–356.

192. Khrennikov, A. (2009). *Contextual Approach to Quantum Formalism* (Springer, Berlin-Heidelberg-New York).

193. Khrennikov, A. (2009). Bell's inequality, Einstein, Podolsky, Rosen arguments and von Neumann's projection postulate. *Laser Phys.* **19**, pp. 346–356.

194. Khrennikov, A. (2009). Demystification of Bell inequality. In Roychoudhuri, Ch., Kracklauer, A. F., Khrennikov, A. (eds.), *The Nature of Light: What are Photons? III, Proc. SPIE* **7421**, art. no. 74210E.

195. Khrennikov, A. (2010). Subquantum detection theory, SDT. *Physica E: Low-Dimensional Systems and Nanostructures* **42**, pp. 287–292.

196. Khrennikov, A. (2010). An analogue of the Heisenberg uncertainty relation in prequantum classical field theory, *Physica Scripta* **81**, art. no. 065001.

197. Khrennikov, A. (2010). Pairwise correlations in a three-partite quantum system from a prequantum random field. *J. Russ. Laser Res.* **31**, pp. 191–200.

198. Khrennikov, A. (2010). Prequantum classical statistical field theory: Schrödinger dynamics of entangled eystems as a classical stochastic process. *Found. Phys.* **40**, pp. 1–13.

199. Khrennikov, A. (2010). Description of composite quantum systems by means of classical random fields. *Found. Phys.* **40**, pp. 1051–1064.

200. Khrennikov, A. (ed.) (2010). *Quantum Theory: Reconsideration of Foundations-5*, Ser. Conference Proceedings **1232** (American Institute of Physics, Melville, NY).

201. Khrennikov, A., Ohya, M., Watanabe, N. (2010). Classical signal model from quantum channels. *J. Russ. Laser Res.* **31**, pp. 462–468.

202. Khrennikov, A. (2010). Quantum correlations from classical Gaussian random variables; fundamental role of vacuum noise. *Fluctuations and Noise Lett.* **9**, pp. 331–341.

203. Khrennikov, A. (2010). Two-particle wave function as an integral operator and the random field approach to quantum correlations. *Theor. Math. Phys.* **164**, pp. 1156–1162.

204. Khrennikov, A. (2010). Quantum correlations and dynamics from classical random fields valued in complex Hilbert spaces. *J. Math. Phys.* **51**, art. no. 082106.

205. Khrennikov, A. (2010). Correlations of components of prequantum field corresponding to biparticle quantum system. *Europhys. Lett.* **90**, art. no. 40004.

206. Khrennikov, A. (2010). *Ubiquitous Quantum Structure: from Psychology to Finances* (Springer, Berlin-Heidelberg-New York).

207. Khrennikov, A. (2010). On the physical basis of theory of "mental waves," *Neuroquantology* **8**, pp. S71–S80.

208. Khrennikov, A. (2011). Quantum-like model of processing of information in the brain based on classical electromagnetic field. *Biosystems* **105**, pp. 250–262.

209. Khrennikov, A. (2011). Violation of Bell's inequality by correlations of classical random signals, quant-ph/1111.1907v1.

210. Khrennikov, A. (2012). Quantum probabilities from a mathematical model of threshold detection of classical random waves, *J. Phys. A Math. Theor.* **45**, art. no. 215301.

211. Khrennikov, A. (2012). Born's rule from measurements of classical signals by threshold detectors which are properly calibrated. *J. Mod. Opt.* **59**, pp. 667–678.

212. Khrennikov, A. (2012). Towards new Grangier type experiments. *Ann. Phys.* **327**, pp. 1786–1802.

213. Khrennikov, A., Nilsson B., Nordebo, S. (2012). Classical signal model reproducing quantum probabilities for single and coincidence detections, *J. Phys. Conf. Ser.* **361**, art. no. 012030.

214. Kim, Y. S., Noz, M. E. (2007). Can one do quantum mechanics without Einstein? In Adenier, G., Fuchs, C., Khrennikov, A. (eds.), *Foundations of Probability and Physics-3*, Ser. Conference Proceedings **889** (American Institute of Physics, Melville, NY), pp. 152–161.

215. Klyshko, D. N. (1993). The Bell and GHZ theorems: a possible three-photon interference experiment and the question of nonlocality. *Phys. Lett. A* **172**, pp. 399–403.

216. Klyshko, D. N. (1998). Basic notions of quantum mechanics from the operational viewpoint. *Physics-Uspekhi: Adv. Phys. Sc.* **41**, pp. 885–922.

217. Kochen, S., Specker, E. (1967). The problem of hidden variables in quantum mechanical systems. *J. Math. Mech.* **17**, pp. 59–87.

218. Kolmogoroff, A. N. (1933). *Grundbegriffe der Wahrscheinlichkeitsrechnung* (Springer Verlag, Berlin); English translation: Kolmogorov, A. N. (1956). *Foundations of Theory of Probability* (Chelsea Publishing Company, New York).
219. Kracklauer, A. F. (2007). What do correlations tell us about photons? In Roychoudhuri, Ch., Kracklauer, A. F., Creath, K. (eds.), *The Nature of Light: What Are Photons? Proc. SPIE* **6664**, art. no. 66640H.
220. Kracklauer, A. F., Rangacharyulu, Ch., Roychoudhuri, Ch., Brooks, H. J., Carroll, J., Khrennikov, A. (2009). Is indivisible single photon really essential for quantum communications, computing and encryption? In Roychoudhuri, Ch., Kracklauer, A. F., Khrennikov, A. (eds.), *The Nature of Light: What are Photons? III, Proc. SPIE* **7421**, art. no. 74210Y.
221. Lamb, W. E., Scully, M. O. (1969). The photoelectric effect without photons. In *Polarization, Matter and Radiation. Jubilee Volume in Honour of Alfred Kasiler* (Press of University de France, Paris), pp. 363–369.
222. Lamb, W. E. (2001). *The Interpretation of Quantum Mechanics* (Rinton Press, Inc., Princeton).
223. Lande, A. (1955). *Foundations of Quantum Theory* (Yale University Press).
224. Lande, A. (1965). *New Foundations of Quantum Mechanics* (Cambridge University Press, Cambridge).
225. Landsman, N. P. (2008). In *Compendium of Quantum Physics* Weinert, F., Hentschel, K., Greenberger, D., Falkenburg, B. (eds.). (Springer, Heidelberg-Berlin-New York).
226. Larsson, J.-Å. (2000). *Quantum Paradoxes, Probability Theory, and Change of Ensemble* (Linköping University Press, Linköping).
227. Larsson, J.-Å. (2003). Bell inequalities for position measurements. In Khrennikov, A. (ed.), *Quantum Theory: Reconsideration of Foundations*, Ser. Math. Modeling **10** (Växjö University Press, Växjö), pp. 353–364.
228. Larsson, J.-Å., Gill, R. (2005). Bell's inequality and the coincidence time loophole. In Adenier, G., Khrennikov, A. (eds.), *Foundations of Probability and Physics-3*, Ser. Conference Proceedings **750** (American Institute of Physics, Melville, NY), pp. 228–235.
229. Lockwood, M. (1992). What Schrödinger have learned from his cat? In Bitbol, M., Darrigol, O. (eds.), *E. Schrödinger: Philosophy and the Birth of Quantum Mechanics* (Editions Frontieres, Gif-sur-Yvette), pp. 380–390.
230. Louisell, H. H. (1973). *Quantum Statistical Properties of Radiation* J. Wiley, New York.

231. Loubenets, E. R. (2004). General framework for the probabilistic description of experiments. In Khrennikov, A. (ed.), *Quantum Theory: Reconsideration of Foundations-2*, Ser. Math. Modeling **10** (Växjö University Press, Växjö), pp. 385–387.

232. Ludwig, G. (1983). *Foundations of Quantum Mechanics* (Springer, Berlin).

233. Mackey, G. W. (1963). *Mathematical Foundations of Quantum Mechanics* (W. A. Benjamin Inc., New York).

234. Mandel, L., Wolf, E. (1995). *Optical Coherence and quantum Optics* (Cambridge University Press, Cambridge).

235. Manko, V. I. (1996). Classical formulation of quantum mechanics. *J. Russ. Laser Res.* **17**, pp. 579–584.

236. Manko, O. V., Manko, V. I. (1996). Classical mechanics is not $h \to 0$ limit of quantum mechanics. *J. Russ. Laser Res.* **25**, pp. 477–492.

237. Manko, V. I., Shchukin, E. V. (2001). A charged particle in an electric field in the probability representation of quantum mechanics. *J. Russ. Laser Res.* **22**, pp. 545–560.

238. Manko, M. A., Manko, V. I., Mendes, V. I. (2006). A probabilistic operator symbol framework for quantum information. *J. Russ. Laser Res.* **27**, pp. 507–532.

239. Margenau, H. (1950). *The Nature of Physical Reality* (McGraw-Hill, New York).

240. Margenau, H. (1963). Measurements in quantum mechanics. *Ann. Phys.* **23**, pp. 469–485.

241. Michielsen, K., De Raedt, H. (2010). Quantum interference with macroscopic object. In Khrennikov, A. (ed.), *Quantum Theory: Reconsiderations of Foundations-5* (American Institute of Physics, Ser. Conference Proceedings **1232**, Melville, NY), pp. 27–32.

242. Mermin, N. D. (2002). Whose knowledge? In Khrennikov, A. (ed.), *Quantum Theory: Reconsideration of Foundations*, Ser. Math. Modeling **2** (Växjö University Press, Växjö), pp. 261–270.

243. Mie, G. (1912). Grundlagen einer Theorie der Materie. *Annalen der Physik* **37**, pp. 511–534.

244. Mie, G. (1912). Grundlagen einer Theorie der Materie. Zweite Mittelung. *Annalen der Physik* **39**, pp. 1–40.

245. Mie, G. (1913). Grundlagen einer Theorie der Materie. Dritte Mittelung. *Annalen der Physik* **40**, pp. 1–66.

246. Morse, P. M., Ingard, K. U. (1968). *Theoretical Acoustics* (Mc Graw-Hill Book Company, New York, St. Louis, San Francisco, Toronto, London, Sydney).

247. Mukunda, N. (2000). The story of the photon. *Resonance* **5**, pp. 35–51.

248. Nelson, E. (1985). *Quantum Fluctuations* (Princeton University Press, Princeton).

249. Nieuwenhuizen, Th. M. (2006). Classical phase space density for relativistic electron. In Adenier, G., Khrennikov, A., Nieuwenhuizen, Th. M. (eds.), *Quantum Theory: Reconsideration of Foundations-3*, Ser. Conference Proceedings **810** (American Institute of Physics, Melville, NY), pp. 198–210.

250. Nieuwenhuizen, Th. M., Mehmani, B., Spicka, V., Aghdami, M. J., Khrennikov, A. (eds.) (2007). *Beyond the Quantum.* (WSP, Singapore).

251. Nyman, P. (2010). On consistency of the quantum-like representation algorithm. *Int. J. Theor. Phys.* **49**, pp. 1–9.

252. Ohya, M., Watanabe, N. (1986). A new treatment of communication processes with Gaussian channels. *Jpn. J. Ind. Appl. Math.* **3**, pp. 197–206.

253. Ohya, M., Volovich, I. (2011). *Mathematical Foundations of Quantum Information and Computation and Its Applications to Nano- and Bio-systems* (WSP, Singapore).

254. Ojima, I. (2011). Micro–macro duality and space–time emergence. In Jaeger, G., Khrennikov, A., Schlosshauer, M., Weihs, G. (eds.), *Advances in Quantum Theory*, Ser. Conference Proceedings **1327** (American Institute of Physics, Melville, NY), pp. 197–206.

255. Ozawa, M. (1983). Conditional expectation and repeated measurement of continuous quantum observables, *Lecture Notes in Math.* **1021**, pp. 518–525.

256. Ozawa, M. (1980). Optimal measurements for general quantum systems. *Rep. Math. Phys.* **18**, pp. 11–28.

257. Ozhigov, Yu. (2011). *Constructive Physics* (Nova Scence Publishers Inc., NY).

258. Pais, A. (1983). *Subtle Is the Lord: The Science and the Life of Albert Einstein* (Oxford University Press, Oxford).

259. Perez-Suarez, M., Santos, D. J. (2002). Quantum mechanics as an information theory: Some further musings on a Fuchsian proposal. In Khrennikov, A. (ed.), *Quantum Theory: Reconsideration of Foundations*, Ser. Math. Modeling **2** (Växjö University Press, Växjö), pp. 469–478.

260. Pitowsky, I. (1989). *Quantum Probability: Quantum Logic* (Springer, Berlin).

261. Planck, M. (1899). Uber irreversible Strahlungsvorgange. *Sitzungsberichte der Koniglich Preussischen Akademie der Wissenschaften zu Berlin* **5**, pp. 440–480.

262. Plotnitsky, A. (2005). "This is an extremely funny thing, something must be hidden behind that": Quantum waves and quantum probability with Erwin Schrödinger. In Adenier, G., Khrennikov, A. (eds.), *Foundations of probability and physics-3*, Ser. Conference Proceedings **750** (American Institute of Physics, Melville, NY), pp. 388–408.

263. Plotnitsky, A. (2006). *Reading Bohr: Physics and Philosophy* (Springer, Dordrecht).

264. Plotnitsky, A. (2009). *Epistemology and Probability: Bohr, Heisenberg, Schrödinger, and the Nature of Quantum-Theoretical Thinking* (Springer, Heidelberg-Berlin-New York).

265. Ritov, S. M. (1966). *Introduction to Statistical Radiophysics* (Nauka, Fizmatlit, Moscow).

266. Robertson, A. P., Robertson, W. J. (1964). *Topological Vector Spaces* (Cambridge University Press, Cambridge).

267. Roychoudhuri, Ch., Creath, K. (eds.) (2005). *The Nature of Light: What Is a Photon? Proc. SPIE* **5866.**

268. Roychoudhuri, Ch., Kracklauer, A. F., Creath, K. (eds.) (2007). *The Nature of Light: What Are Photons? Proc. SPIE* **6664**.

269. Roychoudhuri, Ch., Kracklauer, A. F., Khrennikov, A. (eds.) (2009). *The Nature of Light: What are Photons? III. Proc. SPIE* **7421**.

270. Roychoudhuri, Ch., Kracklauer, A. F., Khrennikov, A. (eds.) (2011). *The Nature of Light: What are Photons? IV, Proc. SPIE* **8121**.

271. Schack, R. (2007). Bayesian versus frequentist predictions in quantum tomography. In Adenier, G., Fuchs, C., Khrennikov, A. (eds.), *Foundations of Probability and Physics-3*, Ser. Conference Proceedings **889** (American Institute of Physics, Melville, NY), pp. 230–234.

272. Schilpp, P. A. (1979). *Albert Einstein: Autobiographical Notes* (Open Court Publishing Company, Chicago).

273. Schrödinger, E. (1935). Die gegenwärtige Situation in der Quantenmechanik. *Naturwissenschaften* **23**, pp. 807–812, 823–828, 844–849.

274. Schrödinger, E. (1984). *E. Schrödinger Gesammelte Abhandlungen* (Wieweg and Son, Wien); especially the paper "What is an elementary particle?" pp. 456–463.

275. Scully, M. O., Zubairy, M. S. (1997). *Quantum Optics* (Cambridge University Press, Cambridge).

276. Shiryaev, A. N. (1989). Kolmogorov: Life and creative activities. *The Annals of Probability* **17**, pp. 866–944.

277. Shiryaev, A. N. (1984). *Probability* (Springer, New York-Berlin-Heidelberg).

278. Sinha, U., Couteau, C., Medendorp, Z., Sollner, I., Laflamme, R., Sorkin, R., Weihs, G. (2009). Testing Born's rule in quantum mechanics with a triple slit experiment. In Accardi, L., Adenier, G., Fuchs, C., Jaeger, G., Khrennikov, A., Larsson, J.-A., Stenholm, S. (eds.), *Foundations of Probability and Physics-5*, Ser. Conference Proceedings, **1101** (American Institute of Physics, Melville, NY), pp. 200–207.

279. Skorohod, A. V. (1974). *Integration in Hilbert Space* (Springer, Berlin).

280. Smolaynov, O. G., Khrennikov, A. Yu. (1985). Central limit theorem for generalized measures on infinite-dimensional space. *Dokl. Akademii Nauk USSR, ser. Mat.* **281**, pp. 279–283.

281. Smolin, L. (2006). Can quantum mechanics be considered as an approximation to another theory? ArXiv: quant-ph/0609109.

282. Sorkin, R. (1994). Quantum mechanics as quantum measure theory. *Mod. Phys. Lett. A* **9**, pp. 3119–3127.

283. Stenholm, S. (2003). Two dogmas of quantum theory. In Khrennikov, A. (ed.), *Quantum Theory: Reconsideration of Foundations-2* Ser. Math. Modeling **10** (Växjö University Press, Växjö), pp. 519–532.

284. Strocchi, F. (1966). Complex coordinates and quantum mechanics. *Rev. Mod. Phys.* **38**, pp. 36–40.

285. Summhammer, J. (1989). Neutron interferometric test of nonergodic interpretation of quantum mechanics. *Il Nuovo Cimeno B* **103**, pp. 265–280.

286. Svozil, K. (2005). On counterfactuals and contextuality. In Adenier, G., Khrennikov, A. (eds.), *Foundations of Probability and Physics-3*, Ser. Conference Proceedings **750** (American Institute of Physics, Melville, NY), pp. 351–360.

287. 't Hooft, G. (1999). Quantum gravity as a dissipative deterministic system. ArXiv: gr-qc/9903084.

288. 't Hooft, G. (2006). The mathematical basis for deterministic quantum mechanics. ArXiv: quant-ph/0604008.

289. 't Hooft, G. (2011). The free-will postulate in quantum mechanics. *Herald Russ. Acad. Sci.* **81**, pp. 907–911; ArXiv: quant-ph/0701097v1.

290. Ventzel, E. (1958). *Theory of Probability* (Fizmatlit, Moscow).

291. Vladimirov, V. S., Volovich, I. V., Zelenov, E. I. (1993). *P-adic Analysis and Mathematical Physics* (WSP, Singapore).

292. Volovich, I. V. (2001). Quantum cryptography in space and Bell's theorem. In Khrennikov, A. (ed.), *Foundations of probability and physics*, QP–PQ: Quantum Prob. White Noise Anal. **13** (WSP, River Edge, NJ), pp. 364–372.

293. Von Mises, R. (1919). Grundlagen der Wahrscheinlichkeitsrechnung. *Math. Z.* **5**, pp. 52–99.

294. Von Mises, R. (1957). *Probability, Statistics and Truth* (Macmillan, London).

295. Von Mises, R. (1964). *The Mathematical Theory of Probability and Statistics* (Academic, London).

296. Von Neuman, J. (1955). *Mathematical Foundations of Quantum Mechanics* (Princeton University Press, Princeton).

297. Vorob'ev, N. N. (1962). Consistent families of measures and their extensions. *Theory of Probability and its Applications* **7**, pp. 147–162.

298. Weihs, G. (2007). A test of Bell's inequality with spacelike separation. In Adenier, G., Fuchs, C., Khrennikov, A. (eds.), *Foundations of Probability and Physics-4*, Ser. Conference Proceedings **889** (American Institute of Physics, Melville, NY), pp. 250–262.

299. Weinberg, S. (1989). Precision tests of quantum mechanics. *Phys. Rev. Lett.* **62**, pp. 485–488.

300. Zeilinger, A. (2010). *Dance of the Photons: From Einstein to Quantum Teleportation* (Farrar, Straus and Giroux, New York).

Index